William B. Ruger Chair of National Security Economics Papers

Number 4

American Foreign Policy: Regional Perspectives

Proceedings
A Workshop Sponsored by the
William B. Ruger Chair of National Security Economics
Newport, Rhode Island
13–15 May 2009

Richmond M. Lloyd, editor
William B. Ruger Chair of National Security Economics

Naval War College
Newport, Rhode Island

The Naval War College expresses appreciation to the Naval War College Foundation, through the generosity of the William B. Ruger Chair of National Security Economics, in the preparation and presentation of this workshop.

Correspondence concerning the Ruger Papers may be addressed to Richmond M. Lloyd, William B. Ruger Chair of National Security Economics, Naval War College, 686 Cushing Road, Newport, RI 02841-1207; by telephone at 401.841.3669; or by e-mail at richmond.lloyd@usnwc.edu. Our website is http://www.usnwc.edu/academics/courses/nsdm/rugerpapers.aspx.

ISBN 978-1-884733-70-3

Printed in the United States of America

Contents

Introduction 1

Workshop Agenda. 3

Executive Summary 7

Opening Remarks 37
Ambassador (Ret.) Mary Ann Peters, Provost, Naval War College

Panel I: A Global Perspective. 39

U.S. Security Policy in a Changing World 41
Dr. Patrick M. Cronin, Director, Institute for National Strategic Studies, National Defense University

The New Foreign Policy Agenda: Energy, the Environment, and the Global Economy 47
Dr. Michael T. Klare, Five College Professor of Peace and World Security Studies, Hampshire College

Panel I Summary of Discussion 55
Moderator: Dr. Thomas R. Fedyszyn, Professor of National Security Affairs, Chair Eurasia Regional Study Group, Naval War College

Panel II: Western Hemisphere 61

Another Chance for U.S. Policy in the Americas 63
Peter Hakim, President, Inter-American Dialogue

U.S.–Latin American Relations 75
Dr. Shannon K. O'Neil, Douglas Dillon Fellow for Latin America Studies, Council on Foreign Relations

The Outlook for U.S. Foreign Policy in Latin America and the Caribbean: The Challenges of Transforming Goodwill into Effective Policy 85
Ambassador (Ret.) Paul D. Taylor, Senior Strategic Researcher, Naval War College

Panel II Summary of Discussion 95
Moderator: Professor Laurence L. McCabe, Associate Professor of National Security Affairs, Chair Latin America Regional Study Group, Naval War College

Panel III: Asia and the Pacific. 99

U.S. Asia-Pacific Strategy in the Obama Administration. . . . 101
Dr. Jonathan D. Pollack, Professor of Asian and Pacific Studies, Chair Asia-Pacific Study Group, Naval War College

The New Security Drama in East Asia: The Responses of U.S. Allies and Security Partners to China's Rise 111
Dr. Evan S. Medeiros, Senior Political Scientist, RAND Corporation

U.S. Foreign Policy and Southeast Asia: From Manifest Destiny to Shared Destiny 125
Dr. Emrys Chew, S. Rajaratnam School of International Studies, Singapore

Panel III Summary of Discussion 141
Moderator: Dr. John F. Garofano, Professor of Strategy and Policy, Jerome Levy Chair of Economic Geography and National Security, Naval War College

Panel IV: South Asia 147

A U.S. Strategy for Pakistan: Future Directions 149
Dr. Daniel Markey, Senior Fellow for India, Pakistan, and South Asia, Council on Foreign Relations

India and the United States: Making the Partnership Strategic . . 157
Ambassador (Ret.) Teresita C. Schaffer, Director, South Asia Program, Center for Strategic and International Studies

Panel IV Summary of Discussion 163
Moderator: Dr. Timothy D. Hoyt, Professor of Strategy and Policy, Co-chair Indian Ocean Regional Study Group, Naval War College

Panel V: The Greater Middle East. 167

New Directions for U.S. Foreign Policy in the Greater Middle East 169
Ambassador (Ret.) David C. Litt, Executive Director, Center for Stabilization and Economic Reconstruction, Institute for Defense Business

Engaging the Muslim World beyond al Qaeda 179
Dr. Marc Lynch, Associate Professor of Political Science and International Affairs, The Elliot School of International Affairs, George Washington University

Why the United States Should Engage Islamists 189
Dr. Heidi E. Lane, Associate Professor of Strategy and Policy, Greater Middle East Area of Study Coordinator, Naval War College

Panel V Summary of Discussion 199
Moderator: Dr. Hayat Alvi-Aziz, Associate Professor of National Security Affairs, Naval War College

Panel VI: Europe and Russia 203

Advancing a Strategy for Constructive Security Engagement: "Resetting" the U.S./NATO Approach toward Russia . . . 205
Dr. Sharyl Cross, Professor and Director of Studies, Program in Advanced Security Studies, George C. Marshall European Center for Security Studies

Resetting U.S.-Russian Relations 223
Dr. R. Craig Nation, Professor of Strategy and Director, Eurasian Studies, U.S. Army War College

Panel VI Summary of Discussion 239
Moderator: Ambassador (Ret.) Mary Ann Peters, Provost, Naval War College

Panel VII: Africa. 247

Africa in U.S. Foreign Policy 249
Ambassador (Ret.) Princeton N. Lyman, Adjunct Senior Fellow for Africa Policy Studies, Council on Foreign Relations

Great Expectations versus Daunting Challenges: Prospects for U.S. Foreign Policy toward Africa during the Obama Administration 261
Dr. Peter J. Schraeder, Professor and Graduate Program Director, Department of Political Science, Loyola University, Chicago

Panel VII Summary of Discussion 273
Moderator: Dr. Stephen A. Emerson, Associate Professor of National Security Affairs, Naval War College

Concluding Remarks 277
Dr. Richmond M. Lloyd, William B. Ruger Chair of National Security Economics, Naval War College

Participant Biographies 279

Introduction

Workshop Focus

The purpose of this workshop is to provide a collegial forum for a small and select group of foreign policy and regional experts to formulate and recommend new directions for American foreign policy for each of the major regions of the world.

Workshop Background

With a new American administration in office, this is an opportune time to assess American foreign policy and to set future directions. What challenges and opportunities will the United States, and its allies and friends, face in the future? What changes should be made to all elements of U.S. foreign policy, including the diplomatic, economic, military, and informational elements? What elements should continue? What are the varying perspectives of nations within the regions concerning U.S. foreign policy? What changes in U.S. foreign policy would they desire? Overall, what new directions for U.S. foreign policy will better support the interests and objectives of the United States, its allies, and its friends?

Workshop Venue and Format

A total of thirty-three individuals participated in this by-invitation-only workshop held at the Naval War College in Newport, Rhode Island. The college and its staff provide a professional environment to facilitate small group workshops in exploring specific issues.

Seventeen panelists prepared and presented papers on topics of their choice within the subject areas of their respective panels. Following a presentation of the papers, all participants engaged in extensive discussion of the papers and of the focus of the panel. All discussions were conducted under a nonattribution policy.

All papers and summaries of working-group discussions (prepared by each panel moderator) are included in this monograph. The monograph is being widely distributed within the national security community and the general public. The monograph is also available electronically at http://www.usnwc.edu/academics/courses/nsdm/rugerpapers.aspx.

William B. Ruger Chair of National Security Economics

The Ruger Chair was established to support research and study on the interrelationships between economics and security. A fundamental premise is that without security it is difficult to have economic prosperity and without prosperity it is difficult to have security.

The intent of this Ruger Chair–sponsored workshop is to support individual research, publication, and a continuing dialogue on matters important to national security economics. It is hoped that research done for this workshop will provide participants with the building blocks for further research and publication.

Agenda

American Foreign Policy: Regional Perspectives

A Workshop Sponsored by the William B. Ruger Chair of National Security Economics Naval War College, Newport, Rhode Island 13–15 May 2009

Wednesday, 13 May 2009

1840 Depart Hotel

1900 Welcome Dinner, Officers' Club, Naval Station Newport

Thursday, 14 May 2009

0730 Depart Hotel

0745 Welcome Breakfast, Decision Support Center

0830 Opening Remarks

Ambassador (Ret.) Mary Ann Peters, Provost, Naval War College

0845 Panel I: A Global Perspective

Dr. Patrick M. Cronin, Director, Institute for National Strategic Studies, National Defense University

Dr. Michael T. Klare, Five College Professor of Peace and World Security Studies, Hampshire College

Moderator: Dr. Thomas R. Fedyszyn, Professor of National Security Affairs, Chair Eurasia Regional Study Group, Naval War College

1015 Break

1030 Panel II: Western Hemisphere

Peter Hakim, President, Inter-American Dialogue

Dr. Shannon K. O'Neil, Douglas Dillon Fellow for Latin America Studies, Council on Foreign Relations

Ambassador (Ret.) Paul D. Taylor, Senior Strategic Researcher, Naval War College

Moderator: Professor Laurence L. McCabe, Associate Professor of National Security Affairs, Chair Latin America Regional Study Group, Naval War College

1200 Lunch, RADM Joseph Strasser Dining Room

1330 Panel III: Asia and the Pacific

Dr. Jonathan D. Pollack, Professor of Asian and Pacific Studies, Chair Asia-Pacific Study Group, Naval War College

Dr. Evan S. Medeiros, Senior Political Scientist, RAND Corporation

Dr. Emrys Chew, S. Rajaratnam School of International Studies, Singapore

Moderator: Dr. John F. Garofano, Professor of Strategy and Policy, Jerome Levy Chair of Economic Geography and National Security, Naval War College

1500 Break

1515 Panel IV: South Asia

Dr. Daniel Markey, Senior Fellow for India, Pakistan, and South Asia, Council on Foreign Relations

Ambassador (Ret.) Teresita C. Schaffer, Director, South Asia Program, Center for Strategic and International Studies

Moderator: Dr. Timothy D. Hoyt, Professor of Strategy and Policy, Co-chair Indian Ocean Regional Study Group, Naval War College

1645 Adjourn

1650 Return to Hotel

1845 Depart Hotel

1900 Dinner, La Forge Casino Restaurant

Friday, 15 May 2009

0730 Depart Hotel

0745 Breakfast, Decision Support Center

0830 Panel V: The Greater Middle East

Ambassador (Ret.) David C. Litt, Executive Director, Center for Stabilization and Economic Reconstruction, Institute for Defense Business

Dr. Marc Lynch, Associate Professor of Political Science and International Affairs, The Elliot School of International Affairs, George Washington University

Dr. Heidi E. Lane, Associate Professor of Strategy and Policy, Greater Middle East Area of Study Coordinator, Naval War College

Moderator: Dr. Hayat Alvi-Aziz, Associate Professor of National Security Affairs, Naval War College

1000 Break

1015 Panel VI: Europe and Russia

Dr. Sharyl Cross, Professor and Director of Studies, Program in Advanced Security Studies, George C. Marshall European Center for Security Studies

Dr. R. Craig Nation, Professor of Strategy and Director, Eurasian Studies, U.S. Army War College

Moderator: Ambassador (Ret.) Mary Ann Peters, Provost, Naval War College

1145 Lunch, RADM Joseph Strasser Dining Room

1300 Panel VII: Africa

Ambassador (Ret.) Princeton N. Lyman, Adjunct Senior Fellow for Africa Policy Studies, Council on Foreign Relations

Dr. Peter J. Schraeder, Professor and Graduate Program Director, Department of Political Science, Loyola University, Chicago

Moderator: Dr. Stephen A. Emerson, Associate Professor of National Security Affairs, Naval War College

1430 Concluding Remarks

Dr. Richmond M. Lloyd, William B. Ruger Chair of National Security Economics, Naval War College

1445 Adjourn

1500 Depart to Airport

Participants:

Dr. Rocky R. Meade, Colonel General Staff, Headquarters, Jamaica Defense Force

Dr. Hussein Solomon, University of Pretoria, Republic of South Africa

Captain David G. Manero, U.S. Navy, Security Cooperation Programs Division, U.S. European Command

Captain (Sel) Victor M. Ott, U.S. Navy, Regions Branch Head, International Engagement Division N52, OPNAV

Ms. Deborah A. Bolton, State Department Advisor to President, Naval War College

Dr. Christopher R. Jasparro, Associate Professor of National Security Affairs, Africa Area of Study Coordinator, Naval War College

Dr. Terence Roehrig, Associate Professor of National Security Affairs, Naval War College

Professor Sean C. Sullivan, Workshop Administrative Assistant, and Assistant Professor, Naval War College

Executive Summary

Workshop Purpose and Organization

The purpose of this workshop is to provide a collegial forum for a small and select group of foreign policy and regional experts to formulate and recommend new directions for American foreign policy for each of the major regions of the world.

With a new American administration in office, this is an opportune time to assess American foreign policy and to set future directions:

1. What challenges and opportunities will the United States, and its allies and friends, face in the future?
2. What changes should be made to all elements of U.S. foreign policy, including the diplomatic, economic, military, and informational elements?
3. What elements should continue?
4. What are the varying perspectives of nations within the region concerning U.S. foreign policy?
5. What changes in U.S. foreign policy would they desire?
6. Overall, what new directions for U.S. foreign policy will better support the interests and objectives of the United States, its allies, and its friends?

A total of thirty-three foreign policy and regional experts participated in the workshop. Seventeen panelists presented papers on seven panels: A Global Perspective, Western Hemisphere, Asia and the Pacific, South Asia, The Greater Middle East, Europe and Russia, and Africa.

This Executive Summary highlights the major ideas presented in each paper and the topics discussed in the follow-on panel discussions. The summary of each paper draws upon extensive verbatim extracts from each author's work with minor paraphrasing and editing.

Panel I: A Global Perspective

U.S. Security Policy in a Changing World
Dr. Patrick M. Cronin, Director, Institute for National Strategic Studies, National Defense University

The New Foreign Policy Agenda: Energy, the Environment, and the Global Economy
Dr. Michael T. Klare, Five College Professor of Peace and World Security Studies, Hampshire College

Moderator: Dr. Thomas R. Fedyszyn, Professor of National Security Affairs, Chair Eurasia Regional Study Group, Naval War College

Patrick Cronin discusses the challenges of crafting U.S. security policy in a changing world. The United States needs a new conceptual pathway for policy makers to begin recalibrating America's security role to reverse what has appeared to be a widening gap between U.S. ends and means, now and in the future. International security requires U.S. active engagement, but the character of that engagement is changing along with the global environment. Worldwide trends suggest that the United States will increasingly have to approach complex challenges and surprises through wider and more effective partnerships and more integrated strategies. This essay begins a discussion on how to think about the complex security environment and how in particular the United States can begin the process of strategic adaptation.

Complexity is the watchword of our century. The overriding message is to emphasize global complexity and America's vital yet limited role in coping with that complexity. Policy makers are only beginning to come to terms with the uncertain, complex world in which we operate. For instance, too little systematic thought has been given to the dynamic interactions between state and nonstate actors, between economics and security, and to comprehend security in the contested global commons of international waters, airspace, space, and cyberspace—to cite only three issue areas. Moreover, to the extent that officials and analysts are able to stay on top of global trends, they also realize that our prescriptions, policies, and strategies tend to lag woefully behind them.

There are at least eight strategic trends shaping both near- and long-term challenges and opportunities:

1. A gradual global redistribution of economic power from the West to the "rest" is under way, and economic power is the bedrock of enduring military and political power.
2. We are on the cusp of but not yet in a multipolar world.
3. The globalization of communications is challenging more than just the virtual foundations of the postmodern information society.
4. Energy and environmental insecurity have reached a tipping point. The era of cheap hydrocarbons and scant ecological regard is finished.
5. Transnational terrorism and stateless actors can inflict unprecedented damage, and we must be on our guard against catastrophic terrorism.
6. September 11 and growing insecurity in Afghanistan and Pakistan remind us of the growing challenge posed by fragile states and "ungoverned" spaces, with the "bottom billion" in some sixty countries left behind in dire poverty.
7. The character of war is changing, forcing low-level uses of force and greater civil-military integration in a renaissance in counterinsurgency and irregular warfare. We must prepare for "hybrid warfare" and hedge against emerging peer competitors.
8. The proliferation of weapons of mass destruction has made it increasingly possible that nuclear or biological weapons may be used in the coming years, including uncontrolled biological agents.

This complex environment poses a potent set of challenges for how the administration of President Barack Obama seeks to exert America's significant yet finite power to safeguard against a diverse set of traditional and modern threats and challenges, while also seizing as many opportunities as possible to build more durable, peaceful, and collaborative solutions for the twenty-first century. In his first months in office, President Obama demonstrated a keen ability to change the basic narrative of the United States, placing it in a far less confrontational stance with most of the world, and showing a willingness to give greater weight to local and multilateral solutions.

There are five pathways to a game-changing strategy for the United States:

1. ***Heal thyself.*** To a remarkable degree, security hinges on America having its house in order.
2. ***Redefine problems.*** Ends should be realistic, recognizing emerging interrelationships among energy, the environment, food, and climate change.
3. ***Surge civilians.*** Complex challenges require a larger whole-of-government team of national security professionals.
4. ***Countermobilize.*** The United States can mobilize emerging power centers into action through bilateral alliances, coalitions of the willing, and multilateral institutions.
5. ***Exercise strategic restraint.*** We cannot afford quagmires that drain resources without providing lasting security. A strong military is the U.S. ace in the hole, but better still are indirect approaches, strategies of leverage, and "smart power."

The need for broader U.S. strategic thinking is obvious. But equally important is the need to mobilize partners, conduct serious planning, integrate a rich variety of disciplines and actors, follow through on implementation, and then assess actions with an appreciation of history. And all of these steps must then, in turn, inform our education and training. There is an obvious case for all-of-government and coalition-based solutions.

The challenges are great, but so are the opportunities. The world is changing, but the United States still has the greatest capacity to cope with these vicissitudes, to lead global responses, and to make the world a safer place.

Michael Klare identifies the key problems that he believes will dominate U.S. foreign policy in the years ahead. Since World War II, American policy makers have been preoccupied with what he calls **the old foreign policy agenda**, which dealt with problems arising from the competitive pursuit of power and the dominance of nation-states within the international system. This agenda was dominated by the "three a's"—allies, adversaries, and armaments—and will continue in the future as long as nation-states vie for power and influence and armed groups pose a security threat.

Nevertheless, he believes increasing attention will be given to the **new foreign policy agenda**: problems arising from transboundary and nonmilitary issues, including global economic disorder, pervasive underdevelopment in the poorest countries, environmental degradation, humanitarian disasters, mass migrations, global pandemics, resource scarcities, and international crime. Given the limits of time, he focuses on the "three e's": energy, the environment, and the world economy.

The economy comes first. The United States is the principal architect and advocate of the liberal economic order so failure of the American model will lead to a decline in American political leadership. And without a healthy world economy, we will lack the resources to deal with the longer list of challenges. Finally, weak or negative economic growth is likely to lead to social unrest, political turmoil, and, potentially, state collapse.

Energy is important because without it the world economy cannot thrive and much of international relations is dominated by imbalances between supply and demand. Global supplies of oil and natural gas have failed to grow in tandem with soaring international demand, much of it coming from expanded economic activity in Asia. Surplus states, including Iran, Russia, and Venezuela, have sought to use their windfall profits to pursue political projects that conflict with U.S. interests. And the vast oil wealth of privileged elites in oil-producing states of Africa, the Middle East, and Asia tends to heighten internal tensions, leading sometimes to ethnic strife, terrorist violence, or separatist conflicts.

The environment will become more important as warming of the planet proceeds. This will pose two challenges: first, addressing the social and political consequences of climate change and, second, engaging in the diplomacy of mitigation.

The three e's are closely interconnected. The world economy cannot function without adequate supplies of affordable energy, and world trade is supported by global logistical networks that are especially dependent on transportation fuels. The consumption of fossil fuels is the leading source of carbon dioxide emissions that are largely responsible for humanity's share of greenhouse gases.

While no clear set of guiding principles has yet to be articulated such as during the Cold War, based on statements from the administration a matrix of principles would include:

(1) *On the global economy*: The United States must exercise leadership within the G-7, the G-8, the G-20, the World Bank and International Monetary Fund, and other such bodies to stabilize the international financial system and promote economic recovery.

(2) *On energy*: The United States must diminish its current dependence on importing petroleum, thereby diminishing the power and influence of oil-exporting countries, and enhance energy security by diversification of supply by types, provider, and means of delivery.

(3) *On the environment*: The United States should play a constructive role in design and implementation of international regimes for the reduction of greenhouse gas emissions.

The three e's will govern America's ties to specific countries, regions, and international bodies. While elements of the old agenda will continue to apply to China, energy and the environment will pose daunting problems. By 2030

China is projected to be the second-biggest importer of oil after the United States. Because of its heavy reliance on coal, China will remain the biggest emitter of carbon dioxide. The best way to deal with this is to engage China diplomatically and to engage in collaborative development of alternative energy sources.

Similarly, while legacy issues of the Cold War still dominate U.S.-Russian relations, energy issues will play an increasingly important role. Russian leaders have sought to harness Russia's massive oil and natural gas reserves and control over pipelines from Central Asia to Western Europe.

In Africa, the new foreign policy agenda is ascendant. Africa is expected to suffer most from the onset of global climate change, producing humanitarian disasters and possibly state collapse. And yet Africa possesses untapped supplies of oil and natural gas that, if developed prudently, could spur development. The three e's will play a pivotal role in Central America, the Caribbean, the Middle East (where water scarcity is a growing concern), Central Asia, and South and Southeast Asia.

Policy makers will need to become familiar with international financial systems, the worldwide energy trade, and the causes and mitigation of global climate change.

Topics discussed included how violent the world will be, how we might prioritize the challenges we face, the elements of a grand strategy, the importance of domestic politics and pressures on foreign policy, especially with economic challenges, and the difficulties of moving from a focus on "a" factors of security to a focus on "e" factors of security.

Panel II: Western Hemisphere

Another Chance for U.S. Policy in the Americas
Peter Hakim, President, Inter-American Dialogue

U.S.–Latin American Relations
Dr. Shannon K. O'Neil, Douglas Dillon Fellow for Latin America Studies, Council on Foreign Relations

The Outlook for U.S. Foreign Policy in Latin America and the Caribbean: The Challenges of Transforming Goodwill into Effective Policy
Ambassador (Ret.) Paul D. Taylor, Senior Strategic Researcher, Naval War College

Moderator: Professor Laurence L. McCabe, Associate Professor of National Security Affairs, Chair Latin America Regional Study Group, Naval War College

Peter Hakim reports that the election of Barack Obama was enthusiastically welcomed throughout Latin America and the Caribbean. Latin Americans want a better relationship with the United States, but they also want Washington to approach the region differently. So far, U.S.–Latin American relations under the Obama administration are off to a good start.

The governments of Latin America are more ambivalent about the role the United States should play in the region, and they are far more independent and assertive than ever before. Brazil is an alternative pole of power with an increasing regional and global profile. Most Latin American countries now have a diversity of international ties. And a few countries, led by Venezuela, have become "adversaries" of the United States.

Hakim provides recommendations to deal with ten critical challenges.

While most nations in the region are today far better prepared to withstand the shocks of the *financial crisis*, the impressive gains made in growth rates, keeping inflation low, building a middle class, and reducing pervasive poverty and inequality could be reversed. The best thing the United States can do is to resolve its own economic problems. In addition, it should reject protectionist measures and mobilize support for increased resources of the IMF and multilateral development banks.

Policy toward Cuba is the issue on which the United States is most out of step with the rest of the region. The administration should start to dismantle restrictions imposed on Cuba. Instead of pursuing a bilateral approach, the United States should stop trying to block other countries and multilateral institutions from doing business with Cuba.

Integration with Mexico presents the toughest challenges and the greatest opportunities. The central challenge is managing the accelerating economic and demographic integration of the two nations. Mexico faces a complex of dangerous security problems, aggravated by the economic recession. The government has waged a fierce military campaign against drug gangs and other organized criminals. The Obama administration should follow through with delivery of promised equipment and aid and intensify efforts to reduce the use of illicit drugs in the United States and the smuggling of weapons to Mexico.

Crime, violence, and drugs beyond Mexico are urgent concerns throughout Latin America and the Caribbean, with many countries seeking U.S. cooperation to deal with problems. Virtually everywhere in the Americas, crime and violence are fueled by illegal drug profits. It is clear U.S. anti-drug efforts are not doing much to reduce either supply or demand. An honest, well-informed, and wide-ranging debate on alternative drug policies is needed.

Reforming U.S. immigration policy is the single most pressing issue for a dozen or more countries in their bilateral relations with the United States. The U.S. immigration system is broken and badly serves U.S. and Latin American interests. Key elements of a new approach include (1) offering sufficient work visas to satisfy U.S. labor market demands, (2) providing legal status (and the opportunity to earn permanent residence and citizenship) to immigrants residing in the United States illegally, and (3) putting in place effective (and humane) incentives and enforcement mechanisms to curb illegal migration.

The *unfinished trade agenda* should be completed, including the ratification of free trade agreements with Colombia and Panama, doing more to mitigate dislocations that free trade can produce, finding ways to implement existing NAFTA provisions without opening the agreement to renegotiation, reinstating trade preferences with Bolivia, and working with Brazil on a negotiating formula for the Doha Round of trade negotiations.

The United States should strengthen *cooperation with Brazil,* whose rapidly escalating regional and global influence represents a pivotal change in inter-American affairs and an encouraging development for the United States. Brazil is a constructive force in hemispheric affairs, leading peacekeeping operations in Haiti and helping to resolve conflicts in South America. Mutual interests overlap in the Doha Round, climate change, environmental protection, and new energy sources.

Hugo Chávez, *the challenge from Venezuela*, has been a polarizing force in inter-American relations since taking office. There is no urgency to engage him, so Washington should keep the Venezuelan leader at a distance and let the situation play itself out.

Democracy is the norm in the Americas today. *Advancing democracy* means more than periodic elections. The fundamental institutions of democracy still perform badly in much of the region. Democratic progress will depend mostly on the governments and citizens of each country. However, the United States should be an advocate for democracy, and this is most effectively achieved multilaterally.

Finally, the new administration has an opportunity to build upon recent inter-American cooperation to establish a long-term, multilateral approach to a *failing Haiti*.

Shannon O'Neil explains that Latin America's strategic, economic, and political importance to the United States is growing because it provides more oil than the Middle East, it is an important source of alternative fuels, it sends more immigrants, it is one of the United States' fastest-growing regional trading partners, it is the largest source of illegal drugs, and nearly all Latin American nations are vibrant, if imperfect, democracies. O'Neil focuses on four main issues: public security, sustainable energy, economic advancement, and hemispheric migration.

Public security is an overriding concern of all governments in the region. In spite of the near absence of cross-border threats, Latin America is the most violent region in the world, with a homicide rate three times the global average. Law enforcement and judicial systems are unable or unwilling to take on powerful criminal organizations. Half of Latin Americans place very little trust in their police and judicial systems. The narcotics trade flourishes in this general atmosphere of impunity and corruption, coupled with difficult socioeconomic conditions and the high demand for narcotics in the United States, Europe, and increasingly in Latin America itself.

The United States should broaden and deepen efforts to assist law enforcement and judicial reform efforts. Long-term solutions will not result from more military or police hardware, but instead must come from the strengthening of law enforcement and judicial institutions, reducing the impunity and corruption that deter citizen involvement and support, and allow crime to thrive. There should be greater focus on vetting, training, and investigation; international cooperation and information sharing; evaluation of programs and practices; and innovative solutions.

The United States needs to improve efforts on its side of the border to combat cartels and organized crime, better control guns, and enforce its own laws. This will require more resources for the Bureau of Alcohol, Tobacco, Firearms, and Explosives. And it means more vigilance on the border, checking vehicles not just coming north but also going south. A second effort should be the strengthening of drug-related anti–money laundering initiatives. Finally, the United States needs to establish a comprehensive drug policy that addresses not only supply but also demand.

Sustainable energy security will come through diversification of energy sources. Canada provides 18 percent of U.S. oil supplies, while Latin America provides 30 percent. Latin America and the Caribbean have the potential to be an important source of natural gas. And Latin America is one of the largest sources of alternative energies. The region will require over $1 trillion in investment in the energy sector over the next twenty-five years in order to meet its own increased energy demand. The United States can help by supplying foreign direct investment financing incentives through multilateral institutions.

Despite recent improvements in economic performance, Latin America lags other developing regions in combating poverty and inequality. Latin America and the Caribbean remains the most unequal region in the world, with 37 percent of the population poor. Income inequalities mirror structural inequalities, particularly in terms of access to health care, education, credit, and economic opportunity. Added to these are the costs of the current economic crisis.

The United States can best help these countries get through the current downturn by first stimulating its own economy. It should go beyond its historically narrow approaches to poverty and inequality alleviation. President Obama's announcement of a Microfinance Growth Fund is an example of a step in the right direction. Our approach to trade agreements should change since existing agreements have not benefited the broader populations. U.S. targeted aid for poverty alleviation has stagnated and needs to be increased. Finally, the United States can assist in the strengthening of public institutions.

Latin America is the largest supplier of U.S. immigrants, legal or not. Immigration has mixed effects for Latin America. U.S.-bound migrants can be a significant portion of a nation's population. They send back nearly $50 billion each year in remittances to Latin America. Yet U.S. opportunities attract the best and the brightest. Meanwhile, the U.S. workforce has become increasingly dependent on Latino workers. The United States needs to negotiate and approve comprehensive immigration reform. And it should pursue policies that promote circular, as opposed to permanent, migration.

Four important bilateral relations will dominate U.S. policy toward the region; two represent strategic partners while two present diplomatic challenges. Relations with Brazil and Mexico are important and strategic. Relations with Cuba offer the opportunity for change, while the relationship with Venezuela is less likely to change.

The biggest shift in relations will result from a change in process rather than substance—valuing ideas from the region and working together in search of multilateral solutions. The best approach will be one designed to enhance partnerships rather than just programs.

Paul Taylor explains that U.S. policy objectives have remained remarkably consistent through changes in parties and administrations with the United States seeking democratic government, good governance, economic growth, and security in the Western Hemisphere. He suggests that despite this broad consensus on policy, it has not been easy for policy makers to formulate and execute policy for four reasons:

- The history of U.S. interventions in Latin America has left a legacy of suspicion that colors the way U.S. actions are perceived and limits the political space in which Latin American leaders can conduct relations with the United States.
- Frequently our possible actions in the region are conditioned profoundly by events in *other* parts of the world.
- Increasingly, events in Latin America involve things that are either not our business or may turn out better if we refrain from engaging on them.
- The instruments needed to achieve many of our objectives and to meet the concerns of Latin Americans themselves often involve agencies in the U.S. government other than the traditional foreign affairs agencies.

Events in other parts of the world affected U.S. policy with respect to immigration reform and a Free Trade Area of the Americas (FTAA). George W. Bush came to office promising to emphasize Latin America. But hope for liberalizing immigration was dashed by 9/11. The U.S. government shifted focus to border security, and the attention of busy policy makers was consumed with the wars in Afghanistan and Iraq. And the U.S. decision to use military force without international approval reawakened our neighbors' worst fears.

Similarly, the failure to come to agreement in the Doha Round of trade negotiations in 2008 put a stop to progress on FTAA negotiations since access to agricultural markets and subsidies to agriculture production were central to both negotiations. The sequence was wrong; without agreement in the Doha Round, the United States could not make concessions in FTAA that countries such as Argentina and Brazil demanded.

There are now events in the region in which the United States is not the central player. The expropriation of foreign investments in Bolivian natural gas affected Brazil as the home country of investors at risk and not the United States. Similarly, the issue of a permanent seat on the United Nations Security Council involves differences of interests among Latin American governments. While Brazil has much in its favor, countries such as Argentina, Chile, and Mexico suggest they would rather have a revolving seat for Latin America.

Some in the United States are bothered by what they see as a worrisome increase in Chinese involvement in the region. China has increased its trade and foreign investments in Latin America and the Caribbean. Yet such purchases have fueled above-average economic growth, an objective that has enjoyed bipartisan support in Washington. Clearly, the United States has large economic relations with China, so on what basis can others be denied economic dealings with China?

Drug-related violence in Mexico is an important issue in which domestic agencies have the lead. President Obama has usefully acknowledged that the United States bears some responsibility, both because U.S. demand for drugs energizes the industry and because many of the powerful guns used in the drug wars were bought in the United States and smuggled south. Traditional foreign affairs agencies have little authority to act against drug demand or control guns, so the task falls to agencies that do not have international relations in their core missions.

The initiative of the U.S. Southern Command to bring representatives of civilian agencies into the organization is a well-intended effort to improve coordination of U.S. activities. It must be exercised with caution, to keep from putting a military face on U.S. civilian activities. A better approach would have the chief of plans and strategy of the Southern Command resident in Washington and assigned concurrently as a deputy assistant secretary of state in the Bureau of Western Hemisphere Affairs.

Fixing the American economy, as Brazil's president Lula stated, is the best thing the United States can do for the region. The negative effects of the global crisis will set back progress on multiple fronts throughout the region. And many of the above constraints will intensify during economic recession, making it more difficult to consider issues such as immigration reform and trade agreements.

The best path is to engage our Latin American and Caribbean neighbors in a dialogue on how to manage the crisis, not in the donor-recipient relationship that has so often characterized the past, but as partners searching for ways to mitigate and contain the costs of the economic crisis. If we can do that, the process will be part of the solution.

Topics discussed included the role of the U.S. military in the Western Hemisphere, strategic intentions of rising powers such as Brazil and Mexico, the effectiveness of the "war on drugs," the role of China in the hemisphere, and U.S. relations with its neighbors to the south.

Panel III: Asia and the Pacific

U.S. Asia-Pacific Strategy in the Obama Administration
Dr. Jonathan D. Pollack, Professor of Asian and Pacific Studies, Chair Asia-Pacific Study Group, Naval War College

The New Security Drama in East Asia: The Responses of U.S. Allies and Security Partners to China's Rise
Dr. Evan S. Medeiros, Senior Political Scientist, RAND Corporation

U.S. Foreign Policy and Southeast Asia: From Manifest Destiny to Shared Destiny
Dr. Emrys Chew, S. Rajaratnam School of International Studies, Singapore

Moderator: Dr. John F. Garofano, Professor of Strategy and Policy, Jerome Levy Chair of Economic Geography and National Security, Naval War College

Jonathan Pollack outlines some prospective choices in the Obama administration's Asia-Pacific strategy. The United States has a clear need to refashion U.S. strategy in a manner that reflects and responds to the Asia-Pacific's dynamism and success. The Asia-Pacific region is more economically vibrant, more politically self-confident, and more militarily robust than at any time in its modern history. Though many states face daunting political, economic, and societal challenges, and numerous states express unease about future power configurations, the region's successes far outweigh its failures and uncertainties. However, there is neither clarity nor closure on the contours of a reconfigured regional order, in regional expectations of the United States, or on long-term U.S. strategy.

Beneath a veneer of regional cooperation and multilateral institution building, the states of Asia and the Pacific continue to enhance their absolute and relative power; strategic trust and longer-term collaborative habits and practices remain a scarce commodity, especially in Northeast Asia. There is also as yet no discernible power equilibrium among Asia's major powers. China, India, and Japan are all seeking to enhance their economic, political, and military weight and diversify their strategic options, even as all hope to build durable relationships with the United States in light of their respective strategic circumstances. Russia also seeks to reassert its claims to major-power status in the region. In addition, divergent national interests and the persistence of historically rooted conflicts continue to complicate the building of a coordinated, consensual strategy. Immediate as well as long-term political and security issues also intrude on strategic calculations. North Korea's nuclear weapons development, for example, is a major regional concern as well as a pressing issue for the future of the nonproliferation regime that cannot be addressed by the United States alone.

The Obama administration seeks to define future U.S. leadership in terms of enhanced international engagement without a threat-dominant major-power rationale, and Asia and the Pacific will be central to this equation. Even as many U.S. policies remain predominantly bilateral in design, a longer-term regional strategy must incorporate China as a full participant at the table. This process must be a two-way street. Long-term stability will require China to fully articulate its security perceptions and expectations, and to mesh Chinese power with larger collaborative ends. Regional powers must also build a deeper understanding of how competing national-level goals, interests, and capabilities intersect and interact.

The challenge and opportunity confronting the United States and the emergent powers of the Asia-Pacific region is to vest regional states (especially the major powers) in an inclusive, future-oriented order. Though bilateral relations will remain a cornerstone of American strategy, they must be embedded in a larger concept of the region's long-term future.

Regional states largely fit into three broad categories: autonomous major powers, long-time allies pursuing an enhanced national identity while seeking assurance from the United States, and local actors prepared to collaborate with the United States and to facilitate complementary political and security goals. (North Korea and Burma remain outside these categories.) In oversimplified terms, the United States can pursue (1) inclusion and integration, (2) prevention and inhibition, or (3) preservation and hedging. These correspond to three impulses underlying these alternative possibilities: an opportunity or incentive-based approach, a threat-driven approach, or an uncertainty-based approach.

The United States needs to weigh much more fully the opportunities and possibilities of Asia's ongoing transformation in conjunction with relevant regional states. Such an approach would entail at least six principal security goals:

- Preventing a strategic breakdown or major regional crisis (e.g., conflicts in Korea, in the Taiwan Strait, or between India and Pakistan);
- Enhancing communication related to potential contingencies that could involve multiple powers, with Korea as the preeminent example;
- Achieving sustainable alliance bargains that move beyond traditional approaches and would entail more meaningful responsibility sharing;
- Simultaneously achieving durable relationships with China and Japan, while facilitating and sustaining a longer-term strategic accommodation between Beijing and Tokyo;
- Undertaking a far more extensive set of exchanges between the U.S. and Chinese militaries, with particular emphasis on China assuming more of a "stakeholder" role in international security;
- Pursuing international arrangements where the United States has either been skittish or oppositional in the past (e.g., accession to the ASEAN Treaty of Amity and Cooperation, and ratification of the UN Convention on the Law of the Sea).

Lasting American engagement is both necessary and possible. Without a concurrent approach that vests the United States and regional states in a compatible vision of the longer-term future, neither America nor the region can expect to ensure mutual security and well-being on which the vital interests of all will depend.

Evan Medeiros analyzes the responses of U.S. allies and security partners to China's rise. China's rise is affecting the perceptions, interests, and policies of all nations throughout East Asia.

To understand and evaluate these evolving dynamics, the RAND Corporation conducted a study of responses of the five U.S. allies in the Asia-Pacific (Japan, South Korea, Thailand, the Philippines, and Australia) and of Singapore, a major security partner. The study sought to answer the following questions: How have these nations responded to China? What forces are driving these reactions? How will the drivers change? What are the implications for American security interests? Responses were analyzed in four areas: domestic politics and public opinion, economic policy, foreign policy, and defense policy.

In contrast to much current research, China's growing presence and interactions with U.S. allies and security partners are not fundamentally transforming the security order in the Asia-Pacific. China is having an influence on these relationships, but these changes are not as rapid or comprehensive as many presume.

First, the foundation of the U.S. alliances in Asia continues to endure. No allies or major security partners see China as a viable strategic alternative to the United States. The United States remains the security partner of choice, largely because it is the one nation seen as possessing the capability and resolve to balance China. Its allies and partners prefer that Washington do the "heavy lifting" of deterring China and, ultimately, preventing Chinese domination of regional affairs. U.S. allies are all intensely pursuing engagement strategies with China, driven principally by an economic logic. They want to benefit from China's large and growing economy. But these goals exist alongside concerns about China's long-term intentions, particularly its military modernization plans.

Second, China is affecting American relationships with its allies and security partners. On the one hand, China's rise makes some U.S. security commitments more relevant. These countries can interact with China more confidently because they know (and Chinese leaders see) that the U.S. commitments to them and to involvement in Asia continue. On the other hand, allies and partners are also positioning themselves to benefit from both the United States and China. This is a recalibration more than a transformation. None of these countries want to choose between the United States and China, and all reject having to make such a choice.

Third, China is undoubtedly gaining influence with U.S. allies and partners—in the defined sense of looming larger in their economic, diplomatic, and defense policies decisions. The key question is how it is manifesting itself in these states' regional behaviors. U.S. allies and partners have become more sensitive to some of China's preferences and interests, especially on China's self-identified "core interests," which now include Taiwan and Tibet.

A related indicator of Chinese influence is that Beijing has been effective at precluding the emergence of "anti-China" containment efforts. China has been effective at accumulating "defensive influence," persuading nations to avoid taking actions China deems to be threatening. There is very little evidence that China has accumulated "offensive influence," in the sense of policies that could effectively degrade or dismantle U.S. alliances or security partnerships.

China's growing presence and interactions do not directly translate into influence. The regional consensus favoring engagement with China has a tentative quality. There is a creeping uncertainty about China's future: some nations fear a weak China, and some fear a strong China. Few are willing to bet their futures on Beijing's assurances about a "peaceful rise." China's large and growing economy is not a geopolitical "tractor beam." And there are nagging concerns about Chinese military modernization. However, there has been a lack of a regional rush, over the last decade, to increase military budgets and modernize conventional forces in response to concerns about China's military.

Medeiros concludes, first, that the United States remains well positioned to achieve its long-standing regional objectives. The United States does not face a crisis of confidence, and the foundations of its influence will endure. The United States needs to improve the legitimacy of its role and the credibility of its commitments in the Asia-Pacific. That effort will require an adaptation to the changing constellation of the equities of U.S. allies and security partners. None want to provoke China or be drawn into a containment effort; none want China to dominate the region; none want the United States to leave or even substantially draw down its presence; and all want China to play a major role in managing regional challenges.

Second, there was no strong correlation between high levels of economic integration with China and accommodation with it. While Japan, Singapore, and Australia all have significant economic relations with China, this is not reflected in their foreign and security policy making in any direct manner.

Finally, domestic politics matters a lot in determining nations' responses to China's rise. For most East Asian states, China's rise generates a variety of contradictory reactions.

Emrys Chew provides a historical study of the distinctive phases and emphases of U.S. foreign policy in Southeast Asia, as well as evolving Southeast Asian perspectives on U.S. foreign policy.

From postcolonial state to global superpower, America's relations with Southeast Asia—as with the rest of the world—have been driven by a peculiar sense of "manifest destiny." Founded upon such transcendent values as "life, liberty, and the pursuit of happiness," the United States as champion of those values in the world has, time and again, rightly or wrongly, made a case for American exceptionalism if not interventionism.

The United States has cast itself uniquely as champion of a new world order built upon universal values of self-determination and human rights. Throughout the history of U.S. foreign policy, however, such notions as American exceptionalism have been manifested unevenly in terms of both the power of America's example and the example of America's power. Woven into U.S. foreign policy tradition are almost contradictory, alternating strands of unilateralism and universalism, liberal as well as fundamentally conservative values, where ideals and national interests intertwine but have not always complemented one another; the Republican administrations have tended to place greater emphasis on military-strategic interests, while Democratic administrations have emphasized human rights.

Chew illustrates by discussing three main phases of U.S. foreign policy in Southeast Asia. First, a period of early adventurism and expansion (1776–1946), culminating in America's colonial policy in the Philippines during the era of Western imperialism, followed by the end of that colonial experiment through America's promotion of national self-determination in the era of world wars and decolonization. Second, a period of anticommunism and ambiguity (1946–1989), where America's containment policy during the Cold War was marked by a certain ambivalence in its support of authoritarian regimes while proclaiming liberal-democratic values in the bid to counter the communists. This period came to be dominated by the imperatives of the "domino" theory and "quagmire" thesis, as Southeast Asia became a critical frontier and the United States was increasingly bogged down by the military-strategic commitments in Vietnam. Finally, a period of unparalleled authority mingled with uncertainty (1989–2009), in which America's post–Cold War global hegemony was challenged in such a manner as to require post-9/11 counterterrorism strategies dealing with the Islamic extremist threat. Southeast Asia, as home to the largest concentration of Muslims in the world, became a crucial frontier once again in America's military-strategic calculations.

Analysts have subdivided the region into three categories. First, nations engaging with China but still placing greater emphasis and faith in their long-term strategic relations with America: Philippines and Singapore. Second, nations charting a middle course between America and China, mainly due to geographical distance from China and unease over pursuing closer strategic relations with America: Indonesia and Malaysia. Third, nations whose security strategies are dominated by their proximity to China: Burma, Thailand, Vietnam, Laos, and Cambodia.

Given the vast economic opportunities, and the need to safeguard the flows as well as fruits of trade and investment in an age of global interdependence, ASEAN countries see the way forward in terms of both regionalism and multilateralism. In ASEAN's view, an expanded, reinforced regional architecture that engages and enmeshes both China and the United States can only be a positive, constructive development.

Whatever their differences of perspective, the nations of Southeast Asia would all prefer a greater measure of clarity and consistency in U.S. foreign policy: less prescriptive, more sensitive. While proclaiming the virtues of liberty and democracy to parts of Southeast Asia, America would do well to remember its patchy historical record of supporting right-wing dictatorships. While championing its notions of good governance and human rights, America could display deeper cross-cultural sensitivity and patience, promoting more constructive diplomatic and developmental approaches over military solutions or economic sanctions.

In conducting its "war on terror," the United States again proved inconsistent with its own principles in dealing with terrorist suspects and political detainees. Vindicating the dreams of the founding fathers at long last, it has taken the almost ironic election of an exceptional man of color to the White House to restore some measure of hope in the promise of America for the rest of the world: inaugurating a new era of internationalism—both responsible and responsive—in which the United States pledges to listen more than dictate; dismantling Guantánamo while engaging with others—especially the Muslim world—on the basis of "mutual interest and mutual respect." There is a sense that manifest destiny has given way to shared destiny.

Finally, the nations of Southeast Asia would prefer a greater degree of commitment and compromise in U.S. foreign policy: less unilateralist, more multilateralist. The United States would do well to commit itself to achieving its foreign policy agenda in partnership with multilateral institutions in the region, paying more attention to the regional agenda rather than resorting to "coalitions of the willing."

Topics discussed included great-power politics; how the United States should engage China; what kind of regional framework would aid in dealing with North Korea; Russia, Japan, Australia's defense policies; and specific U.S. policy options.

Panel IV: South Asia

A U.S. Strategy for Pakistan: Future Directions
Dr. Daniel Markey, Senior Fellow for India, Pakistan, and South Asia, Council on Foreign Relations

India and the United States: Making the Partnership Strategic
Ambassador (Ret.) Teresita C. Schaffer, Director, South Asia Program, Center for Strategic and International Studies

Moderator: Dr. Timothy D. Hoyt, Professor of Strategy and Policy, Co-chair Indian Ocean Regional Study Group, Naval War College

Dan Markey proposes a new U.S. strategy for Pakistan. U.S. interests will be best served by recognizing (1) that Pakistan, not Afghanistan, poses the paramount challenge to American security; (2) that building a strong partnership with Pakistan while working to transform perceptions of the strategic environment in South Asia holds the greatest potential for sustainable U.S. security; and (3) that even under the best of circumstances, success in the region will take a long time and may prove extremely costly.

Pakistan has a much larger population than Afghanistan. By nearly all accounts, Taliban and al Qaeda leadership find sanctuary in Pakistan, and Pakistan's security apparatus has long supported domestic Islamist militant groups as an asymmetric means to achieve strategic equilibrium with India. It has a history of alternating authoritarian military and largely ineffectual and corrupt civilian rule. Hollowed-out state institutions, inadequate civilian control over the military, and an unsettled ideological debate about its own strategic interests leave Pakistan unable, and perhaps unwilling, to fulfill Washington's expectations of it as a partner against Islamist militancy. Pakistan's internal instability and the geographic proximity of al Qaeda to Pakistan's nuclear arsenal drastically raise the strategic stakes.

Markey explains why two credible alternative strategies—containment and coercion—are unlikely to succeed over the long term. Instead, he argues that a better long-term strategy to advance U.S. goals in Pakistan should be based on the twin pillars of induced bilateral partnership and the reshaping of Pakistani perceptions of the regional security environment. Washington should seek to induce, rather than coerce, allies and partners within Pakistan's civilian political leadership, military, and wider public. American security and development assistance should be used to strengthen these elements of Pakistani society. Too often lost in the discussion is the fact that the Taliban and other extremist groups do not represent the goals and aspirations of the people, the vast majority of whom prefer to live in a peaceful, prosperous, and moderate country. Washington should therefore pursue policies that will be perceived in Pakistan as supporting these goals. Along these lines, the United States should help to build a stronger Pakistani civilian administration, capable of delivering law and order; enhance the educational and economic prospects for millions of young Pakistanis; and train and equip effective counterinsurgency forces within the military. The United States should patiently but persistently seek the most efficient means to deliver its assistance, as waste and corruption will undermine trust on both sides. On the civilian side, one way to improve transparency and accountability would be to create a multilateral trust fund, possibly administered by the World Bank, which could work with the United States, other donors, Pakistan's government, and nongovernmental groups to identify, formulate, and implement effective assistance.

As the second pillar of its effort, the United States should work to reshape the strategic environment as understood by the Pakistanis. U.S. assistance, military operations, and diplomacy should be employed to create new incentives that will convince fence-sitters within Pakistan's political and military leadership of the benefits of working with the United States, and the costs inherent to opposing American efforts. Washington should work to create conditions that diminish the political and military uncertainty along Pakistan's borders. Washington should make a clear commitment to regional stability, demonstrating that militancy will be defeated decisively in Afghanistan and lending under-the-radar support to a normalization of relations between India and Pakistan. The United States should devote greater, sustained resources to the fight in Afghanistan, thus eliminating the incentives for Pakistan to hedge its bets and support Taliban and other anti-Kabul factions. At the same time, the United States should approach New Delhi and Islamabad to encourage a return to dialogue and normalization of Indo-Pak relations.

These steps represent only the very beginning of a much longer, exceedingly complicated, and costly process. This process should begin quickly. The many forces undermining Pakistan's stability are now ascendant if they are not yet dominant. The United States can reverse this momentum, not by seeking to contain the threat from a distance or by leveling coercive threats, but by cultivating and empowering a wide range of Pakistani partners.

Teresita Schaffer describes how the United States and India have transformed their relationship into a serious bilateral engagement based on a growing array of common strategic and economic interests. She argues that the partnership is serious; that it is not yet strategic; and that in order to become strategic, it needs to be reinvented.

The new Indian foreign policy is based upon a new calculus of India's power with its economic power central, the United States becoming India's most important external friend, relations with China and East Asia becoming much more important, and Indians remaining committed to strategic autonomy—the concept that India must neither allow anyone else to dictate its foreign policy, nor permit such an impression to be created.

India's national security policy is based upon preeminence in South Asia, countering any major threat that intrudes into this space, and deterring major threats from beyond. The inner ring of its security environment is a dangerous place, with the greatest security challenges stemming from internal insurgencies connected to problems in their neighbors' territories, with Pakistan being the prime example. The outer circle of its security environment extends from the Middle East to Malacca with India's goal being to protect its lifelines for trade, investment, and energy. Here maritime and littoral security is supported by naval and air power. India sees the Indian Ocean as a single strategic environment with its navy as the dominant littoral force, and India does not want to see its primacy challenged. Unlike in the past, India now sees U.S. presence in the Indian Ocean as benign. Finally, India's long-term challenge to the east is China, still regarded as a strategic rival, but India does hope to build a peaceful and profitable relationship.

Schaffer explains the successful building of bilateral relations between the United States and India. The achievements include a very dramatic expansion of economic ties resulting from India's higher economic growth and greater integration with the global economy. Military cooperation is almost unrecognizable compared to the 1990s. Given common interests, there has been a sharp increase in joint exercises. Finally, the most dramatic accomplishment was the Indian-U.S. agreement on civil nuclear cooperation. Thus, these three areas—economic ties, defense relations, and nuclear and high-tech trade—represent extraordinary achievements that transform the political landscape between India and the United States.

The building of this relationship did relatively little, however, to move the partnership on to a regional or global stage. For this to happen, both nations will have to address the question of what kind of partnership they would like to have and can sustain. The time is ripe for the partnership to go global. Both nations should begin dialogue on global issues such as global warming, nonproliferation, and financial reform. Three rules should govern: candor, inclusiveness, and no surprises. The end result will not be an alliance nor a seamless agreement, but rather an expanding agenda of selective cooperation, and a relationship that can drive the increasingly important Asian balance of power.

Topics discussed included U.S.-Pakistani relations, the challenges of partnering with Pakistan, elements of a U.S. strategy toward Pakistan, and evolving U.S.-Indian relations and future prospects.

Panel V: The Greater Middle East

New Directions for U.S. Foreign Policy in the Greater Middle East
Ambassador (Ret.) David C. Litt, Executive Director, Center for Stabilization and Economic Reconstruction, Institute for Defense Business

Engaging the Muslim World beyond al Qaeda
Dr. Marc Lynch, Associate Professor of Political Science and International Affairs, The Elliot School of International Affairs, George Washington University

Why the United States Should Engage Islamists
Dr. Heidi E. Lane, Associate Professor of Strategy and Policy, Greater Middle East Area of Study Coordinator, Naval War College

Moderator: Dr. Hayat Alvi-Aziz, Associate Professor of National Security Affairs, Naval War College

David Litt argues for both continuity and significant change in U.S. foreign policy in the Greater Middle East: continuity in addressing regional conflict (albeit more effectively) and change in the attention and resources the U.S. government devotes to diplomacy, especially public diplomacy, and socioeconomic development. We must rebuild America's image and reputation. We must shift the center of gravity of the instruments of national power away from nearly exclusive reliance on the military toward enhancing our civilian prowess in foreign affairs, especially diplomacy and development.

Rebuilding America's credibility requires first and foremost changes in policies and attitudes on the part of the government, but intensive public diplomacy throughout the region must accompany those changes. Public diplomacy begins with the president, who has already launched the campaign calling on the Muslim world to give the United States another look. Other senior officials must likewise inform regional audiences on a regular and sustained basis that a new approach is under way. The messages must be clear and consistent. The U.S. government respects Islam and the peoples and cultures of the Middle East. We will listen more and we will not view "success" only in terms we define. We will be—and be perceived as—a wise and generous nation that uses all instruments of its national power judiciously and to good effect. Moreover, we should eschew the reputation of a go-it-alone Goliath. Only collaborative efforts of many nations can resolve the tribulations of the Greater Middle East.

No improvement in our image or our relationships will gloss over ineffectual policies. The administration's next immediate challenges will involve the simmering conflicts, principally (but not exclusively) Iraq, Iran, and the Israeli-Palestinian conflict. Iraq's reconciliation process is underdeveloped at best. True reconciliation must be deeply rooted, and take place above the level of governmental or political party leadership, involving the nation's ethnic and sectarian leaderships on a societal level. American diplomacy and efforts to support good governance, transparency, and accountability in Iraq must not diminish. The State Department and USAID run important programs to build capacity in the Iraqi government at both the national and local levels. The perennial struggle to build a just, comprehensive, and durable peace between Israel and all of its Arab neighbors continues—and that is the good news. The bad news is that sustainable progress seems even more elusive today. The appointment of former senator George Mitchell as special envoy for the Middle East went a long way to enhancing the government's credibility. Finally,

America will wrestle with the conundrum of Iran. Even before the flawed election in June 2009, the U.S. administration, with a few deft strokes, had altered the political landscape for shaping America's relations with Iran. The United States has yet to announce, however, any new strategic policy toward Iran, preferring to watch and wait. Nevertheless, it is not likely that the overall goals—especially with respect to independent Iranian control over the nuclear fuel cycle and its pursuit of a nuclear weapons program—will change in any substantial way. The path to persuading Iran diplomatically to make the best policy choices with regard to a peaceful nuclear energy program lies in unambiguous international unanimity over those choices.

America is not prepared to undertake these missions given its current foreign affairs infrastructure. We have significantly cut the corps of U.S. Foreign Service diplomats and development officers over the past two decades. We have an opportunity to resource foreign policy properly, by ramping up and retooling the civilian side of our capabilities, while resetting and reorienting military power into its appropriate channels. Even with the passage of the needed authorizations and appropriations, much time and effort will be required to retool and redevelop the experience and expertise that we will need in the State Department and USAID. A preferred outcome would be an interagency determination of what current and future skills are needed, and where they should reside. Our newly bulked-up cadre of skilled international affairs officers across many agencies must then receive the education and training to perform effectively in the *comprehensive* approach to foreign policy issues.

In sum, we must continue our intense pursuit of peaceful resolution of the critical, "front-page" issues, especially the Israeli-Palestinian dispute, Iraq, and Iran. However, we must also recognize that viewing the region only as a function of these specific issues will ignore the many sources and drivers of conflict there. A "crisis-focused" approach overlooks the broader impediments to economic, social, and political development that will sustain stability, growth, and global engagement for the region. The best way for the United States to contribute to this latter vision is to recruit, hire, and train the experts needed to perform inherently governmental functions in foreign affairs; to re-create the ground-level, person-to-person programs that enhance America's credibility; and to integrate all instruments of national power in effective pursuit of the U.S. national interest.

Marc Lynch states that President Obama's call in Turkey for an engagement with the Muslim world beyond al Qaeda offers the prospect for a dramatic and long-overdue shift in the American approach to the "war of ideas" that since 9/11 has occupied a central place in American national security policy. But the nature, objective, and appropriate means of this "war of ideas" remain ill-defined and poorly conceptualized. The Bush administration came to define the "war of ideas" primarily in terms of counterterrorism, counter-radicalization, and "combating violent extremism." While this does indeed represent an important component of any serious approach to the problem of al Qaeda, it represents a dangerously narrow focus for American engagement with the Islamic world. The focus on violent extremism privileges and reinforces al Qaeda's conception of the nature of the confrontation, ironically when al Qaeda is weaker than it has ever been as a political force in the Arab world. Al Qaeda should be marginalized, recognized for the radical fringe movement that it is, and not allowed to dominate our vital dialogue with the mainstream of the Arab and Muslim worlds.

While al Qaeda is relatively weak today, the spirit of "resistance" to American hegemony—a mass-based, political resistance rather than a fringe, religious radicalism—is strong and rising. The focus of U.S. engagement must be to reframe and transcend the binary oppositions that fuel the appeal of the advocates of resistance. That means focusing far less on al Qaeda or upon grand ideological rhetoric, and more on the practical issues related to core diplomatic agendas: building broad support for American foreign policy goals, establishing long-term foundations of trust and mutual respect, supporting engagement with potential adversaries, and addressing the political issues that provide sustenance for the rhetoric of resistance. The key is to disaggregate rather than aggregate, to split the problem rather than lump it together into a single threat, and to deny the adversary the advantage of being viewed as the primary alternative to the United States.

There are three different conceptions of the "war of ideas," all of which move beyond the traditional conception of public diplomacy as explaining American policy and values to foreign publics. The first, which emerged powerfully in the later portion of the Bush administration, is rooted in counterterrorism, and involves a narrow campaign to marginalize al Qaeda and delegitimize violent extremism. The second is the vastly more ambitious campaign to spread liberal values through the Islamic world, bringing about fundamental changes in Arab and Muslim political cultures and promoting Western civilization. Third, and by far the most important, is broad-based engagement across the mainstream of Arab and Muslim political society with the goal of explaining American policies, building support where possible while building networks and relationships of mutual respect. This should involve sustained and productive dialogue with those with whom we disagree, whether states or publics, secularists or Islamists. It also requires serious changes in policy to give substance to the dialogues—full engagement on the Israeli-Palestinian conflict, careful management of the withdrawal from Iraq, and openings to Syria, Iran, and others.

The dominant approach to the war of ideas after 9/11 took the form of "lumping": conceptualizing the Islamist threat as a single, undifferentiated challenge. This approach squanders the opportunities to divide and conquer, obscures very real and crucially significant differences among movements, magnifies the challenges, and unintentionally strengthens the hand of our most radical adversaries. The second approach, which has gained increasing acceptance, might be called "splitting"—seeing the internal differences among Islamist groups as extremely significant both analytically and politically. The Iraq experience showed graphically the tactical value of careful exploration of the lines of division within Islamist movements. Years of undifferentiated warfare against an insurgency seen as monolithic only strengthened that insurgency, while the decision to work with the "Awakenings" and to cooperate with "former" insurgents proved far more effective.

How can Obama defuse the "resistance" discourse and genuinely transform the political contours of America's engagement with the Middle East? To change attitudes will require a new form of engagement that adopts a genuinely different approach, as outlined in this paper. It is here where the Obama administration has made a strong beginning: high-level engagement on the Israeli-Palestinian peace process, announcement of withdrawal from Iraq, direct outreach to the Islamic world, the closure of Guantánamo, and renunciation of torture.

Heidi Lane provides five good reasons why it is time for the United States to launch an initiative aimed at direct and focused engagement with Islamist organizations. Although such a proposition may seem counterintuitive and at odds with some aspects of our national security policies, opening channels of dialogue with Islamist organizations would have far-reaching benefits for U.S. objectives in the Middle East region.

The first reason is that the entrance of the new American administration onto the world stage has produced both real and imagined "windows of opportunity" for reshaping the image of the United States, especially in areas of "soft power" such as public diplomacy. Both at home and abroad, there exists a hopeful expectation that the new American administration will make a clear break with the policies carried out under the former administration.

The second reason that we should engage Islamist organizations stems from the fact that the American public is better prepared to take this step than ever before. The American public has struggled mightily (along with our government) to grasp what separates those who advocate political Islam from those who advocate terrorism, where the term "moderate" begins and ends, and how societal ills such as poverty, lack of political rights, and failed states are believed to contribute directly to the growth of violence carried out under the banner of religious extremism.

The third reason that the United States should engage Islamists has to do with the political attitudes and perceptions of those who make up the proverbial "Arab street." President Obama and the new administration have already made a valuable initial investment in recasting America's image in the region. This was evident in the president's June 2009 speech in Cairo, but such efforts must be ongoing if they are ultimately to make a difference. The United States could continue this positive trend by publicly addressing the failure of its democratization policies over the past decade. The regrettable outcome for the United States is that the policies associated with the war on terror paradoxically undermined those of the Freedom Agenda.

The fourth reason that the United States should engage Islamists is because the rapid growth and evolution of Arab satellite television now offer a public and transparent medium that is viewed as increasingly credible by its broad and diverse consumer base. Any program of public diplomacy must take into account this meaningful change in the way that information is disseminated in the region.

The fifth reason that the United States should engage Islamists is because both moderate and radical Islamist organizations have begun to show signs of failure. Moderates have seldom succeeded in breaking into formal political circles through "playing by the rules" of their respective governments. For the most liberally oriented Islamist organizations, this means that they lose disenchanted members of their prospective support bases to other, often more radical organizations.

On the other side of the spectrum are radical Islamist organizations. Contrary to popular myth and propaganda disseminated widely by regional governments themselves, most of these groups do not possess the popular support to lead a revolution against or undertake the violent overthrow of their respective regimes. Most would be unprepared to govern even if they could somehow grab hold of the reins of power.

That leaves those Islamists and their supporters who are somewhere in the middle between participating and excluded, moderate and radical, manageable and unmanageable. It is precisely this opaque center that makes up a growing and dynamic, but unknown, percentage of Islamist activism. They guarantee themselves popular support in places where government services, public order and security, and other basic necessities are denied or become contested. It is in precisely these environments, such as Iraq and Pakistan, where Islamist successes are based on one part "popularity" and one part coercion and fear. In these "tipping states" a proactive program of public diplomacy by the United States may prove a most valuable investment.

One important caveat must be heavily emphasized here: This program of engagement with Islamist organizations should never be interpreted as a change in the fundamental principles that the United States has upheld for itself. It should be made abundantly clear to our own public and others that the United States continues to view terrorism and the use of indiscriminate violence against civilians (no matter what the political cause) as abhorrent.

The United States should pioneer a program of direct and focused engagement with Islamists as an outgrowth of a new commitment to "soft power"—investing in the future of the region by engaging Islamists entails less risk than does allowing this opportunity to slip away for yet another generation.

Topics discussed included strengthening U.S. public diplomacy in the region, strategic relationships in the region and the need for collaborative efforts, resolving the Arab-Israeli conflict, stabilizing Iraq and Afghanistan, and dealing with Iran and its nuclear ambitions.

Panel VI: Europe and Russia

Advancing a Strategy for Constructive Security Engagement: "Resetting" the U.S./NATO Approach toward Russia

Dr. Sharyl Cross, Professor and Director of Studies, Program in Advanced Security Studies, George C. Marshall European Center for Security Studies

Resetting U.S.-Russian Relations
Dr. R. Craig Nation, Professor of Strategy and Director, Eurasian Studies, U.S. Army War College

Moderator: Ambassador (Ret.) Mary Ann Peters, Provost, Naval War College

Sharyl Cross proposes a strategy for constructive security engagement to "reset" the U.S./NATO approach toward Russia. The United States and NATO will undoubtedly require Russia's cooperation for meeting the priority transnational security challenges of the twenty-first century. The Obama administration must not only foster an improved climate with Moscow, but also manage the often divergent perspectives among NATO member nations. At least for the foreseeable future, the United States and NATO will encounter significant obstacles attempting to build a relationship with the Russian Federation on the basis of common values. The intention of the new U.S. administration to "reset" the relationship with Russia should not be dismissed as a "naïve hope," but as recognition that the United States and Russia share vital interests. We should promote cooperation in areas of common interests, and offer a unified Western voice in seeking to resolve differences on issues where serious conflicts of values/interests emerge.

Russia's cooperation is needed for almost every priority security concern for the United States and NATO nations. Both the Obama and Medvedev administrations recognize continued progress in arms control and counterproliferation of weapons of mass destruction (WMD) requires cooperation. Russia's support is crucial in preventing the advancement of nuclear programs in Iran and North Korea, for strengthening the Nuclear Non-proliferation Treaty, and arresting further proliferation. Both nations share varying commitment to countering the ambitions of al Qaeda and its affiliates, and recognize the threat posed by the potential access of terrorist groups to WMD. The United States, NATO, and Russia share the common objective to prevent the Taliban from returning to power in Afghanistan, and they share interests in ensuring the stability of nuclear-armed Pakistan. The United States/NATO and Russia would all stand to benefit by fostering stable energy regimes providing reliable access and markets for Europe/Eurasia. Finally, ensuring reliable communication channels and clear understanding of objectives on the part of all actors is essential for avoiding misperception that holds the potential for escalating into a crisis situation between Russia and the West.

The United States and Euro-Atlantic community should work cooperatively to forge a constructive, consistent, and coherent strategy toward Russia on the basis of long-term strategic vision. Greater attention must be devoted to fostering transatlantic unity in building strategy toward Russia. Over the past decade, Putin and Medvedev have sought to divide the United States and European nations. A resurgent Cold War is not inevitable. The United States/NATO and Russia recognize the imperative need for cooperation. Reversing the deterioration in Russia's relationship with the West must be among the highest priorities for the new U.S. administration. This will require sustained commitment and patience, and it will not be easy.

Bold new initiatives from the West will be needed. The United States and European partners must demonstrate a willingness to work with Russia on equal terms. Resuming the NATO-Russia Council and seeking renewed military-to-military ties should be first priority. Medvedev has proposed a new European security architecture to encompass Vancouver to Vladivostok. While the suggestion to replace NATO is unrealistic, the current system is obviously not working from Moscow's perspective, and Western nations should be willing to discuss these objections and explore viable alternatives. With respect to NATO enlargement, the actual prospects for early NATO membership of Ukraine and Georgia are unlikely, though eventual membership should remain open. Instead of fueling the prevailing perception among several post-Soviet nations that they must choose between either Russia or the West, the United States/European nations should attempt to cultivate a more constructive climate emphasizing the importance of contributions that both Russia and the West can bring to these societies.

A logical area to "reset" the U.S.-Russian relationship is in arms control and weapons proliferation. The initial steps taken by the presidents of the United States and Russia in April 2009 to resume cooperation in arms control are encouraging. Another more promising area is the issue of missile defense.

The United States/NATO will need to engage the Russian leadership more effectively in the energy security area. Charges that Russia monopolizes energy supply lines for political purposes must be addressed, and developing a sustainable energy security regime for Europe/Eurasia will be important for the coming decades. Similar discussions should be initiated with Russia to tackle the transnational threats of terrorism, crime, drug trade, human trafficking, piracy, climate change, poverty, disease/pandemics, and other issues.

The infusion of expertise, technology, investment, and human support provided by the Western private sector will continue to be important to Russia. The United States and European partners should support Russia's progress toward membership in the World Trade Organization and remove outdated trade restrictions.

Finally, while it is difficult to move beyond the stereotypes of the Cold War, these perceptual legacies must be overcome in order to build a more constructive U.S./NATO-Russian relationship.

R. Craig Nation explores the challenges of "resetting" U.S.-Russian relations. If the current attempt at normalization is to succeed where others have failed it will need to rest upon a more sophisticated understanding of the larger trends working to shape U.S.-Russian relations both for good and for ill.

The promise of renewed strategic partnership after 9/11 proved to be a mirage. Russian-American cooperation in Afghanistan and elsewhere was short-lived, while sources of disagreement and confrontation (NATO enlargement, the American military presence in Central Asia, frozen conflicts on the territory of the former USSR, the Kosovo question, access to oil and natural gas resources of the Caspian basin, the premises of Russian governance, U.S. military action in Iraq, the missile defense program in central Europe, etc.) proliferated.

The failure to reestablish a meaningful Russian-American partnership in the wake of the 9/11 attacks was to some extent the result of choices made by leaders on both sides, for whom national priorities took precedence over a bilateral relationship whose relevance seemed to lie in the past. But the sources of strategic rivalry between Russia and the West also have a structural foundation. Russia and the United States are different kinds of polities with distinct strategic cultures and sometimes sharply contrasting priorities. A history of rivalry has led to the accumulation of considerable mistrust. The systemic forces at work include a resurgent Russia, an authoritarian drift, a possible decline of the West, a possible Moscow-Beijing axis, and the geopolitics of Eurasia.

What needs to be done? Recognizing the considerable weight of inherited hostility and the magnitude of the task ahead is a good starting point. Chronic criticism of Russia's domestic regime based upon a purported clash of values has become unhelpful. To make democratization on Western terms a litmus test for political reliability is not sound policy. The United States does not apply such a litmus test to its relations with other key international actors.

Russia and America share a large number of coinciding interests that provide a strong foundation for practical cooperation. Together Russia and America control over 90 percent of the world's nuclear weapons. They have a powerful vested interest in regulating strategic competition, maintaining transparency, and sustaining a minimal deterrent posture. Renegotiating the START Accord is hopefully a step toward a more comprehensive rethinking of respective nuclear postures and doctrines. Russian sensitivities to the U.S. agenda for deploying missile defense systems in Poland and the Czech Republic should be respected. Ideally this agenda would be rethought and recast in such a way as to engage Russia as a partner in the context of a larger, reanimated arms control dialogue. Both parties likewise have a related interest in blocking the proliferation of weapons of mass destruction and reinforcing a robust nonproliferation regime.

Washington and Moscow share a common perception of the threat of catastrophic terrorism emerging from radical jihadist movements. Russia's geopolitical situation and security assets position it to play the role of a useful ally in a "long war" against terrorism.

As the world's largest consumer and producer of energy resources, respectively, the United States and Russia should have a vested interest in regulating global energy markets to their mutual advantage. Stability of supply and stability of demand are not mutually exclusive categories.

Confronting the global economic crisis is of interest to both nations. Russia's economic fortunes have an important influence on systemic stability and it possesses important economic assets. The United States should eliminate the Jackson-Vanik Amendment, a Cold War relic, and use its influence to bring Russia into the World Trade Organization.

Finding a way to escape from the zero-sum logic of U.S.-Russian competition in the new Eurasia is an essential basis for meaningful rapprochement. NATO enlargement needs to be slowed, frozen, or if possible put off given Russian sensitivities of NATO membership for Ukraine or Georgia. The best way to work around contrasting perceptions that make such issues so difficult to resolve cooperatively would be to reanimate NATO-Russian dialogue and reconstruct a more robust and dynamic NATO-Russia Council.

Finally, recasting U.S.-Russian relations will have to be inspired by a new and expanded conception of security itself. The most pressing security concerns no longer relate to the threat of armed aggression by neighbors. Rather they are economic instability, social inequities, environmental degradation, depletion of energy sources, pandemic disease, mass casualty terrorism, the proliferation of weapons of mass destruction, and many other real and present dangers and challenges that can only effectively be addressed through cooperative initiatives on a global scale. Taking advantage of a relatively benign security environment and bringing Russia, China, India, and other emerging powers out from the cold and into a functional great-power concert represents a strategic opportunity. Fixing what is broken in the U.S.-Russian relationship would be one good place to start.

Topics discussed included Russia and energy, the geopolitics of energy, how to reset U.S./NATO-Russian relations, how Russia could assist with the Iranian issue, and a comparison of relations with China and with Russia.

Panel VII: Africa

Africa in U.S. Foreign Policy
Ambassador (Ret.) Princeton N. Lyman, Adjunct Senior Fellow for Africa Policy Studies, Council on Foreign Relations

Great Expectations versus Daunting Challenges: Prospects for U.S. Foreign Policy toward Africa during the Obama Administration
Dr. Peter J. Schraeder, Professor and Graduate Program Director, Department of Political Science, Loyola University, Chicago

Moderator: Dr. Stephen A. Emerson, Associate Professor of National Security Affairs, Naval War College

Princeton Lyman explains that for a long time, Africa was seen as outside the strategic interests of the United States, and nowhere more so than within the U.S. military. As a Department of Defense (DoD) official, charged with Africa policy, put it in the 1980s, "DOD sees Africa as a place to fly over, not stop there." This perspective was reinforced by the U.S. military intervention in Somalia in 1992–1993. In foreign policy circles, Africa was seen as largely a humanitarian interest, especially following the Cold War. U.S. assistance to Africa during the 1990s drifted downward.

This situation changed in the years since 2000. The European Command (EUCOM) began to focus more attention on Africa, into which it had been drawn for several humanitarian or rescue missions, and saw in the weakness and vulnerability of African states a long-term strategic threat. In 2006, a Council on Foreign Relations report, *More Than Humanitarianism: A Strategic U.S. Approach toward Africa*, called attention to Africa's growing importance as an oil exporter, its importance in the global war on terror, the costly series of conflicts there, its central role in the fight against HIV/AIDS, its importance as a voting bloc in international organizations, and China's growing activities in Africa.

President Bush during this period began placing greater emphasis on Africa. Much of it was a continuation of the humanitarian focus. Altogether Bush more than tripled U.S. aid to Africa after 2001 to $6 billion by 2007 with a promise to raise it to $9 billion by 2010. In the wake of 9/11, the Bush administration created the Combined Joint Task Force/Horn of Africa, and stationed some 1200 American service personnel in Djibouti. Bush proclaimed a $100 million counterterrorism program for East Africa. EUCOM initiated the Pan-Sahel Initiative, a small counterterrorism training program, an initiative that would grow into the much larger Trans-Sahara Counterterrorism Partnership. Finally, the Bush administration created a single African Combatant Command, AFRICOM.

Today Africa provides some 24 percent of U.S. oil exports, most of which comes from the states in the Gulf of Guinea. West African crude oil is attractive in that it is low in sulfur, and Africa is one of the few areas that encourage private investment. China has made its major Africa energy forays in Sudan and Angola. Some fear that China intends to lock up supply, but the evidence is not convincing. China has obtained several oil blocs in West Africa, but they are neither promising nor economical, and China lacks offshore technology. The primary problem for U.S. energy interests is not China, but instability in Africa.

International terrorism raised its ugly head in Africa in 1998, with the bombing of American embassies in Kenya and Tanzania. It is now clear that al Qaeda cells operate down the east coast of Africa and in some parts of southern Africa. At present, Somalia's anarchy and violence continue and the radicalism of insurgents has deepened. U.S. attention has been paid to the threat from al Qaeda in the Maghreb operating in and out of Algeria. The bottom line is that bringing terrorism under control, both that which is internationally linked and that which is indigenous, rests more with the underlying weaknesses and vulnerabilities of African states than with improving the purely security capacities of these governments.

Conflicts have been enormously costly in lives, rape, pillage, destruction of property, and undermining development progress. The war in the Democratic Republic of Congo alone has cost directly and indirectly nearly 5 million lives. In Sudan, the conflict in Darfur, labeled a genocide by the United States, continues after six years. Overall American policy has been to foster African peacekeeping and conflict resolution capacity.

A new and growing threat is narcotics trafficking from South America through West Africa to Europe. While the drugs are destined for Europe, criminality undermines efforts at peace and development. The potential alliance of drug syndicates and terrorist elements is worrisome.

Finally, Africa is the epicenter of the HIV/AIDS pandemic, on which the United States has spent $18 billion over the past year and for which Congress has authorized $48 billion for the next five years. Unfortunately, the rate of infections continues to outpace the number of new persons being treated.

The Obama administration has the advantage of building on the Bush administration's increased attention and resources earmarked for Africa. Without adding too much new funding, the United States could meet Bush's pledge to reach $9 billion in assistance by 2010. The Obama administration has already appointed several high-level officials with strong Africa credentials.

Africa registers today as a factor in energy security, terrorism, international crime, conflict, global health, and trade. The challenge for the United States now is to translate that recognition into the long-range, capacity- and institution-building programs that will enable Africa to become a more effective and reliable partner in all the areas of mutual concern.

Peter Schraeder states that Barack Obama's historic election as the forty-fourth president of the United States has raised extraordinary expectations among both Africans and Africanists as concerns the future of U.S. foreign policy toward Africa. If history is our guide, however, Africa will remain the region of least concern, as the Obama administration by necessity focuses on domestic issues and other regions of perceived greater importance.

Five sets of constraints may limit the maneuverability of the Obama administration, potentially reinforcing continuity in U.S. foreign policy toward Africa: (1) crisis in the U.S. economy, (2) inheritance of a residual fear of 9/11, (3) historic White House neglect of the African continent, (4) historic congressional neglect of the African continent, and (5) bureaucratic influence in the policy-making process.

President Obama's number one priority is responding to the crisis in the U.S. economy. This domestic crisis will clearly consume a significant portion of the Obama administration's first two years in office, leaving little time relatively speaking for foreign initiatives and especially those targeted toward Africa. Moreover, the financial requirements of dealing with the crisis will limit resources to fund new initiatives in Africa.

The African dimension of the residual fear of 9/11 is that the administration has inherited a series of national security initiatives, often critiqued as the "militarization of U.S. foreign policy toward Africa," that a president seeking to avoid missteps on the path to reelection may be hesitant to dismantle.

Historically presidents have lacked knowledge of Africa, tended to view Africa as the responsibility of European former colonial powers, delegated responsibility for foreign policy for areas considered marginal, and felt the necessity to focus on domestic priorities. Even if we recognize that Barack Obama is different, an Obama White House will be consumed by foreign policy issues in other regions perceived of greater importance.

Like their White House counterparts, members of Congress historically have neglected Africa relative to other regions. As a result, membership on the Africa subcommittees is among the least desired congressional positions in both houses of Congress.

The net result of White House and congressional neglect of Africa is that U.S. foreign policy toward Africa remains largely delegated to the high-level bureaucrats and political appointees. The effect of this "bureaucratic influence" is that it fosters continuation of established policies.

There are six trends in an Obama administration's foreign policy toward Africa: (1) campaign and presidential appointments demonstrating a heightened interest in Africa; (2) Obama's governing ideology: pragmatic traditional realism with strong tendencies toward liberal internationalism; (3) sober reckoning as concerns Africa programs that will require additional U.S. financial resources; (4) cautious approach to conflict resolution that will nonetheless use the White House as a bully pulpit; (5) heightened focus on socioeconomic and development issues; and (6) unclear change in the democratic deficit in U.S. foreign policy toward Africa.

Obama's early appointments include an unprecedented number of close advisers with impressive backgrounds in Africa. Three characteristics in common to this team include early opposition to the Iraq war, a tendency toward liberal internationalism, and an emphasis on the use of "soft power." Interestingly, this team demonstrates Obama's reliance on those with military experience with implications concerning the current overwhelming influence of strategic/military initiatives in Africa.

President Obama is not a typical liberal, but rather a pragmatic traditional realist with strong tendencies toward liberal internationalism sometimes referred to as an "optimistic realist" or a "realistic optimist."

President Obama has proposed several new foreign aid initiatives and the expansion of existing ones for Africa. However, there clearly has been a sobering reckoning, especially in light of the economic meltdown in the U.S. economy.

President Obama has underscored his administration's intention to take a more proactive approach to African conflict resolution, and has specifically cited ongoing conflicts in Darfur, Zimbabwe, the eastern Congo, the Niger Delta, and Somalia. It is unclear if foreign policy challenges and priorities elsewhere will permit the degree of high-level White House attention necessary for effective conflict resolution.

The Obama administration has pledged to reverse the recent trends in decreased development assistance to African nations by strengthening a variety of nonmilitary programs. The most noteworthy and far-reaching proposal is to remake and restore the United States Agency for International Development.

Finally, an overriding preoccupation with terrorist threats led the Bush administration to overlook authoritarian excesses of African regimes. The key dilemma for pro-democracy activists: it is unclear whether Obama's governing ideology will lead to significant change in the democratic deficit in U.S. foreign policy in Africa.

Topics discussed included potential changes in U.S. foreign policy toward Africa, the role of trade and investment, the role of short-term aid versus long-term development assistance, and U.S. objectives and likely nature of American engagement in Africa.

Opening Remarks

Ambassador (Ret.) Mary Ann Peters
Provost, Naval War College

Good morning ladies and gentlemen. I hope you all had a good night's rest, because we're going to keep you very busy for the next couple of days.

Again, I want to welcome you to the third research workshop sponsored by the William B. Ruger Chair of National Security Economics here at the Naval War College.

The very first workshop explored the link between maritime strategy and economic prosperity and security. Its purpose was to contribute to the development of the new maritime strategy that has since been released by the U.S. Navy as *A Cooperative Strategy for 21st Century Seapower,* and if you are not familiar with it, I commend it to you. In my view it's a very forward-looking strategy.

The second Ruger Workshop was devoted to defense strategy and forces and addressed the links among security challenges, defense resources, and risks.

So as you can see, this particular workshop is perhaps even more ambitious than past workshops because it addresses the broader topic of American foreign policy.

Our purpose *this* year is to provide a collegial forum for you, a select group of foreign policy scholars and practitioners, to formulate and recommend new directions for American foreign policy, for each of the regions of the world.

Of course, we are using regions only as an organizing principle, recognizing fully that the nation-state is still the primary actor on the world stage, and that to some extent the regions we talk about are just lines on our own organization charts.

With the new administration just over one hundred days in office, we see this workshop as a chance to look at American foreign policy and consider the challenges and opportunities that the United States faces now and into the future.

To do that we—by "we" I mean "you"—are going to examine the interests and perspectives of nations and other actors in each region. You are, we hope, going to take a dispassionate look at how U.S. interests can best be served around the globe.

We expect that, in some areas, you may recommend significant shifts in priorities for the new administration. Of course, in other areas you may argue for continuity.

As experts in national security and foreign policy, we look forward to hearing your ideas—and to hearing them challenged.

As you can see by the venue we have chosen, we hope that the dialogue among this very impressive group of people will generate new insights or reaffirm enduring themes into the foreign policy process.

I want to especially thank you for your extensive efforts in researching and preparing your papers. And for taking time from your busy schedules to join us in Newport. I've had a chance to read your papers and I really have been very impressed by the perspectives and insights that you bring to us here at the workshop.

As I believe you know, your papers and discussions will be published in a monograph, and it will be widely distributed throughout the national security community and the general public. And we do expect that your work will be used in our curriculum here.

The Ruger Workshop is part of a robust program of regional studies and in-the-field engagement programs to support our foreign and defense policies.

In recent years, the College has added significant regional area studies to our core curriculum. And we provide students the opportunity to further concentrate on area studies in their elective program.

We have organized some of our faculty along six regional studies groups that focus on research, teaching, and engagement or outreach programs. The newest one is the Indian Ocean Regional Studies Group.

The leaders of these regional study groups are here with us for this workshop.

The College's engagement—or outreach—programs include visits to and from governments, sister war colleges, universities, and think tanks in countries around the world. Increasingly, College faculty members are being asked to assist with curriculum development for professional military institutions around the globe.

Our study groups sponsor regional workshops on contemporary issues.

The College sponsors major events such as the International Seapower Symposium, which will bring heads of navies and coast guards from over ninety nations to the College in October. The last symposium, held here in 2007, discussed ways to strengthen maritime partnerships, reflecting the new maritime strategy.

At the College we will host the Secretary of the Navy's Current Strategy Forum this June. It will bring over one thousand participants to discuss security challenges for the United States. The theme is "Strategic Opportunities: Challenging the Paradigm." Among our speakers at the forum will be Greg Mortenson, the author of *Three Cups of Tea,* and Anne Marie Slaughter, the head of Policy Planning at the State Department.

But the Ruger Workshop is one of those events we prize most because it gives us—and you—a chance to step back and refresh our thinking on the critical issues of the day. So, we are very grateful for your help in that endeavor. Thank you.

And now it is my pleasure to turn the podium over to our Ruger chair and the organizer of this conference, Professor Rich Lloyd.

Panel I

A Global Perspective

Dr. Patrick M. Cronin
Director, Institute for National Strategic Studies, National Defense University

Dr. Michael T. Klare
Five College Professor of Peace and World Security Studies, Hampshire College

Moderator:
Dr. Thomas R. Fedyszyn
Professor of National Security Affairs, Chair Eurasia Regional Study Group, Naval War College

U.S. Security Policy in a Changing World[1]

Dr. Patrick M. Cronin
Director, Institute for National Strategic Studies
National Defense University

Although the United States cannot afford to be the world's exclusive security guarantor, the world is ill-prepared for U.S. retrenchment. The United States needs a new conceptual pathway for policy makers to begin recalibrating America's security role to reverse what has appeared to be a widening gap between U.S. ends and means, now and in the future. International security requires U.S. active engagement, but the character of that engagement is changing along with the global environment. Worldwide trends suggest that the United States will increasingly have to approach complex challenges and surprises through wider and more effective partnerships and more integrated strategies. This essay begins a discussion on how to think about the complex security environment and how in particular the United States can begin the process of strategic adaptation.

Complexity is the watchword of our century. This assessment should be a healthy reminder of just how complex—and dangerous—a world we live in. That complexity was encapsulated by the Greek poet Archilochus, who said that the fox knows many things but the hedgehog has only one big idea. During the previous administration, the United States conflated security issues under the umbrella of a "global war on terror" and focused on a single big idea. Thus, a central idea, if not an organizing principle, is that the United States will have to be as clever as the fox, keeping its eye on multiple challenges and taking care not to exert its finite resources on any single problem. Preparing for and dealing with such profound complexity requires particular capabilities, approaches, and proclivities: cultural, developmental, experiential, technical, organizational, political, and operational. These attributes can be selected, cultivated, and enhanced, and it seems that they will have to be if we are to survive, let alone succeed.

We need to bridge the gap between theory and praxis. The overriding message is to emphasize global complexity and America's vital yet limited role in coping with that complexity. Some critics will hew to a traditional view of security and the world, claiming that the threats are far more straightforward and the world quite predictable. Indeed, the world of tomorrow will carry on with a great deal of continuity. The gist of my research is that policy makers are only beginning to come to terms with the uncertain, complex world in which we operate. For instance, too little systematic thought has been given to the dynamic interactions between state and nonstate actors, between economics and security, and to comprehend security in the contested global commons of international waters, airspace, space, and cyberspace—to cite only three issue areas. Moreover, to the extent that officials and analysts are able to stay on top of global trends, they also realize that our prescriptions, policies, and strategies tend to lag woefully behind them.

Today's world is marked by the uneasy coexistence between traditional geopolitics and ever-widening globalization. A fundamental question is how the United States can best use its essential and yet insufficient influence in a world marked by both rising state power centers and the devolution of power into the hands of more nonstate actors. Clearly there is no simple prescription for the problem of how the United States can best exert its influence in this dynamic security landscape. Even so, the breadth of threats, challenges, and opportunities that may surface in the coming years will require a comprehensive approach that utilizes the full continuum of power—be it hard, soft, smart, dumb, or fuzzy. Complexity should not be an excuse for ignoring clear, urgent, and obvious dangers, but responses to those threats must better assess the side effects and opportunity costs of neglecting the full array of challenges confronting the United States and the world. In short, there is no substitute for making conscious choices within a grand strategic perspective: the world cannot afford for us to be narrow, near-sighted, or parochial.

Safeguarding U.S. national interests and global security is complex and uncertain today and is only likely to become more so tomorrow. First, we need to appreciate global trends. Second, we need a global analysis of the world's seven regions, to consider important developments in their distinctive neighborhoods. Finally, we need an examination of prospective U.S. contributions, military capabilities and force structure, national security organization, alliances and partnerships, and strategies.

I would contend that there are at least eight strategic trends shaping both near- and long-term challenges and opportunities. Economic and political

Figure 1
Eight Global National Security Challenges: A Sobering Agenda

1. A gradual global redistribution of economic power from the West to the "rest" is under way, and economic power is the bedrock of enduring military and political power.
2. We are on the cusp of but not yet in a multipolar world.
3. The globalization of communications is challenging more than just the virtual foundations of the postmodern information society.
4. Energy and environmental insecurity have reached a tipping point. The era of cheap hydrocarbons and scant ecological regard is finished.
5. Transnational terrorism and stateless actors can inflict unprecedented damage, and we must be on our guard against catastrophic terrorism.
6. September 11 and growing insecurity in Afghanistan and Pakistan remind us of the growing challenges posed by fragile states and "ungoverned" spaces, with the "bottom billion" in some sixty countries left behind in dire poverty.
7. The character of war is changing, forcing low-level uses of force and greater civil-military integration in a renaissance in counterinsurgency and irregular warfare. We must prepare for "hybrid warfare" and hedge against emerging peer competitors.

8. The proliferation of weapons of mass destruction has made it increasingly possible that nuclear or biological weapons may be used in the coming years, including uncontrolled biological agents.

    National Security Reform: A Global Roundtable

Source: Institute for National Strategic Studies, "Adapting to Eight Global Challenges," in *Global Strategic Assessment 2009: American's Security Role in a Changing World,* available at http://www.ndu.edu/inss/index.cfm?secID=8&pageID=126&type=section.

power is shifting, technology is altering political and social patterns of behavior, energy and the environment are looming as larger long-term drivers of security than in the past, permanent fragile states and nonstate actors are creating new dimensions to what had once been seen by many as a big-power chessboard, and the proliferation of weapons and hybrid warfare are likely to change the character of conflict in the future. The world seems stuck in a constant tussle between geopolitics and globalization, between classic state-power contests for competition and cooperation and emerging dynamics in which the good and ill effects of globalization take on heightened importance. Policy makers will have to seek the best balance between these traditional and emerging forces.

In addition to these global trends, I would argue that a survey of the seven regions of the world highlights the rich and distinctive issues, uncertainties, competitions, and partnerships that characterize each region of the world. Trends may be global, but they affect and shape each region in different ways. Moreover, each region appears to have largely local domestic and regional concerns, even while increasingly intersecting with other regions and global security issues. As for which countries will contribute to regional and international security, there is an obvious gap between the array of challenges transcending narrow national interests and the level of contributions most countries are making. Again, clearly policy makers will have to find a balance between local and regional priorities on the one hand and more global and transnational issues on the other.

This complex environment poses a potent set of challenges for how the administration of President Barack Obama seeks to exert America's significant yet finite power to safeguard against a diverse set of traditional and modern threats and challenges, while also seizing as many opportunities as possible to build more durable, peaceful, and collaborative solutions for the twenty-first century. In his first months in office, President Obama demonstrated a keen ability to change the basic narrative of the United States, placing it in a far less confrontational stance with most of the world, and showing a willingness to give greater weight to local and multilateral solutions.

We know from the first months of the Obama administration's tenure that in many ways the United States has turned the page on its style and narrative in many parts of the world. At the same time, it should be obvious that while diplomacy and rhetoric can provide an important new beginning, the hard work of seeking security, building support, and implementing whole-of-government solutions across a vast number of complex challenges is a neverending business. The administration has not only embraced the "three d's" of diplomacy, development, and defense, but has also recognized that many broad security issues are interwoven with the "three e's" of economics, energy, and the environment. Other issues, such as democracy and human rights, cannot be divorced from security, whether concerning the future course of Iran or the difficulties democracies have in waging protracted counterinsurgencies without losing popular support or straying from democratic values.

Figure 2
Five Pathways to a Game-Changing Strategy for the United States

1. ***Heal thyself.*** To a remarkable degree, security hinges on America having its house in order.
2. ***Redefine problems.*** Ends should be realistic, recognizing emerging interrelationships among energy, the environment, food, and climate change.
3. ***Surge civilians.*** Complex challenges require a larger whole-of-government team of national security professionals.
4. ***Countermobilize.*** The United States can mobilize emerging power centers into action through bilateral alliances, coalitions of the willing, and multilateral institutions.
5. ***Exercise strategic restraint.*** We cannot afford quagmires that drain resources without providing lasting security. A strong military is the U.S. ace in the hole, but better still are indirect approaches, strategies of leverage, and "smart power."

Albert Einstein once said that given an hour to save the world, he would devote fifty-nine minutes to thinking about the problem and one minute to resolving it. A global strategic assessment provides a purposefully broad point of departure for many national security functions: subsequent analysis, interagency coordination, policy derivation, coalition-building, reorganization, long-range planning, and operations. The need for broader U.S. strategic thinking is obvious to me and to my colleagues at the Institute for National Strategic Studies (INSS) at the National Defense University. But equally important is the need to mobilize partners, conduct serious planning, integrate a rich variety of disciplines and actors, follow through on implementation, and then assess actions with an appreciation of history. And all of these steps must then, in turn, inform our education and training. There is an obvious case for all-of-government and coalition-based solutions.

This should be a familiar process: on the modern battlefields of Iraq and Afghanistan, a deep and sober understanding of what U.S. and coalition forces faced had to emerge before any hope of a comprehensive and successful strategy was possible. We need to get this strategic learning process off the battlefield to the maximum extent possible and appropriate.

The challenges are great, but so are the opportunities. The world is changing, but the United States still has the greatest capacity to cope with these vicissitudes, to lead global responses, and to make the world a safer place. Many of the trends are positive, and the contributions of issues as diverse as the information revolution and advances in the life sciences are bringing greater overall good than ill to humankind. Even so, it would be a dereliction of duty to avoid difficult questions about better ways to manage the challenges emerging even from positive trends.

In addition to the elaborate interrelatedness of international security, we need to be reminded of the enduring realities of American power. There is nothing permanent about the U.S. global security role, and there are no guarantees in international security, but no other nation has America's unique attributes: a global zeal to make the world a better place, potent expeditionary forces to project power on all continents and oceans, a large and open economy, and a diverse and ever changing society built on freedom and the rule of law. As the nation is refocusing its foreign policy on diplomatic rather than military capabilities, the fact remains that formidable military power has supercharged our diplomacy and remains key to providing the Obama administration with far more purchase than other countries. Whether through settled or ad hoc collective security arrangements, no other country appears ready to mobilize its instruments of power to address threats posed by state and nonstate actors. Even as American power measured as a percentage of the global economy has declined, its comparative advantage in terms of hard military power has expanded.

There remains a good deal of optimism that problems can be resolved or at least better managed; that a more humble America that is more sensitive to diverse views from around the world is ready to work together with others; and that for America's relative decline in perceived and actual influence, perhaps, there is every reason to believe that the United States will remain a powerful and unique contributor—only one, to be sure—to global security.

Note

1. Portions of this article are drawn from the author's introduction to the Institute for National Strategic Studies' *Global Strategic Assessment 2009: America's Security Role in a Changing World*, available at http://www.ndu.edu/inss/index.cfm?secID=8&pageID=126&type=section.

The New Foreign Policy Agenda: Energy, the Environment, and the Global Economy

Dr. Michael T. Klare
Five College Professor of Peace and World Security Studies
Hampshire College

Perhaps never before have American policy makers had to confront such a wide array of global challenges simultaneously. In his first 100 days in office, President Obama was compelled to address such far-ranging issues as the global economic meltdown, the wars in Iraq and Afghanistan, the deteriorating security environment in Pakistan, rising drug violence in Mexico, the resumption of nuclear weapons production by North Korea, and a global flu pandemic. And this list does not include initiatives by the White House to "reset" U.S. relations with Russia, open a dialogue with Iran, and open the door to fresh contacts with Cuba. Under these circumstances, it is exceedingly difficult for anyone to devise a grand scheme to describe the overriding themes and concerns that are likely to dominate American foreign policy in the years ahead, given the multitude of known and unknown challenges facing this country. Patrick Cronin and his colleagues at the Institute for National Strategic Studies (INSS) of the National Defense University (NDU) and Mathew Burrows and his colleagues at the National Intelligence Council (NIC) have made a heroic attempt to address these multiple factors and to put them into some sort of rational framework in two remarkable documents, the NDU's *Global Strategic Assessment 2009: America's Security Role in a Changing World*[1] and the NIC's *Global Trends 2025: A Transformed World*.[2] I have not attempted to duplicate such an effort—no individual can—but rather to draw on their impressive work in identifying a number of key problems that I believe will dominate U.S. foreign policy in the years to come.

First, some broad generalizations. For most of the past sixty-five years, since the end of World War II, American policy makers have largely been preoccupied with what I would call the **old foreign policy agenda**: problems arising from the competitive pursuit of power and dominance by nation-states within the international system. Typically, these include problems of power blocs and alliance relations, arms balances and arms transfers, threats and counterthreats, interventions, crises, and wars. One could say that this foreign policy agenda was dominated by the "three a's": allies, adversaries, and armaments. These three agenda items were almost totally dominant during the Cold War era, when NATO affairs, the U.S.-Soviet rivalry, so-called proxy wars in the developing world, and arms control were the overarching concerns of American foreign policy. A similar outlook prevailed in the immediate post–Cold War era, with the focus of attention shifted to the threat of "rogue states" equipped with or seeking weapons of mass destruction, and, after 9-11, to state sponsors of terrorism.

What I call the old foreign policy agenda certainly retains substantial relevance today and will continue to do so for so long as nation-states vie for power and influence in the international system and armed groups of whatever sort pose a security threat to this country. Certainly Iran and North Korea continue to represent a significant problem for American policy makers as they have in the past, while arms control talks with Russia have again gained the spotlight. The war in Iraq may be winding down, but the war in Afghanistan is entering a new, more vigorous phase. The rise of China will involve many of the same sorts of challenges that accompanied the rise of Germany, Japan, and the Soviet Union in the twentieth century. So the sorts of problems embedded in the old foreign policy agenda will not disappear in the years to come, though they may appear in new forms.

Nevertheless, I believe that we will see ever-increasing attention devoted to what might be termed the **new foreign policy agenda**: problems arising from transboundary and nonmilitary issues, including global economic disorder, pervasive underdevelopment in the poorest countries, environmental degradation, humanitarian disasters, mass migrations, global pandemics, resource scarcities, and international crime. We have already seen how these issues intruded into U.S. policy making during the first 100 days of the Obama administration, with the enormous attention devoted to the global economic crisis and the global A(H1N1) influenza epidemic. Moreover, during her first trips abroad as Secretary of State, Hillary Clinton addressed issues of energy and the environment in China and drug violence and immigration in Mexico. I believe that any realistic projection of where American foreign policy is headed would have to conclude that these issues will increasingly dominate the global landscape upon which American policy makers will be forced to tread.

If time and space were not an issue, it would be useful to examine all of the items on the new foreign policy agenda. Given practical limitations, however, I will focus on three of them—what I call the "three e's": energy, the environment, and the world economy. As I see it, these three closely related factors will play an ever more significant role in shaping U.S. foreign policy in the years to come. But before I discuss the implications of each for American policy makers, why pick these three above all others in the list of items on the new foreign policy agenda?

The **economy** comes first, I believe, because the United States has been the principal architect and advocate of a liberal economic order, and so any sustained downturn in the global economy will be interpreted not only as an indictment of the American economic model but will also lead to a decline in American political leadership. Without a healthy world economy, moreover, there is little hope of overcoming such American objectives as the elimination of persistent underdevelopment, world hunger, illiteracy, soil degradation, deforestation, poverty-related disease, and a host of other critical problems. On the other side of the coin, anemic or negative economic growth, as we are experiencing today, is likely to produce social unrest, political turmoil, and, in the worst case, state collapse. This could prove enormously challenging for U.S.

policy makers if it threatens the survival of governments in states of importance to this country, as appears to be the case in Pakistan.

Energy is important because the world economy cannot thrive without an adequate and affordable supply of it and because so much of international relations has come to be dominated by problems arising from imbalances between supply and demand. When supplies of energy are sufficient and affordable, issues of supply and demand tend to recede from international consciousness, but when shortages arise or prices soar, these issues acquire great significance. This has been the case in recent years as global supplies of oil and natural gas have failed to grow in tandem with soaring international demand, much of it generated by the vast expansion of economic activity in Asia. As demand overtakes supply, prices rise, and this creates a host of political as well as economic problems. Some states with a surplus of oil and gas for export, including Iran, Russia, and Venezuela, have sought to use their windfall profits to pursue political projects that conflict with American interests. The acquisition of vast oil wealth by privileged elites in the oil-producing states of Africa, the Middle East, and Asia has also tended to heighten internal tensions in these countries, in some cases precipitating ethnic strife, terrorist violence, or separatist conflicts; these conflicts, in turn, have occasionally prompted direct or indirect involvement by the major powers, including the United States, China, and Russia. Given that the imbalance between demand and supply is likely to persist in the years ahead, problems of this sort are likely to recur.

The **environment** is destined to become an ever more important problem in U.S. foreign policy as the warming of the planet proceeds and the resulting climatic effects gain momentum. This is not the place to debate the science of climate change or the likely timetable of its most severe effects. If we rely on the most authoritative report to date, the Fourth Assessment Report of the Intergovernmental Panel on Climate Change (2007), we can assume that the planet is already experiencing the earliest impacts and these impacts will grow in severity as we move further into the future. These findings pose two sorts of challenges for American foreign policy: first, addressing the social and political *consequences* of climate change; and second, engaging in the *diplomacy of mitigation*. With respect to consequences, we will be looking at massive humanitarian disasters, akin to that now unfolding in Darfur (which some analysts believe is partly caused by desertification in the Sahel region of Africa), as well as continuous waves of migration, much of it illegal. In extreme cases, there is a potential for some combination of water scarcity, mass starvation, and coastal inundation leading to social chaos and state failure.[3] With respect to diplomacy, we can expect many rounds of negotiation over successive pacts aimed at reducing the emission of greenhouse gases. These, and other environmentally related matters, will, I believe, consume an ever-increasing share of the time, attention, and ingenuity of American policy makers.

As indicated, the three e's are closely interconnected. The global economy cannot function effectively without an adequate supply of affordable energy, while global logistical networks—the very sinews of globalization—are especially dependent on an abundant supply of transportation fuels. Petroleum is also the world's leading trade commodity, and its ownership, extraction, refining, and distribution constitute a major factor in the economies of numerous countries, including some of America's major allies. It is the consumption of fossil fuels, moreover, that is the leading source of the carbon dioxide emissions that are largely responsible for humanity's share of the greenhouse gases now accumulating in the atmosphere, raising planetary temperatures. Any effort to reduce CO_2 emissions will, therefore, require a complete transformation of the global energy system, and this, in turn, has vast economic implications. These, and other such interrelationships, ensure that the three e's will figure ever more prominently in American foreign policy.

The Three E's in Practice

When American foreign policy was largely governed by the three a's, the priorities for American policy makers were usually straightforward: to assemble a constellation of strong, reliable, and pro-American allies; to isolate, contain, and weaken the nation's adversaries; and to maintain a balance of military power that favored the United States and its allies vis-à-vis the Soviet Union and its allies, while mitigating the dangers inherent in the nuclear arms race. For the most part, these principles could be and were applied to the everyday policy matters that arose throughout the Cold War era and, in modified form, in the post–Cold War struggle to isolate the so-called rogue states.

As the three e's grow in importance, they, too, will begin to set the priorities for American foreign policy. At present, no clear set of guiding principles has been articulated comparable to those that governed U.S. policy during the Cold War era, but we can begin to see the outlines of such a matrix. Based on the statements of President Obama and his senior foreign policy aides during the first months of the new administration, it could be said that this matrix encompasses the following broad principles:

1. *On the global economy:* The United States must exercise leadership within the G-7, the G-8, the G-20, the governing boards of the World Bank and International Monetary Fund, and other such multilateral bodies to ensure the stability of the international financial system, and, so far as possible, to promote economic recovery and prevent any future contractions caused by inadequate governmental oversight of financial transactions. Beyond this, Washington must contribute to the economic health of particular states of concern (e.g., Pakistan) whose collapse or decay would threaten the vital interests of the United States.

2. *On energy:* The United States should take vigorous action to diminish its current dependence on imported petroleum, thereby diminishing the power and influence of the oil-exporting countries—especially those with hostile intents toward this country. In addition, Washington should enhance the "energy security" of this country and its allies by promoting the diversification of energy supplies by type, provider, and means of delivery.

3. *On the environment:* The United States should play a constructive role in the design and implementation of international regimes for the reduction of greenhouse-gas emissions. In addition, this country

should cooperate with others in the development of climate-friendly alternative fuels and transportation systems.

Much of this remains to be formulated, but the groundwork has already been laid. Just as was true of the three a's, moreover, the three e's will govern America's ties to particular countries, regions, and international bodies.

Take China, for example. As noted earlier, elements of the old agenda still apply. The United States must worry about the Chinese medium-range ballistic missiles aimed at Taiwan and the modernization of the People's Liberation Army (PLA). China's growing investment in the PLA Air Force and Navy, and the lack of transparency in the allocation of this funding, are natural sources of concern. But of equal concern to U.S. policy makers are issues arising from China's behavior in the economic, energy, and environmental realms.

For the most part, American analysts have praised China's handling of the current economic crisis. But some economists—and some members of Congress—have complained that China has kept its currency artificially low with respect to the dollar in order to facilitate Chinese exports to the United States and has otherwise engaged in unfair trading practices. Chinese authorities have also refused to eliminate various rules and regulations that impede American investment in China. As economic concerns loom ever more prominently in Sino-American relations, these are the sorts of issues that will consume more and more of the time and attention of U.S. diplomats and policy makers.

Energy and the environment pose equally daunting problems. According to the U.S. Department of Energy (DOE), China's need for imported oil is projected to jump from 3.0 million barrels per day in 2005 to 11.6 million barrels in 2030, making it the second-biggest importer after the United States.[4] Given trends in the production and export of oil, this will mean that the United States and China will be competing for access to relatively limited sources of supply in a dozen or so countries located in Africa, the former Soviet Union, Latin America, and the Middle East. Because both Washington and Beijing have often chosen to cement their ties to favored suppliers in these areas through military means—arms transfers, military training, participation in joint exercises, and so on—the potential exists for geopolitical rivalry and arms competition. China's cooperation with Russia in efforts to dislodge the United States from Central Asia is one expression of this phenomenon. Managing this new source of friction in Sino-American relations will pose a new challenge to American policy makers.[5]

Finally, there is the environmental dimension of the U.S.-Chinese relationship—potentially the most challenging of all. Because of its heavy reliance on coal, China is destined to remain the world's biggest emitter of carbon dioxide for the foreseeable future. According to the DoE, China's emissions of CO_2 will jump from 5.3 billion metric tons in 2005 to 12.0 billion tons in 2030, at which time its CO_2 emissions will account for 28 percent of the world total, nearly twice as much as those of the United States, the next-biggest emitter.[6] If these projections prove accurate, it will be virtually impossible to avert the most severe effects of global warming, even if other countries, including the United States, make considerable progress in reducing *their* CO_2 emissions. It is essential, therefore, that the United States persuade China to curb its use of coal, or to use coal in a climate-friendly manner; the best way to accomplish this, I suspect, will be to engage China in extensive diplomacy aimed at the collaborative development of alternative energy sources.

As in the case of China, we can see how economic, energy, and environmental issues arise in U.S. ties with other major players on the world stage. Russia is another important case in point. Admittedly, much of the current agenda in U.S.-Russian relations is dominated by legacy issues from the Cold War, including the status of nuclear arms control treaties and the ongoing safety of nuclear materials, along with more recent Cold War–like irritants, such as Russian opposition to NATO expansion and the deployment of U.S. missile interceptors in Poland and the Czech Republic. Nonetheless, energy issues have come to play an ever more important role in Russia's ties with the West and with the former Soviet republics on its periphery. In a concerted drive to restore Russia's status as a major regional power, Russian leaders led by Vladimir Putin and Dmitri Medvedev have sought to harness their country's massive oil and natural reserves—and its control over pipelines stretching from Central Asia to Western Europe—to become the world's first "energy superpower." This, in turn, will pose ongoing challenges to U.S. policy makers as they seek to ensure the political and economic independence of the former Soviet republics and to promote energy security within NATO.[7]

For some countries and areas of the world, the old foreign policy agenda still holds considerable sway. America's ties with the two Koreas, for example, are still dominated by a schism that dates back to the 1950–1953 war and to North Korea's pursuit of nuclear weapons. In many areas, however, the new foreign policy agenda is almost totally ascendant. In Africa, for example, problems of energy, the environment, and the economy reign supreme. On one hand, Africa is expected to suffer more from the onset of global climate change than any other region, producing massive humanitarian disasters and possibly widespread state collapse. On the other hand, Africa possesses vast untapped supplies of oil and natural gas that could, if developed in a prudent manner, generate the capital to spur development and allow African states to overcome problems of underdevelopment, global warming, hunger, AIDS, and other afflictions. The three e's are also likely to play an increasingly pivotal role in U.S. ties to Central America and the Caribbean, the Middle East (where water scarcity is becoming a matter of growing concern), Central Asia, and South and Southeast Asia.

In conclusion, it appears that American policy makers and diplomats will have to become increasingly cognizant of the role of the three e's in international affairs. Whereas effective policy making once required a thorough knowledge of NATO politics, Soviet affairs, and the nuclear arms race, policy makers must now become familiar with the international financial system, the worldwide energy trade, and, more and more, the causes and mitigation of global climate change. Fortunately, many senior officials—beginning with the president and secretary of state—seem aware of this need, but much more effort will be required to educate American policy makers to successfully address this complex new agenda.

Notes

1. Institute for National Security Studies, National Defense University (NDU), *Global Strategic Assessment 2009: America's Security Role in a Changing World* (Washington, DC: NDU Press, 2009). Preview edition available online at http://www.ndu.edu/inss.
2. National Intelligence Council, *Global Trends 2025: A Transformed World* (Washington, DC: U.S. Government Printing Office, 2008).
3. For discussion of these dangers, see statement of Thomas Fingar, Chairman of the NIC, before the House Permanent Select Committee on Intelligence, 110th Cong., 2nd Sess., June 25, 2008.
4. U.S. Department of Energy, Energy Information Administration (EIA), *International Energy Outlook 2008* (Washington, DC: DoE/EIA, 2008), Tables A5, G1, pp. 100, 209.
5. For background and discussion, see Michael T. Klare, *Rising Powers, Shrinking Planet: The New Geopolitics of Energy* (New York: Metropolitan Books, 2008), esp. chaps. 2, 3, 6, 7, and 8.
6. DoE/EIA, *International Energy Outlook 2008*, Table A10, p. 105.
7. For background and discussion, see Klare, *Rising Powers, Shrinking Planet*, pp. 88–145.

Panel I: A Global Perspective

Summary of Discussion

Dr. Thomas R. Fedyszyn
Professor of National Security Affairs
Chair Eurasia Regional Study Group
Naval War College

The Ruger Chair Workshop's goal of developing a better appreciation for American foreign policy must necessarily begin with a better understanding of the world we face at the start of the Obama administration. It is imperative that this appreciation be objective in that it reflects more than just the views of the Washington decision-making elite.

The first presentation, by Patrick Cronin, highlighted the National Defense University's forthcoming *Global Strategic Assessment*, which will attempt to achieve this goal, integrating 125 contributions from American and international scholars. Owing to the complexity of *how today's world works*, the editors will identify the broad trends of the international system, then allow them to be put through regional filters in an attempt to develop prescriptive thinking for the United States. This must be the foundation for any effort aimed at developing grand strategy. A crucial observation is that the world is being driven by two sets of forces. The first can loosely be termed "globalization": those transnational forces whose actors can be anything from individuals up to nongovernmental organizations (NGOs) such as Hezbollah, which have access to destructive capabilities and may be supported by nation-states. The second is the traditional forces of geopolitics, whose actors remain nation-states. Both sets of forces, and their mutual interaction, must be understood prior to the successful determination of American foreign policy.

The eight principal national security challenges facing America are the following:

1. A gradual global redistribution of economic power from the West to the "rest" is under way, and economic power is the bedrock of enduring military and political power.
2. We are on the cusp of but not yet in a multipolar word.
3. The globalization of communications is challenging more than just the virtual foundations of the postmodern information society.
4. Energy and environmental insecurity have reached a tipping point. The era of cheap hydrocarbons and scant ecological regard is finished.
5. Transnational terrorism and stateless actors can inflict unprecedented damage, and we must be on our guard against catastrophic terrorism.
6. September 11 and growing insecurity in Afghanistan and Pakistan remind us of the growing challenges posed by fragile states and "ungoverned" spaces, with the "bottom billion" in some sixty countries left behind in dire poverty.

7. The character of war is changing, forcing low-level uses of force and greater civil-military integration in a renaissance in counterinsurgency and irregular warfare. We must prepare for "hybrid warfare" and hedge against emerging peer competitors.
8. The proliferation of weapons of mass destruction has made it increasingly possible that nuclear or biological weapons may be used in the coming years, including uncontrolled biological agents.

The principal failing of the United States has been its inability to incorporate nonmilitary factors into long-term strategic planning. We are superb at dissecting and solving problems that are amenable to a military solution. Our strategic frameworks should also include diplomatic, economic, and social perspectives as we propose solutions. Then, we must fight the human inclination toward instant gratification and realize that some issues are resolved only in the course of generations.

There are five pathways to get to a game-changing strategy for the United States:

1. ***Heal thyself.*** To a remarkable degree, security hinges on America having its house in order.
2. ***Redefine problems.*** Ends should be realistic, recognizing emerging interrelationships among energy, the environment, food, and climate change.
3. ***Surge civilians.*** Complex challenges require a larger whole-of-government team of national security professionals.
4. ***Countermobilize.*** The United States can mobilize emerging power centers into action through bilateral alliances, coalitions of the willing, and multilateral institutions.
5. ***Exercise strategic restraint.*** We cannot afford quagmires that drain resources without providing lasting security. A strong military is the U.S. ace in the hole, but better still are indirect approaches, strategies of leverage, and "smart power."

Can this administration change the way we develop foreign policy as well as mobilize the world behind us? Only if the United States doesn't spend its power with reckless abandon and it better takes into account factors beyond the military that must shape policy. The principal nonmilitary factor to be underscored in the globalized world is economics. Economic factors must be systematically integrated into American security planning. Our principal agencies entrusted with this responsibility have been woefully inadequate at injecting the economic element into American grand strategy. Economics is crucial in the development of policy toward fragile and failing states—areas that will be central to the determination of our forthcoming foreign policy.

Michael Klare, in a second presentation outlining a new foreign policy agenda for the next administration, agreed completely with the need to highlight economics and added two other factors beginning with "e": energy and environment. These "three e's" are closely interconnected and form the new basis around which we should build foreign policy. By contrast, we had been living in an "a-dominated" foreign policy world: arms races, alliances, adversaries, and arms control.

The United States is viewed by the entire world as the principal architect of the world's liberal economic order and will therefore receive most of the blame for the global recession. This American failure will undoubtedly result in the decline of American political power and prestige. Since political power and influence remain, to some extent, a zero-sum game, nations like China will gain in importance and prominence.

However, despite the damage done to the economies of the world's great powers, the developing nations will face an even worse prospect because of today's economic conditions. Their recession will last longer and be more profound for their citizenry. These economic disruptions are likely to manifest themselves as regime-threatening instability, civil war, and general disorder. While many possible examples come to mind, Pakistan may be the worst possible situation. Here we have a nuclear-capable nation with anemic economic growth and increasing hunger, poverty, and lawlessness, where the local citizenry is more likely to accept the role of antigovernment forces such as the Taliban.

Energy challenges run parallel with those in the economic realm, and are almost as important. The global economy cannot function without the global flow of petroleum resources. All critical industries are affected by perturbations in the flow of oil. The price of food—which can be the source of domestic riots—is also related to energy supplies. The need to ensure both energy sufficiency and security has driven American foreign policy and attendant military forces to seek access to remote spots in the world like sub-Saharan Africa. More important, the safe delivery of Middle East oil has largely been responsible for the creation of the Central Command and was a crucial factor in the U.S. determination to fight two wars in this theater. This suggests that all oil-producing regions might eventually entice America to get more assertive, leading to the need to expand the Carter Doctrine, which declared that the safe delivery of the Middle East oil supply was a vital national interest. Should the Caspian Sea region assume such prominence, Russia would undoubtedly consider this a belligerent American posture.

Environmental concerns—principally climate change—must be viewed from two perspectives in foreign policy. First, the United States must develop international teams able to respond to a wide variety of catastrophes. These events will easily overwhelm the ability of poor states to respond. While African states are most likely to be devastated by climatic issues such as drought and floods, the rising sea levels are already threatening low-lying nations in the Pacific and Indian oceans. These forces likely lead to anarchy, state collapse, and almost certainly huge refugee flows. Second, the United States must be the vanguard of the diplomatic initiative to develop more aggressive climate treaties, even though it will take generations to undo the damage already done. Along these lines, there is little reason to expect support from either China or India, since they consider industrial growth their ticket to prosperity.

The question period began with a discussion of just how violent the world will be, taking into account all these new forces of globalization and the three e's of the economy, energy, and the environment. Respondents allowed for the possibility that the world may indeed be less violent but still be more precarious.

The new transnational forces of migration, energy insecurity, nuclear proliferation, and terrorism are still not sufficiently understood such that we can understand how dangerous they are. Further, the intensity and duration of the global recession will define the severity and likelihood of violence in many of the lesser-developed countries. As global prosperity diminishes, so rises the level of violence.

Another questioner asked how we might begin to prioritize the challenges we face in the future, both by issue and by region. Does anything dominate the debate and the intellectual market share of the discussion? Since so much has passed out of pure governmental control and out of bureaucratic domains, we should be skeptical and humble about what nations can do to address these uncertainties. The United States has always been exemplary when it responded to threats and crises, yet many of today's issues, such as the environment, may never have a catastrophic event around which to mobilize. Does this leave us with no ability to rise to this challenge? Mobilizing and organizing American power is rarely done right in a preventative way. Discussants concurred that the U.S. government was not adept at translating its response into genuine strategy, policy, implementation, and follow-up. However, there was a sense that the Obama administration had managed to "take a play out of the Chinese playbook" and managed to "lower the temperature" as security challenges are treated not as issues imperiling the existence of the state, but rather as items on which compromises can be stricken. Our new willingness to talk with all parties has delivered the United States from its status as "Great Satan" in Iran, even if we are a still a ways from settling many of our issues. A discussant suggested that the Iranian elections (held on June 13) could be materially affected by Obama's presence in that students and intellectuals will find it easier to mobilize against the government. That is, Iranians are now suggesting that America is changing. The Obama administration (through Director of Intelligence retired admiral Dennis Blair) is painting the international security environment as one replete with opportunities, not only threats and challenges. Ostensibly negative trends like resource limitations can have the benign effect of assisting the worldwide move toward improving the environment and developing energy self-sufficiency.

A participant proposed that the United States should embark on a grand strategy driven by the theory of positive control, based on the writings of a former naval strategist. It is crucial that we first identify which of the trends noted earlier can be manipulated to our benefit, then focus on them. Some potential areas of control are redistribution of economic power, development of a multipolar world, countering the proliferation of the weapons of mass destruction, and, finally, energy and environmental security. Further, if the principal tool used is diplomacy, America will be more likely to engage in a positive way.

A questioner noted that the discussion of economic issues cannot take place without regard to domestic pressures, the principal obstacle in the development of grand strategy in any of the three-e areas of economy, energy, or environment. This will be a difficult sell to the American people, but the reality is that America must be sold on the importance of coming to terms with climate change. Only then will we have a place in the international debate.

The Kyoto Agreement couldn't be sold to Congress, but polling data suggest that the American people are now out in front of their legislators and are willing to pay the required costs. Of course, the lobbyist efforts in Congress to ensure that the burden does not fall too heavily on one industry or district will never go away. A congressman from Detroit will have his carpet worn out by auto industry lobbyists. A respondent concurred with this observation and argued for more American restraint and "offshore balancing" as our preferred approaches. The respondent criticized those who felt that "boots on the ground" was the key to solving all foreign policy challenges. This is commonly the root, not the solution, of the problem.

A questioner added that the domestic element of foreign policy also manifests itself in "protectionist" policies in trade, especially driven by those espousing that the goal of foreign policy is the preservation of prosperity of the United States. There will be serious domestic opposition to any trade agreements not catering to domestic industries. A participant responded that the United States has always been on the side of "open trading" or reduced domestic tariffs. However, domestic opposition to freer trade will put at risk growth opportunities for the world economy. In particular, the participant felt that the forces of nationalism and protectionism could do irreparable harm, noting that the Doha Round had already been put in abeyance. It is too difficult to forecast the specific degree to which these tendencies will hurt the recovery of the international economic system.

A participant noted that there will always be a tension as we try to put behind ourselves the "a" factors of security (arms races, alliances, arms control, adversaries) and move to the "e" factors (energy, environment, economy). It is easy to say that the world is moving in the direction of these transnational forces, but they inevitably spiral back to traditional geopolitical competitions. The participant offered Caspian Sea energy resources as a good example where the e's rapidly evolved into a traditional a-style controversy, noting Russia's belligerence and adventurism. Another participant concurred, citing a recent European Union (EU) conference on alternative sources of energy supplies. The EU now has to begin to focus its attention on new regions of the world (e.g., Africa) to solve its energy shortages and environmental concerns. There is no desire in Brussels to require traditional military responses, but European capitals are wary of the possibility. In fact, energy concerns will likely lead us back into geopolitical rivalries, alliances, and regional arms races, even as we hope to evolve to a new style of cooperative foreign policy.

A questioner was more optimistic about the prospects for the evolution of American foreign policy in the underdeveloped world as well as the growth of prosperity and democracy. For example, in Zimbabwe, the Internet is helping to create a genuine political debate in the country. It is also instrumental in the diffusion of power "eastward." As the United States develops its foreign policy, it must understand the limits of military power. Since complex transnational challenges will not be solved by America alone, we must instinctively look to regional powers like Russia or China with whom we can make collective responses. The questioner saw a positive and constructive example in the global response to piracy, in which all the world's major powers are working together for a common cause. A discussant noted, however, that Beijing will undoubtedly protect Chinese shipping with the presence of its destroyers in the Gulf of Aden. However, it has no intention of affording indirect protection to global shipping. This is an area where we cannot count on regional powers to adopt an attitude that they must maintain the global commons as we have done. While this is a positive trend and a step in the right direction, we are still in a geopolitical, nationalistic world.

A questioner averred that the United States preferred a unipolar world and could operate most effectively within it. The questioner asked how the United States was reacting to the growth of other power centers along with the growing powers of nongovernmental organizations. A respondent held that we never lived in a unipolar world. This has always been a fiction that many believed simply because one power left the scene in 1991. At best, we had the world characterized by Joseph Nye as a chessboard at three levels. In the military sphere, the United States was clearly dominant, but was merely a leader in the other forms of power (economic and diplomatic). Over the last decade, America has been severely challenged in its attempt to translate one type of power into the next. The respondent agreed that the world is multipolar and other power sources are capable of providing the same public goods the United States has attempted to deliver—even if we haven't always done so well—for the last generation. The idealistic spirit of America is having a hard time accepting this devolution. However, until the other regional powers show more responsibility (in maybe two, four, or eight years), we may continue in this role by default. This is counterproductive and could yield a backlash in many regions where local cultures might feel trampled by American force. For American foreign policy to succeed around the world, we must not create the conditions that instigate locals to mobilize against American power.

Another participant held that the "bottom billion" has not profited from the same globalizing tendencies that have improved the economic conditions of most of the world population. Today's global economic recession will undoubtedly hurt the world's poor much more than the affluent and will push hundreds of millions more into the bottom tier. Failing an exuberant comeback in the world economy, people will become not only poorer, but also more bitter and disaffected toward the United States. This resentment will enable nongovernmental actors to recruit and train terrorists, criminal syndicates, and pirates to harm American interests abroad. This should be reason enough to ensure that American policy to recover from the recession should take into account the economic interests of the developing world.

Panel II

Western Hemisphere

Peter Hakim
President, Inter-American Dialogue

Dr. Shannon K. O'Neil
Douglas Dillon Fellow for Latin America Studies, Council on Foreign Relations

Ambassador (Ret.) Paul D. Taylor
Senior Strategic Researcher, Naval War College

Moderator:
Professor Laurence L. McCabe
Associate Professor of National Security Affairs, Chair Latin America Regional Study Group, Naval War College

Another Chance for U.S. Policy in the Americas

Peter Hakim
President, Inter-American Dialogue

The election of Barack Obama was enthusiastically welcomed throughout Latin America and the Caribbean. The choice of an African American leader committed to universal values revealed the vitality of U.S. democracy to many in the region who had become skeptical. By their spirited reaction to the new president, Latin Americans have made clear that they want a better relationship with the United States, but they also want Washington to approach the region differently.

But the Obama administration faces powerful constraints to its policy in Latin America. By necessity, U.S. priorities are directed elsewhere, and the nation's resources are stretched thin. The country now confronts its worst economic crisis in seventy years and, at the same time, is fighting two overseas wars.

U.S. hemispheric policy is also circumscribed by Latin America's ambivalence about the role the United States should play in regional affairs. Latin American governments are today far more independent and assertive than ever. Brazil has become an alternative pole of power in the hemisphere, with a steadily increasing regional and global profile. A few countries, led by Venezuela, have become adversaries of the United States. Most Latin American nations have developed a diversity of international ties, and many advocate new hemispheric arrangements that would diminish Washington's regional influence. These are all elements of continuing, longer-term trends in inter-American affairs that the United States cannot reverse. American interests will be best served by adjusting U.S. policy approaches to the growing independence, confidence, and competence of Latin American nations.

Although the upsurge in anti-American sentiment, which followed the U.S. invasion of Iraq, abated in the final years of the Bush administration, Latin Americans were still alienated by Washington's unilateralism, excessive reliance on military force, and disregard for international rules and institutions. U.S. credibility was badly damaged as well by several regrettable policy choices in the region—Washington's inattention to Argentina's impending economic collapse in 2001; its uncompromising approaches to Cuba; its quick praise for the 2002 coup against Hugo Chávez, which was reversed a day later; the rigidity of U.S. antidrug policies; and plans to construct a "wall" on the U.S.-Mexican border. The U.S. financial meltdown, which has put Latin America's recent impressive social and economic progress at risk, is now a fresh source of resentment.

The Obama administration's preparation for and participation in the 2009 Summit of the Americas in mid-April turned out to be a good period for U.S.–Latin American relations. The new administration set in motion a series of measures that were applauded by nearly every country of the region. In his personal interactions with the Latin American and Caribbean leaders assembled at the Summit, President Obama succeeded in building a new measure of trust and moderating the hostile attitudes of the several countries.

Washington today has a new chance in the Americas. But Latin Americans will need to be convinced that the United States can be counted on as a dependable partner and responsible neighbor. The new administration needs to align U.S. policies with the changed conditions of Latin America and put hemispheric relations on a new, more cooperative course. This is a time to solve problems, reduce discord and friction, and take advantage of opportunities for joint action. Ten critical challenges need to be addressed:

1. The financial crisis
2. Policy toward Cuba
3. Integration with Mexico
4. Crime, violence, and drugs beyond Mexico
5. Immigration policy
6. The unfinished trade agenda
7. Cooperation with Brazil
8. Venezuela and its allies
9. Democracy's problems
10. Failing Haiti.

Blunting the Impact of the Financial Crisis

The United States has plunged into its deepest economic downturn since the 1930s, and the recession may last for another year or more. The economic progress and social welfare of every Latin American nation are today at risk. To be sure, nearly every country is far better prepared today to withstand these external shocks, but Latin American growth will plummet to zero or worse for the region as a whole. Several countries will fall deeply into recession. The region's impressive advances in recent years—in lifting growth rates, keeping inflation low, building a significant middle class, and reducing pervasive poverty and inequality—could be reversed.

An extended period of economic hardship could produce dramatic political shifts as well. In some countries, the crisis will provoke anger toward governing authorities and institutions. Politics may be further polarized, and political stability put at risk. Popular frustration may lead to diminished support for democracy and markets—and it may lead to a new rejection of the United States. How the crisis is managed by Washington and the governments of Latin America and the Caribbean could well shape inter-American relations for years to come.

By insisting that the most important thing the United States can do for Latin America is resolve its own economic problems, Brazilian president Luiz Inácio Lula da Silva made it clear how vital the U.S. economy is to the region. More than any event in recent memory, the U.S. recession has highlighted the deep interdependence of Latin America and the United States. The slumping U.S.

economy has led to declining investments, remittance transfers, and other capital flows to Latin America; decreasing export volumes and lower prices for the region's products; and sharply diminished access to international credit.

Washington can also help Latin America by rejecting protectionist measures that raise barriers to imports, favor U.S. manufacturers, subsidize exports, restrict U.S. investments overseas, or impose new limits on immigration or remittances. Such measures—which would include the "buy American" provisions proposed for the U.S. stimulus package—would deepen Latin America's economic woes and delay recovery.

The United States can also help Latin America weather the crisis by working to mobilize support for an expansion of the resources and programs of the International Monetary Fund (IMF) and multilateral development banks. Except for Chile, which set aside a sizable portion of its copper revenues when prices were high, Latin American governments have very limited capacity to finance countercyclical fiscal policies on their own. Unlike the United States, Europe, or China, they simply do not have the savings, access to credit, or room in their budgets to pay for stimulus programs or new social benefits. Nearly everywhere in Latin America, economic stimulus efforts will require support from multilateral institutions.

The Obama administration has supported the position of most Latin American countries in strongly endorsing expanded resources for these institutions (including the Inter-American Development Bank, World Bank, IMF, and Andean Development Corporation) along with greater flexibility in their programs and increased participation of emerging market nations in their governance. Latin America's poorest countries, including Haiti, Bolivia, Honduras, Nicaragua, and Guyana, will probably require increased aid flows to sustain viable economies and protect their vulnerable populations. To be sure, the "buy American" provision in the Obama stimulus package troubled many Latin American countries as a protectionist measure, but the administration was able to soften what had initially been a far more damaging provision.

The Group of 20 (G-20), made up of the world's leading economies, is responsible for harmonizing global responses to the financial crisis. Its five participants from the Americas—Argentina, Brazil, Canada, Mexico, and the United States—have not yet done much to establish mechanisms to exchange information, analysis, and ideas on the hemispheric dimensions of the crisis or seek to coordinate policy approaches. This would be a helpful contribution to hemispheric cooperation, on economic and other issues.

Opening to Cuba

Cuba is not an urgent concern for the United States. But there is no other issue on which Washington is so out of step with the rest of the region. Nothing would better demonstrate the new administration's intention to pursue a fresh approach to Latin America than making a quick start to dismantle the web of restrictions that the United States has imposed on Cuba.

The Obama administration has wisely started to fashion a new Cuba policy by taking the very modest step of scrapping all restrictions on family travel and remittances to the island. These measures have majority support in the Cuban

American community, and will help to defuse the community's long history of bitter opposition to changes in U.S. strategy toward Cuba.

President Obama has called for a reciprocal step from the Cuban authorities in Havana. But, instead of pursuing only a bilateral approach, Washington should consider pushing some of the burden elsewhere. For example, the United States could stop trying to block other countries and multilateral institutions from doing business with Cuba. It should, once and for all, end its efforts to keep the Organization of American States (OAS) and multilateral development banks from engaging Cuba—and, instead, encourage these agencies and other nations to facilitate Cuba's reintegration into hemispheric affairs, and assist the island's transition toward an open economy and political system.

President Obama could also transfer some of the burden from the U.S. government to local communities and private citizens. He could make it easier for academic, cultural, and athletic exchanges with Cuba by relaxing the bureaucratic obstacles and overdrawn restrictions that now apply. The United States could allow an expansion of its already vigorous agricultural trade with the island.

Ideally, Washington should seek to establish a wide-ranging dialogue between U.S. and Cuban authorities (as it did earlier with Vietnam) that would set the two countries on a course toward normal diplomatic and commercial ties. There should, however, be no question about Washington's support for Cuba's advancing toward free expression and association, the rule of law, respect for human rights, and competitive elections. But a democratic society in Cuba should be the objective of U.S. engagement, not a precondition.

Mexico: Working with a Troubled Partner

Mexico will present the new U.S. administration with both its toughest challenges and greatest opportunities for productive cooperation. For both countries, sustained cooperation on an array of bilateral, regional, and global issues is essential. The two nations share a 2,000-mile border that is crossed some 250 to 300 million times per year. Mexico sends upward of 80 percent of its exports to U.S. markets and is the United States' third-largest trading partner, after Canada and China. It is also overwhelmingly the largest source of immigrants to the United States—both legal and illegal. Over the longer run, the central challenge is managing the accelerating economic and demographic integration of the two nations.

Today, Mexico faces a complex of dangerous security problems, aggravated by economic recession (which the swine flu has made worse). If the country's security conditions deteriorate further, Mexico could become one of Washington's most troublesome foreign policy tests. For the past two years, the Mexican government has waged a fierce military campaign against drug gangs and other organized criminals, who have been terrorizing many parts of the country. The violence has continued to escalate, wreaking havoc on public safety and the rule of law; undermining the credibility of the nation's army, police, and justice systems; and, in some areas, undercutting the governmental authority.

The threat should not be exaggerated, however. Neither Mexico's democracy nor its key institutions are imperiled by criminal activity. Mexico is not in danger of becoming a failed state. Alarmist interpretations of developments in Mexico—which have been advanced by U.S. military and intelligence agencies—may well lead to misguided policy prescriptions and make it harder for the United States and Mexico to cooperate on security matters. Mexico is going through an extremely difficult period, but the problems require careful and intelligent assessment, not fearmongering.

In its first three months, the Obama administration has been highly attentive to Mexico. Prior to his inauguration, President Obama met with President Felipe Calderón in Washington, and visited Mexico on his way to the Summit. Three other cabinet members, including Hillary Clinton, also traveled to Mexico in March and April. Mexican officials were pleased to hear Obama and his advisers emphasize that U.S. drug consumption and arm sales made the United States co-responsible for the unsavory developments in Mexico, that Washington would speed up the delivery of promised equipment and other aid, and that the United States would intensify its efforts to reduce both the use of illicit drugs and the smuggling of weapons to Mexico. The administration has been reluctant, however, to battle the powerful U.S. pro-gun lobby to stop legal sale of assault weapons.

Confronting Crime, Violence, and Drugs

Violent crime and drug trafficking are not only a Mexican problem. They have emerged as an urgent concern in nearly every country in Latin America and the Caribbean—and many governments are seeking U.S. cooperation to deal with them.

With U.S. funding of $5 billion over the last nine years, the Colombian government has succeeded in gaining more effective control over its territory and reducing armed violence from guerrillas and paramilitary forces. The country has had less success in battling illicit drugs, or addressing human rights concerns. Colombia's advances have been mainly due to the strengthened authority and competence of public institutions, not to the disruption of the still-flourishing drug trade.

Colombia's striking progress may now mean that U.S. military aid can be reduced without risking the country's security. But there remains an urgent need for continuing support to help Colombia advance human rights and humanitarian goals and pursue an expanded social agenda. These have suffered neglect during the years of warfare.

Under the Merida Initiative, modest aid is available to Central American and Caribbean governments. But public institutions in these regions are weaker and more vulnerable than in Mexico—and in greater danger of being overwhelmed by criminal activity. Their reliance on U.S. trade, investment, tourism, and remittances puts their economies at grave risk. Waging a successful fight against crime will require additional support from the United States. Washington could also contribute by intensifying its efforts to control rampant arms smuggling and by reviewing the practice of deporting convicted felons to their countries of origin, where they are often recruited into vicious street gangs.

Virtually everywhere in the Americas, crime and violence are fueled by illegal drug profits. It is painfully clear that U.S. antidrug efforts are not doing much either to cut supply or reduce demand. U.S. consumption of cocaine decreased from its peak in the early 1970s, but has remained essentially stable at a rate three times that of Europe. On the supply side, eradication and interdiction—the two pillars of the U.S. antidrug battle—have lost credibility in most quarters. From time to time, progress is made in one or another country, but production and trafficking are then quickly shifted elsewhere. And Latin American countries are no longer just suppliers or transit points; they have become major consumers of drugs.

Sadly, Washington has not learned much from its more-than-twenty-year war against drugs. Latin American governments resent Washington's inflexible approach to fighting drugs. They are frustrated by the unwillingness of U.S. official agencies and political leaders to question current strategies or consider alternatives. For a decade or more, policy debates and discussions on the issues and approaches have been muted. U.S. programs have not been rigorously scrutinized or evaluated.

What is needed is an honest, well-informed, and wide-ranging exploration and debate on alternative drug policies. That will require a major hemisphere-wide initiative to collect the statistics and conduct the research, evaluation, and experimentation needed to diagnose the problems, assess current policies, and test new proposals. Washington should relinquish its dominant, often suffocating, role in shaping counternarcotics efforts in the hemisphere and cooperate with other governments to develop fresh ideas and strategies.

Reforming Immigration Policy

For a dozen or more countries in Latin America and the Caribbean, U.S. immigration policy is the single most pressing issue in their bilateral relations with Washington. How illegal migrants are treated in the United States has become a politically heated issue across the region, while the rising number of deportations from the United States contributes to criminal violence in many places. Remittances from U.S.-resident family members have become a vital source of income for millions in Latin America. The nearly $70 billion transferred annually is also critical to many economies, particularly as other sources of capital are drying up. Within the United States, immigration policy has divided the American people and provoked bitter debates that are often offensive to the migrants and their countries of origin.

The U.S. immigration system is broken. It badly serves U.S. and Latin American interests, and it has become a constant source of friction between Washington and governments in the region. Even though the issues are contentious, considerable agreement has emerged on the key elements of a new approach to immigration. They include (1) offering sufficient work visas to satisfy U.S. labor market demands; (2) providing legal status (and the opportunity to earn permanent residence and citizenship) to migrants residing in the United States illegally; and (3) putting in place effective (and humane) incentives and enforcement mechanisms to curb illegal migration. This, in fact, was the core of the comprehensive reform proposed by President Bush and rejected by the U.S. Senate in 2007.

Despite considerable political risks, President Obama has declared his intention to push forward with immigration reform this year. He faces the same difficulty as his predecessor—translating agreed-upon guidelines into policies and laws that are politically viable in the United States and supported by Latin American governments. The economic security concerns of many U.S. workers also have to be confronted. Their worries about losing jobs to immigrants or seeing health and education services deteriorate may be exaggerated, but if they are not effectively addressed, they will stand in the way of reform.

The Obama administration's decision to suspend mass raids targeting illegal immigrants—which presented a disturbing image of discrimination and abuse—has been welcomed in Latin America. The administration has not, however, discontinued construction of the fence along the U.S. border with Mexico, which is also a highly charged symbol of disrespect.

Completing the Unfinished Trade Agenda

Hemisphere-wide free trade should be a critical long-term goal for the nations of the Americas, but not much progress is likely until the global economic crisis abates. Washington should now concentrate on completing the unfinished agenda left by President George W. Bush.

- The new administration needs to give more attention to the signed, but as yet unratified, Colombia and Panama free trade agreements. If it decides to ignore them or defer action into the future, the United States would justifiably be seen as backtracking on its commitments to the two countries. Securing ratification of agreements (which in the case of Colombia will require negotiation of an amendment on human rights) would help to reassure Latin American governments that Washington is a reliable partner. The Colombia accord has been a casualty of the partisan rancor in U.S. politics. Its approval would show that the Obama government is committed to working across those divisions.
- In recent years, public support for free trade has diminished sharply in the Americas. There is little prospect for significant new trade agreements until ordinary citizens in the United States and elsewhere regain confidence that trade deals boost growth, employment, and living standards—not push down wages and export jobs. Washington and other governments must do more to mitigate the dislocations that free trade can produce. Increasing the economic security of U.S. workers is the basis for building any bipartisan coalition for freer trade.
- It seems increasingly clear that, campaign rhetoric aside, President Obama is not planning to open the North American Free Trade Agreement for renegotiation. Washington needs to find a way to implement existing provisions that remain blocked, such as allowing Mexican-owned trucks to haul goods into the United States. The recent suspension by Congress of a pilot program on trucking was a step

backward; Mexico responded, as it was entitled to do, by raising tariffs on a range of U.S. products.

- The United States should reinstate trade preferences for Bolivia. These were discontinued when the Bolivian government expelled the U.S. ambassador and suspended U.S. counternarcotics programs. Bolivia's actions may have demanded a response from Washington. But ending the trade preferences, which could eliminate upward of one hundred thousand jobs in Bolivia, was viewed as too harsh a penalty by most Latin American governments and many in the U.S. Congress.
- Prospects are limited for progress in the Doha Round of multilateral trade negotiations. Still, the United States should seek agreement with Brazil on a negotiating formula that would be acceptable to other participating nations. Brazil is now one of the most influential participants in the Doha talks and shares many U.S. objectives.

Cooperating with Brazil

The United States needs the cooperation of Brazil and Mexico to deal with almost every challenge it faces in this hemisphere—and with many global concerns as well.

Brazil's rapidly escalating regional and global influence represents a pivotal change in inter-American affairs—and an encouraging development for the United States. To be sure, the two countries are at odds on many issues. Still, Washington has maintained warm ties with the Lula government and has considered Brazil a constructive force in hemispheric affairs in recent years. Brazil has led peacekeeping operations in Haiti for the past four years and has helped resolve some highly charged conflicts in South America.

Neither Brazil nor the United States is yet ready for a broad, long-term partnership. The two nations should, however, be able to cooperate effectively and consistently on specific issues of mutual concern.

For example, Brazil and the United States share overlapping interests in the Doha negotiating round and should be able to work together more actively to secure a favorable outcome. The potential for productive collaboration may be even greater in the areas of climate change, environmental protection, and new energy sources. Since Obama became president, these issues have become prominent on Washington's global agenda—substantially expanding the opportunities for U.S. cooperation with Brazil and other countries. The United States should explore more active policy coordination with Brazil in other areas as well—for instance, in nuclear nonproliferation, battling racial and ethnic discrimination, and the reform of multilateral institutions.

By arranging for an early visit to Washington by Brazilian president Lula da Silva and promising a return visit to Brasília in short order, President Obama has made plain his interest in building a close and cooperative relationship with this increasingly influential country. The easy relationship and broad agreement of the two leaders at the G-20 and the Summit of the Americas was also encouraging.

The Challenge from Venezuela

Washington has been most brazenly challenged in this hemisphere by the government of Hugo Chávez in Venezuela and its allies in Bolivia, Cuba, Nicaragua, and, to a lesser extent, Ecuador and Honduras.

The Venezuelan leader has been a polarizing force in inter-American relations since he assumed office in 1999. He has fueled internal conflicts in several Andean countries, disrupted the operations of regional institutions, and developed close ties to U.S. adversaries worldwide. Although his anti-U.S. alliance currently incorporates only four or five of the weakest and least stable countries in the hemisphere, many other countries look to Venezuela for financial support.

President Chávez should be less troublesome in the coming period. The depressed price of oil, which accounts for 90 percent of Venezuela's exports and half of the government's expenditures, will cost him political support at home and reduce his regional influence. Moreover, his economic dependence on the U.S. market reveals the emptiness of his threats to curb oil exports to the United States. There is no urgency for the new U.S. administration to engage him or his government or to remake U.S. policy toward Venezuela. Washington should keep the Venezuelan leader at a distance and let the situation play itself out over the coming months.

At the Summit, President Obama, probably more than he would have liked, appeared ready to engage Hugo Chávez and perhaps pursue a détente with him. Most of Latin America would welcome this, but the Obama administration should proceed cautiously with the unpredictable and intemperate Venezuelan president.

Advancing Democracy

Democracy is the norm in the Americas today. Of the hemisphere's thirty-five countries, only Cuba is ruled by unelected leaders. Since 1976, the military has assumed power from a civilian government only once, in Haiti in 1991. Across Latin America, elections are the only path to political power, and nearly every election in recent years has been judged free and fair. But democratic politics means more than periodic elections. The fundamental institutions of democracy—political parties, legislatures, courts, electoral systems, and the press—still perform badly in much of the region and, in many, they have little public credibility. Political corruption is widespread.

Polarizing frictions have arisen in several countries between traditional political forces and newly enfranchised groups that want a larger share of power and changes in the rules of politics. Clearly, the widening of political participation to previously excluded groups—Afro-descendants, indigenous communities, younger and lower-income voters, and women—adds to the vibrancy of democracy in Latin America. So does the increasing attention to social justice and the delivery of public services. But these new political groups are placing mounting demands on already overstretched and poorly financed governments—and they have little patience with slow-moving bureaucracies and legislatures. The strains are obvious in many countries, and in some, the legitimacy of the political system has come under challenge.

The Inter-American Democratic Charter, approved by every elected government in the Americas in 2001, was designed to accomplish two objectives: to codify the rules of democratic practice and strengthen the ability of the hemisphere's governments to defend democracy collectively. But since the Charter came into force, the insufficient trust among signatory countries has prevented cooperative action on such a highly sensitive issue as democracy.

Latin American governments in recent years have not viewed the United States as helpful in defending or advancing democracy—or in resolving conflicts in the region. Washington's involvement has often been seen as an irritant and, at times, counterproductive. On issues of democracy, the rule of law, and human rights, Washington's credibility suffered from the war in Iraq and more generally from how it has managed its battle against terrorism.

Democratic progress in Latin America and the Caribbean will depend mostly on the government and citizens of each country. However, the United States can and should be an advocate for democracy in the region. Important issues and consequences are at stake. The new administration must first align U.S. rhetoric and practice on matters of democracy and human rights, in inter-American affairs and globally. It should also begin routinely to consult and cooperate with Latin American governments on political issues in the region—and trust their judgments and be prepared to defer to them. Latin Americans often want to deal with these issues on their own, and have the ability to do so.

President Obama has mostly raised democracy and human rights issues in his discussion of U.S. Cuba policy. But he has spoken very little about the sharp deterioration of democratic practice in several countries of Latin America, particularly Venezuela and Nicaragua.

Washington's advocacy of democracy is most credible and productive when it is carried out multilaterally. As the hemisphere's principal multilateral forum, the OAS should have the lead role in promoting and safeguarding democratic politics in Latin America and the Caribbean. Some governments have proposed to de-emphasize the OAS and shift its authority to recently formed Latin American and Caribbean organizations. These newer groupings have shown they can contribute in important ways. Still, the OAS is the only institution that has the legal authority and broad legitimacy to represent the hemisphere's governments and act regionally. It is also the only continuing forum in which Latin American and Caribbean governments can collectively engage the United States.

Failing Haiti

Haiti is the hemisphere's only failed or nearly failed state. In the past few years, international cooperation has contributed to some modest progress in Haiti. UN troops, mostly from Latin America and led by Brazil, have helped to maintain order and security. A freely elected government is functioning, although with its capacity and authority limited. The economy remains in deep distress. With several recent devastating hurricanes, high food prices (which provoked massive riots and the ouster of the prime minister last year), and the U.S. recession, the economy of Haiti may further unravel and its 8 million people become even more desperate.

No U.S. administration in memory has done enough to assist Haiti in a sustained way. The new administration has an opportunity to build on recent inter-American cooperation (including Canada's priority attention to Haiti) and establish a long-term, multilateral approach to Haiti's improvement. As a start, the Obama administration should take two quick measures to help the country during this period of extreme hardship: suspend the deportation of undocumented Haitian migrants and refugees, and encourage the multilateral banks to forgive Haiti's debt obligations. These issues were discussed during the April visit of Secretary of State Clinton, which, itself, was an encouraging initiative that suggested Washington was preparing to take a more aggressive role in Haiti.

A Second Chance

President Obama's election gives the United States another chance in Latin America. Conditions today appear less favorable than they were two decades ago. Throughout the region, there is a widespread sense of disappointment with the United States. Latin American governments are distrustful of Washington, and many doubt that it can be counted on as a reliable partner. They are increasingly acting independently of the United States and choosing their own courses.

Yet, paradoxically, this increasing assertiveness and independence may open the way for a healthier and more productive relationship with Washington—a relationship that will genuinely reflect the interests and preferences of both the United States and the region.

The vast majority of Latin American governments want good relations with Washington. They know that the United States is vital to their interests. They want strong trade links and other economic ties, and they want to work with the United States to solve other problems. And they are hopeful that the Obama presidency will bring needed changes in inter-American relations.

In its first months, the new administration has taken important initiatives on most of the key challenges it confronts in Latin America. The measures have been modest, not dramatic. But, aside from trade-related issues, they all move U.S. policy in the right direction and have been welcomed in Latin America—along with the changes in tone and texture that President Obama has brought to U.S. regional diplomacy. It is a good start toward a more respectful and co-operative U.S. relationship with Latin America and the Caribbean.

U.S.–Latin American Relations

Dr. Shannon K. O'Neil
Douglas Dillon Fellow for Latin America Studies
Council on Foreign Relations

Latin America is becoming strategically, economically, and politically more important to the United States than ever before. The region provides more oil to the United States than the headline-grabbing Middle East, and is also an important source of alternative fuels. It sends more immigrants to the United States—both documented and undocumented—than any other region in the world. Latin America is one of the United States' fastest-growing regional trading partners, with over a half trillion dollars' worth of goods exchanged annually. It is also the largest source of illegal drugs. No less important, nearly all Latin American nations are now vibrant, if imperfect, democracies. These transnational connections formed among individuals, communities, and economies constitute de facto U.S.–Latin American integration, outpacing formal government-to-government interactions.

The Obama administration enters at a time when most countries in the region feel neglected. This offers a distinct opportunity to reframe and reshape particular bilateral relations, as well as U.S.–Latin American ties more generally. But it also leaves the administration open to disappointment. Realistically, much less change in U.S.–Latin American relations will occur than many hope—or that some fear. But there is a potential for a real shift—not as much in substance, but in process.

Talking about U.S.–Latin American relations is often to use a misnomer. U.S. policy is increasingly wide-ranging and diverse. It is less and less U.S.–Latin American policy than U.S.-Brazilian, U.S.-Ecuadoran, U.S.-Guatemalan, or U.S.-Chilean policy. That said, there are issues and concerns relevant to and in fact requiring the cooperation of a significant number of Latin American countries. These main issues include public security, sustainable energy, economic advancement, and hemispheric migration. How these themes are addressed will have significant ramifications for the region, for the United States, and for U.S.–Latin American relations.

Public Security

An overriding concern of all governments in the region is public insecurity. In spite of the near absence of cross-border threats, Latin America is now the most violent region in the world. Its homicide rate is three times the global average. More than four out of every ten killings by gunfire globally occur in the region, even though Latin America contains only 10 percent of the world's population. The cost of this bloodshed is staggering—the Inter-American Development Bank (IDB) puts it at 14 percent of gross domestic product (GDP). In many nations in the region, people identify security and crime as the most important problem their countries face, ahead of other hot button issues such as the economy.[1] Much of this violence stems from the spread of local and transnational criminal networks and gangs involved in illicit activities ranging from petty crime to smuggling, kidnapping, and drug trafficking.

In many nations, law enforcement and judicial systems are unable or unwilling to take on these powerful criminal organizations, allowing a wide range of illegal activities to thrive. Prominent studies in Mexico show that over 95 percent of crimes go unpunished, and estimate that some 75 percent of crimes are not even reported. Polls show that up to a half of Latin Americans place very little trust in their police and judicial systems. According to Transparency International's Global Corruption Barometer, 10 percent of Latin Americans consistently report paying bribes in the past month. In many Latin American countries, a vicious cycle of corruption and weak state capacity—particularly in law enforcement and judicial institutions—helps drive violence and crime.

The narcotics trade flourishes in this general atmosphere of impunity and corruption, coupled with difficult socioeconomic conditions and high demand for narcotics in the United States, Europe, and increasingly in Latin America itself. With world cocaine sales alone worth $60 billion, the lucrative nature of these illicit markets breeds insecurity and violence, as contracts are often enforced or rewritten in blood.

A more recent concern for the United States and Central American countries in particular is the growth of transnational gangs. The two most prominent are the Mara Salvatrucha 13 and Mara 18, which have somewhere between fifty thousand and one hundred thousand members residing in the Central American nations (particularly in El Salvador, Guatemala, and Honduras) and in the United States. By many accounts these gangs originated in the United States in migrant neighborhoods, and then spread to Central America with the deportation of non-citizen criminals from the United States. There is substantial evidence of cross-border consultation among gang members, as well as movement back and forth between countries.

Drug trafficking, gangs, and organized crime networks in the region are complex issues with no easy solutions. Nevertheless, Latin American security is of crucial concern to the United States, due to the interconnectedness of our societies and economies. The associated violence limits economic growth and opportunities, and diminishes the quality of life for all.

To support Latin American countries better in their quest to improve citizen security, the United States should broaden and deepen efforts to assist law enforcement and judicial reform efforts in Latin America. Long-term sustainable solutions to these criminal threats will not result from increased military or police hardware, but instead must come from the strengthening of law enforcement and judicial institutions, reducing the impunity and corruption that deter citizen involvement and support, and that allow crime to thrive. This means a greater focus on technical issues of vetting, training, and investigation, helping to build the capacity of local institutions in receptive countries. It should also mean facilitating greater international cooperation and information sharing, working with governments and civil society organizations to evaluate programs and practices in other countries, and promoting innovative solutions to security problems. While this is occurring in many areas, particularly with respect to gangs in Central America, expanding efforts throughout the region is crucial.

The United States should also improve efforts on its side of the border to combat drug cartels and organized crime. First, it can better control the flow of guns into Latin America. More than 90 percent of the thousands of traceable illegal firearms confiscated in Mexico every year are traced back to the United States. To help stem the flow of guns south, the United States needs to enforce its own laws that prohibit the sale of firearms to foreign nationals or straw buyers (those who purchase guns for others), and that prohibit the export of guns to countries where they are prohibited by law. In practice, this means more resources for the Bureau of Alcohol, Tobacco, Firearms, and Explosives (ATF) to fulfill its job. It also means more vigilance at the U.S.-Mexican border, checking vehicles not just coming north but also going south.

Second, the United States can strengthen its current drug-related anti–money laundering initiatives. The Drug Enforcement Administration estimates that Mexican and Colombian drug-trafficking organizations launder some $15–$25 billion every year—most of it by shipping bulk cash across the southern U.S. border. Creating a system similar to the multiagency Foreign Terrorist Asset Tracking Group—formed in the wake of the terrorist attacks of September 11, 2001, to go after terrorist money—for narcotics-based money laundering will help in the dissemination of diplomatic, regulatory, and law enforcement information crucial to cut the flow of money that funds illegal activities and fuels violence throughout the hemisphere. Here, too, greater vigilance over cargoes heading south across the U.S.-Mexican border would also help.

Finally, the United States should establish a comprehensive drug policy that addresses not just supply but also demand. This means more resources and support for rehabilitation and prevention programs, helping those hooked on drugs get off of them, and encouraging those not yet involved never to start. These efforts must involve not just the United States, but cooperation with other drug consumer countries, most notably those in the European Union, which now receive nearly half of the cocaine produced in the Andean region.

Sustainable Energy

Energy security and environmental sustainability are closely linked priorities for the United States and Latin American nations. For the United States, greater and more sustainable energy security will come through the diversification of energy sources rather than through domestic production.

The hemisphere's energy matrices are already fairly closely knit. Canada is the United States' largest provider of oil, sending almost 2 million barrels of oil a day to the United States and constituting 18 percent of U.S. supplies. Latin America alone provides 30 percent of U.S. oil—more than any other region of the world, including the Middle East. After Canada, Saudi Arabia and Mexico both send about 1.2 million barrels per day to the United States, while Venezuela (the fourth major source) provides roughly a million barrels a day.[2] While a relatively stable energy region, declining Mexican production and difficult relations with Venezuela may force a shift in U.S. oil sourcing.

Latin America and the Caribbean also have the potential to be an important source of natural gas. Trinidad and Tobago is already the largest provider of liquefied natural gas (LNG) to the United States, and stands to benefit from projected increases in demand. Peru is taking advantage of the growing need throughout the hemisphere, and particularly in the United States, by investing in its LNG facilities for exports. Other countries, notably Bolivia and Argentina, have large supplies of natural gas, though current political decisions discourage investment in exploration and production.

Latin America is one of the largest sources of alternative energies. Brazil produces nearly 40 percent of the world's supply of ethanol, and is a leader in biofuel technology. The region has extensive hydroelectric resources, and is beginning to develop wind and solar energies. There has also been a resurgent interest in nuclear power. As the United States redefines its energy matrix and mix, Latin America provides ample opportunities for cooperation and integration, as well as some significant challenges.

A linchpin for increasing energy supply and market access in Latin America is infrastructure. The region requires over $1 trillion in investment in the energy sector over the next twenty-five years in order to meet its own increased energy demand, according to the International Energy Agency. A recent report commissioned by the IDB from Garten Rothkopf predicted that in the next five years between $50 billion and $100 billion in energy investments will be needed, most of it to connect the 40 million people living without electricity in the region. The report also points out Latin America and the Caribbean have lagged behind the global pace of adoption of new renewable sources (as opposed to ethanol), with geothermal, wind, and solar energy accounting for less than 1 percent of total production (compared to 2.5 percent globally).[3]

The United States can help meet this demand by supplying foreign direct investment financing incentives through multilateral institutions. Helping the region's governments develop renewable-friendly regulation, supporting innovation and technology, improving access to finance for energy projects, and promoting an analysis of the potential for carbon markets in the region would all be steps in the right direction.

Building on basic infrastructure investments, regional energy cooperation should be extended to the development and distribution of the necessary technology to make alternative energies commercially viable and the production of hydrocarbons more sustainable. The creation of an energy and environment research initiative, to develop and distribute energy-focused technologies cooperatively between governments and in partnership with the private sector, would encourage more information sharing and the quicker diffusion of promising approaches. It would also give many countries a stake in the transition to foster sustainable energy sources in the long term.

Economic Ties

Latin America's economic performance in the first years of the twenty-first century was quite good. The region as a whole grew over 5 percent per year for the past five years—marking its strongest economic expansion since the 1970s. Inflation has averaged just 7 percent since 2000, a remarkable achievement for a region known for hyperinflation. Sound fiscal management has left nearly every government with a strong fiscal position.

Despite these improvements, Latin America still lags other developing regions in combating poverty and inequality. Approximately 37 percent of the population remains poor, and 22 percent of people in Latin America and the Caribbean still live on less than two dollars a day.[4] Furthermore, Latin America and the Caribbean remains the most unequal region of the world. Income inequalities mirror structural inequalities, particularly in terms of access to health care, education, credit, and economic opportunity. Unequal distribution of health care is leading to the revival of diseases such as yellow fever, dengue, pertussis, and measles, particularly among the lower socioeconomic sectors of society. Economically, formal credit reaches only 3 percent of households in Latin America and the Caribbean, limiting their ability to create and fund their own economic opportunities.[5] Some 50 percent of the labor force in the region holds informal sector jobs. While these activities can help alleviate immediate poverty, they provide on average much lower salaries than formal jobs, afford no access to state services such as social security, health care, or disability insurance, and they undermine the economic base of Latin America's governments and the effectiveness of state institutions.

Such extreme and persistent hardship for large segments of these populations has a variety of negative ramifications for Latin America and for the United States. Not only do studies show that these inequalities hamper worker productivity, social mobility, and overall economic growth,[6] but these conditions could also threaten to undermine the region's broader economic and political stability.

There have been some successes in reducing poverty and inequality levels in recent years, many coming from within the region. The combination of sustained economic growth, sound macroeconomic policies, and expansion of the social safety net in Chile has lowered poverty rates there from 39 percent in 1990 to the current 14 percent (approximating U.S. levels). Other countries, such as Brazil and Mexico, have made headway as well by controlling inflation and by instituting conditional cash transfer programs to poor families. Brazil's *Bolsa Familia* now reaches 46 million people while Mexico's *Oportunidades* reaches 25 million—covering nearly 75 percent of the poor in those two countries. Nevertheless, as the stubbornly high rates attest, more needs to be done to assist those still without adequate resources and access.

Added to these issues are the costs of the current economic crisis. While many initially hoped Latin America would be spared a significant downturn, it too is now caught up in the global decline. The International Monetary Fund World Economic Outlook, published in April 2009, suggests that Latin American economies will contract by 1.5 percent in 2009 before recovering in 2010. Mexico and Venezuela will be the hardest hit. According to the IDB, sluggish economic growth in Latin America and the Caribbean could send between 2.8 million and 12.7 million people into poverty in the next two years.

The United States can best help these countries get through the current downturn by first stimulating its own economy. But it can go further, moving beyond the historically narrow U.S. policy approaches to poverty and inequality alleviation in the region. President Barack Obama's announcement of a Microfinance Growth Fund for the Western Hemisphere at the Fifth Summit of the Americas is certainly a step in the right direction, although the initial capital of $100 million and goal of $250 million fall below expectations. Fulfilling the region's presidents' agreement to recapitalize the IDB—in order to ensure steady flows of capital for the region's entrepreneurs—also represents an important marker for economic cooperation and recovery.

Trade agreements, while successfully increasing macroeconomic growth, have not benefited the broader populations as much as initially hoped, whether in terms of reducing poverty levels or widening access to economic opportunities. This stems in part from the lack of complementary domestic reforms to enable broader gains. But it also results from the nature of U.S. free trade agreements, which often stress U.S. priorities such as investment and services rather than broader and more balanced economic exchange. The United States should promote more open trade in areas of Latin American comparative advantage, including the liberalization of textile and agricultural policies in the United States, thereby leveling the playing field with "freer and fairer" trade.

U.S. targeted aid for poverty alleviation has stagnated at approximately $600 million a year, meaning that in real terms it now represents a third of what it was in the 1980s. In terms of targeted assistance, the United States should fully fund the Millennium Challenge Account and complement this program with new initiatives that reach the poor regions of large middle-income countries—such as Brazil and Mexico—where some 50 percent of the region's poor now live. It should support the expansion of microenterprise and small-business financing through multilateral institutions, the private sector, and nonprofit institutions. Financial inclusion can go a long way in helping the poor to accumulate assets, manage risk, and leverage their entrepreneurial skills for the betterment of their and their families' economic situations.

Finally, the United States can better assist efforts to reduce poverty and inequality by helping Latin American governments strengthen their public institutions. On a basic level, this means support for and assistance in restructuring tax systems and building the infrastructure and judicial capacity necessary to increase tax collection. Currently, Latin American tax systems rely on regressive value-added taxes, and governments on average collect just 17 percent of GDP in taxes, compared to 35 percent in Organization for Economic Cooperation and Development countries. Greater fiscal resources—collected in more efficient and egalitarian ways—will help Latin American governments build the capacity necessary to address these challenges themselves.

Addressing the high rates of poverty and inequality represents a critical challenge for most governments in the region, and is one in which the United States can only play a supporting role. Yet the failure to improve the lives of millions of citizens throughout Latin America will continue to undercut both domestic and international U.S. policy objectives, and so these fundamental issues must be at the center of U.S. foreign policy toward the region.

Migration

A final broad issue for U.S.–Latin American relations is that of immigration. While this is often considered solely a domestic concern, the United States and Latin America are inextricably linked through their populations. Latin America is currently the largest supplier of U.S. immigrants, legal or not. Some 18 million Latin American citizens now live in the United States. Latinos—comprising Latin Americans and their descendents—total nearly 50 million, or 15 percent of the U.S. population. The pace of migration—driven by the lack of economic opportunity at home as well as the demand in many U.S. economic sectors—has accelerated in the last twenty years. Latinos account for half of U.S. population growth since 2000.[7] While the recent economic downturn has reduced the level of migration, it has not led to a significant return by immigrants, and the pace of U.S.-bound migration is expected to rise again when the U.S. economy recovers.

Immigration has mixed effects for Latin America. U.S.-bound migrants make up substantial portions of particular countries' populations. For instance, some 10 percent of the Mexican and the El Salvadoran populations now reside in the United States. Migrants send back nearly $50 billion each year in remittances to Latin America, benefiting families by alleviating poverty, funding schooling, and in some cases providing capital for local investments and businesses. Yet U.S.-based opportunities also attract many of the best and brightest from Latin American nations. Surveys indicate that one in four Latin Americans wishes to leave his or her country, and those that identify themselves as talented are significantly more likely to want to emigrate.[8] Their absence limits the spillover effects of their productivity for Latin American economies and societies.

Meanwhile, the U.S. workforce has become increasingly dependent on Latino workers. They represent 41 percent of the total employment in farming, fishing, and forestry; 25 percent in construction; and 28 percent in cleaning and maintenance.[9]

This population has varying effects in the United States as well. Most studies show a net national benefit to the United States from migration, including greater flexibility among workers, lower prices for labor-intensive goods and services to U.S. consumers, and the benefits of innovation and new technologies (migrants are granted one out of every four new U.S. patents). At the same time migration—and particularly unauthorized migration—has its costs, pressuring local education and health systems, depressing local wages, and risking U.S. security with millions forced to live on the margins.

Even though Latin American immigrants have been hit disproportionally hard by the economic crisis,[10] as the U.S. economy grows, and as the baby boomer generation begins to retire, migrants will be increasingly needed to meet economic demands. As a result, the disconnect between labor supply in Latin America and job demand in the United States means continued migration, despite restrictive U.S. immigration laws.

The United States needs to negotiate and approve comprehensive immigration reform. This will not only benefit the United States domestically, but will also boost its standing in the region. This reform should improve border security and management; regularize the status of the unauthorized workforce already here; ensure employer security, verification, and responsibility; and expand a flexible worker program to meet changing U.S. economic demands. In addition, the United States should pursue bilateral or multilateral immigration agreements with the most important sending nations. These will allow closer monitoring and control over the flow of individuals, improving the safety of both the United States and the individuals and families on the move. It will also assuage the hard feelings of many in the region, who find the current unilateral U.S. approach counterproductive and discriminatory.

Finally, the United States should pursue policies that promote circular, as opposed to permanent, migration. This includes policies that allow workers to maintain ties with their home communities by returning home in the off-season—without worrying about the return trip. It also means providing portability in terms of health and pension benefits, so workers retain earned rights if and when they leave the United States upon retirement. It could also mean creating programs to provide information and assistance to expatriated workers about projects, jobs, and opportunities within their home countries to encourage their return. These types of policy changes will benefit the United States—providing the workers necessary to keep its economy growing—and will also benefit Latin American nations, encouraging migrants to return with newfound skills and knowledge to invest in their own economies and societies.

Prioritizing Strategic Bilateral Relationships

Even as these overriding themes constitute important anchors for U.S. policy in the hemisphere, particular bilateral relations will dominate U.S. policy toward the region. Of the four most important for the tenor of U.S. relations as a whole, two represent strategic partnerships while two present diplomatic challenges.

As the fourth-largest democracy and ninth-largest economy in the world, Brazil is a crucial actor not only in Latin America but globally. The United States should build on recent collaboration on ethanol to develop a broader diplomatic partnership incorporating a wide range of bilateral, regional, and global issues, including energy and climate change, security, and world trade.

Mexico too remains a key ally and an important harbinger of U.S. well-being due to the deep economic, social, cultural, and security ties between the two bordering nations. U.S. relations must focus on more than border security, addressing issues of economic exchange, human flows, border infrastructure, and a shared environment.

U.S. relations with Cuba offer the biggest opportunity for change. After nearly 50 years, the role of the U.S. embargo is being severely questioned on both sides of the political aisle in Washington, and among the Cuban American population. The Obama administration has taken a few small steps, including more open travel for Cuban Americans, allowing the flow of remittances, and lifting bans on telecommunications investments. Real change, though, still remains difficult, requiring the political capital to overcome both political inertia and vocal (if decreasing) supporters of the embargo. It also requires some reciprocation from the Cuban government. In the end a shift is already occurring, even if just a return to the openness of a decade ago.

The United States' relationship with Venezuela is less likely to change. While personal animosities between the presidents may end, Venezuelan president Hugo Chávez's suspicions about and rhetoric against the United States will most likely persist, as will U.S. worries about the erosion of Venezuelan democracy. Despite the diplomatic tensions between the United States and Venezuela, trade relations have continued and will continue unabated, with the United States remaining the largest consumer of Venezuelan oil.

A New Approach

The biggest shift in U.S.–Latin American relations will result from a change in process rather than substance. The Obama administration has already hinted at this by meeting with regional institutions such as CARICOM and UNASUR. Valuing ideas from the region and working together in the search for multilateral solutions to shared concerns represent a shift from traditional U.S. policy. The best approach going forward will be one designed to enhance partnerships rather than just programs. This change is crucial in achieving the ambitious goals of strengthening Latin American nations' public institutions and their democracies—both necessary to take on the serious issues facing the region.

Improving the lives of Latin Americans will require long-term efforts on the part of many participants—most importantly Latin American governments and societies themselves. But there is a significant supporting role the United States can play. In doing so, the United States should work in partnership with Latin American governments bilaterally and multilaterally through organizations such as the World Bank, the Inter-American Development Bank, the International Monetary Fund, the International Finance Corporation, and the Organization of American States. It should also continue to work closely with civil society organizations and domestic and international businesses to create more inclusive economic, social, and political opportunities for Latin American countries and their citizens. By expanding its policy framework to incorporate issues of security, sustainable energy, economic advancement, and migration, as well as concentrating on strategic regional partnerships in areas of mutual concerns, the United States can best promote its own interests in enhancing stability, security, and prosperity at home and throughout the hemisphere.

Notes

1. Latinobarometro, "Informe 2008," 2008, http://www.latinobarometro.org/docs/INFORME_LATINOBAROMETRO_2008.pdf.
2. "Crude Oil and Total Petroleum Imports Top 15 Countries: February 2009 Import Highlights: April 14, 2009," Energy Information Administration, http://www.eia.doe.gov/pub/oil_gas/petroleum/data_publications/company_level_imports/current/import.html.
3. Garten Rothkopf, *A Blueprint for Green Energy in the Americas 2009* (2009), available at http://www.gartenrothkopf.com/publications.asp.
4. Economic Commission for Latin America and the Caribbean, "Cepalstat: Latin America and the Caribbean Statistics," http://websie.eclac.cl/sisgen/ConsultaIntegrada.asp (accessed April 2009).
5. Luis Tejerina, César Bouillon, and Edgardo Demaestri, eds., *Financial Services and Poverty Reduction in Latin America and the Caribbean* (Washington, DC: Inter-American Development Bank, 2006), available at http://www.iadb.org/sds/doc/POV-FSPovertyReduction.pdf.

6. For instance, see Guillermo E. Perry et al., *Poverty Reduction and Growth: Virtuous and Vicious Cycles* (Washington, DC: World Bank, 2006), available at http://siteresources.worldbank.org/EXTLACOFFICEOFCE/Resources/870892-1139877599088/virtuous_circles1_complete.pdf.
7. "Latinos Account for Half of U.S. Population Growth since 2000," Pew Hispanic Center, October 23, 2008, http://pewhispanic.org/reports/report.php?ReportID=96.
8. Jesus Rios and Steve Crabtree, "One in Four Latin Americans Wishes to Emigrate," Gallup, January 21, 2008, http://www.gallup.com/poll/103837/One-Four-Latin-Americans-Wishes-Emigrate.aspx; and Gerver Torres and Brett Pelham, "Is All the Talent Leaving Latin America?" Gallup, September 26, 2008, http://www.gallup.com/poll/110743/All-Talent-Leaving-Latin-America.aspx.
9. Rakesh Kochhar, "Latino Labor Report, 2006: Strong Gains in Employment," Pew Hispanic Center, September 27, 2006, available at http://pewhispanic.org/files/reports/70.pdf.
10. While the unemployment rate for the general population increased to 6.6 percent by the end of 2008, the unemployment rate for foreign-born Latinos rose to 8 percent. Rakesh Kochhar, "Unemployment Rose Sharply among Latino Immigrants in 2008," Pew Hispanic Center, February 12, 2009, http://pewhispanic.org/reports/report.php?ReportID=102.

The Outlook for U.S. Foreign Policy in Latin America and the Caribbean: The Challenges of Transforming Goodwill into Effective Policy

Ambassador (Ret.) Paul D. Taylor
Senior Strategic Researcher
Naval War College

President Obama got off to a good start in relations with Latin America and the Caribbean. His early meetings with counterparts in Mexico and Canada produced positive results. Obama both signaled a willingness to abandon his campaign suggestions that the North American Free Trade Agreement (NAFTA) would have to be renegotiated and showed an eagerness to engage in businesslike management of the increasingly complex interactions among the three countries participating in NAFTA, which has produced the world's largest economic bloc. On his first trip ever to Latin America, his engagement with other leaders of the hemisphere at the summit in Trinidad and Tobago stressed listening over preaching. Demonstrating the sales axiom that the ear is the most important organ in a salesperson's body, the president set a positive tone. That new tone alone can bear fruit. By taking the wind of out of the sails of Hugo Chávez, Obama may have made it harder for the Venezuelan to advance his career by baiting the United States. Signaling flexibility on Cuba before the summit seems to have played well in other countries of the region and put the Castro brothers on the defensive even if many Latin American presidents might have wished that the U.S. government had gone further to dismantle the trappings of Cold War relations with Cuba.

The question now is what next. We try hard at the Naval War College to focus our thinking on strategy by relating actions to policy objectives. Goodwill alone is not much of a policy objective. If it can be employed to help us achieve other ends, though, then it is an asset to be included among the available tools of statecraft.

Usually, at the Naval War College we start our analysis by looking at national interests and policy objectives, then policy instruments. Thereafter, we try to link everything together with a strategy. Over recent years, U.S. policy objectives in Latin America have remained remarkably consistent even through changes of parties in the presidency. Whether Democrat or Republican, presidents have sought democratic government, good governance (usually employed as a politically correct reference to eliminating high rates of corruption), economic growth, and security in the hemisphere. Despite a broadly bipartisan consensus on policy since the end of the wars in Central America, it has not been easy for U.S. policy makers to formulate and execute policy toward the Americas for at least four reasons:

- The history of U.S. interventions in Latin America has left a legacy of suspicion that colors the way U.S. actions are perceived and limits the political space in which Latin American leaders can conduct relations with the United States.
- Frequently our possible actions in Latin America and the Caribbean are conditioned profoundly by events in *other* parts of the world.
- Increasingly, events in Latin America involve things that are either not our business or may turn out better if we refrain from engaging on them.
- The instruments needed to achieve many of our objectives and to meet the concerns of Latins themselves often involve agencies in the U.S. government other than the traditional foreign affairs agencies.

A History of Intervention

On the first point, John H. Coatsworth has chronicled at least forty-one times from 1898 to 1994 in which the U.S. government intervened successfully to change governments in Latin America.[1] His time frame and definition left out the most momentous U.S. intervention of all, in Mexico from 1835 to 1853, which was vividly recalled by former president Carlos Salinas de Gortari in an interview for PBS with David Frost on the eve of congressional ratification of the North American Free Trade Agreement. Frost alluded to criticism of the draft treaty for not taking adequate account of labor and environmental protections. He asked Salinas whether Mexico would be willing to renegotiate the treaty if the U.S. Congress rejected it in the present form. President Salinas recalled that Mexico had lost more than half of its territory to the United States through a war of intervention. He said, "This was something many Americans do not remember but every Mexican will talk to you about." If the treaty was rejected, Salinas added, an entire generation would have to pass before the opportunity to conclude such a sweeping agreement would arise again.[2] His point underscored the results of two quite different approaches to teaching the history of the hemisphere north and south of the Rio Grande. While Latin America is by no means monolithic on this point, there are few pockets in the region where at least some fear of U.S. dominance is not felt.

Events in Other Parts of the World

Think back to the beginning of the previous administration, in January 2001. President George W. Bush came to office promising to emphasize Latin America. That commitment may have been typical campaign rhetoric, but it was probably more. As a border-state governor who spoke some Spanish and had a Hispanic sister-in-law, Bush also knew Vicente Fox personally. Their vision for immigration reform, based on recognition of the need of the U.S. economy for foreign labor and the desire of the governments of Mexico and Central America that their emigrating citizens be treated with respect, offered better-than-usual prospects for progress.

Immigration Reform

This hope was dashed by the attacks of 9/11. Instead of liberalizing immigration, the U.S. government swung sharply into a focus on border security. Washington became obsessed with wars, first in Afghanistan and then Iraq. Suddenly the scarcest commodity in Washington, the attention of busy policy makers, was consumed with little room left to focus on relations with Latin America and the Caribbean. Washington chose to wage war in Iraq despite opposition in the region. The decision to apply military force without international approval reawakened our neighbors' worst fears and set back their relations with Washington.

Free Trade Area of the Americas

The centerpiece of Bush policy in the hemisphere, negotiation of a Free Trade Area of the Americas—FTAA—was placed in jeopardy by yet another extraregional development: the failure of the Doha Round of trade negotiations in 2008. Center-left governments had come to office in several Latin American countries in part because the opportunities of political mobilization had outstripped the distribution of the benefits of economic liberalization. Thus it was always going to be difficult for certain countries—Argentina and Brazil are prime examples—to manage the adjustments that would be required by any agreement to liberalize trade. Brazil quite understandably demanded improved access to the U.S. market for its agricultural exports and a reduction in U.S. subsidies to agricultural production as an essential counterpart to lowering its barriers to the importation of manufactured goods. For its part, the United States could not make concessions on agriculture unless they could be matched by similar moves in Japan and the European Union, its major export markets. So the sequencing of events was immutable; agriculture had to be resolved in the Doha Round before it could be addressed in the Western Hemisphere. Doha's failure doomed the FTAA. Brazil tried at the eleventh hour to keep the Doha Round alive when it sided with the United States and other industrialized countries against the intransigent positions of China and India, which derailed the talks.

We do not know what developments lurk outside the hemisphere to take our focus off our neighbors in the future, but they are almost certain to come. Latin Americans tend to view U.S. attention to their region as either too little or too much. For their part, the U.S. public and government have a habit of thinking about Latin America only when they perceive a threat to U.S. security emanating from the region, as when Fidel Castro was trying to export revolution or wars in Central America threatened to spread.

Events in the Region in Which the United States Is Not a Central Player

Several recent developments in Latin America serve to remind us that the United States is not the dominant player in the hemisphere that we once thought we were.

Expropriation of Investments in Bolivian Natural Gas

When Evo Morales early in his Bolivian presidency nationalized foreign investments in natural gas he refreshingly revealed that this time the home country of the investors most at risk was Brazil, not the United States. Unlike the situation several decades earlier, his expropriation would not put the U.S. government at the center of a dispute over nationalization of mineral investments. And it also made the much-heralded "lurch to the left" in South America less ominous than some alarmists had predicted. The leaders of countries with center-left governments, stretching from Venezuela to Ecuador, Bolivia, Chile, Argentina, and Uruguay, may have similar personal stories, forged in opposition to extra-constitutional or insensitive governments, but their countries have particular interests and those interests sometimes diverge.

Permanent Seat on the Security Council

Differences in interests can explain the failure of Latin American governments to come to consensus on the matter of a new permanent seat on the UN Security Council. Brazil has been the Latin American country most often mentioned as a prime candidate for a permanent seat, albeit without the right of a veto, in an expanded Council. Brazil has a number of good arguments in its favor: it has the region's largest population, geographic size, and economy and has served as a participant in UN peace operations. Nevertheless, Argentina, Chile, and Mexico have raised objections to selecting Brazil to occupy a permanent seat, suggesting that they would rather see an additional rotating seat for Latin America. The U.S. government has wisely concluded that little would be gained by favoring any country in the region until a consensus candidate can be identified.

Increasing Chinese Involvement in Latin America

Some U.S. observers, particularly those who worry about security issues, have been bothered by what they see as an alarming increase in Chinese involvement in the region. Clearly, China has greatly increased its trade with many Latin American countries. Similarly, the People's Republic has committed to expanding investments, even though they are relatively slow to be realized. I believe that there are several reasons why the United States stands in fact to benefit from this Chinese activity and should hope that the current downturn in the world economy will not seriously interrupt it. First, my colleague Lyle J. Goldstein, director of the Chinese Maritime Studies Institute of the Naval War College, working with counterparts in Argentina, found little evidence that the Chinese interest in Latin America went beyond legitimate commercial activity.[3] In addition, Chinese purchases have fueled several years of above-average growth in Latin America, an objective that has enjoyed bipartisan support in Washington over the years. Politically, if the United States has found its relations with Latin America complicated by the perception that it is the hegemon bent on dominating the hemisphere, then a greater presence by extra-hemispheric players can place that issue in a better perspective. Finally, consider the oft-repeated threat of Hugo Chávez that he will cut off supplies of oil to the United States, Venezuela's biggest customer. Aside from some refineries along the Gulf coast that are engineered to process the heavy crude that

Venezuela produces, the U.S. economy would not suffer if China's modest purchases expanded to absorb all of Venezuela's petroleum exports. As long as Venezuela maintains its level of production, international markets will clear with little disruption. The United States could make up any loss of Venezuelan product elsewhere. Only by cutting the level, not the destination, of exports could Chávez use his oil as a weapon, but he cannot afford to cut exports and lose revenue for any extended period. Even if he could, the price increases that step would cause would hit consumers everywhere, not just in the United States.

Latin American "Mega-summit"

Michael Shifter and Daniel Joyce's analysis in the February 2009 issue of *Current History* of the Latin American "mega-summit" last December, to which the United States was not invited, provided an even-handed examination of the implications for the United States.[4] Some have seen that meeting as a repudiation of U.S. leadership and a call to arms for an aggressive response. Why not view it instead as a natural consequence of things we have always wished and then decide how to move on? After all, we have long favored governments in this hemisphere that are democratically elected and whose countries are economically prosperous and generally self-reliant. This is a reason that trade liberalization as a strategy to foster economic growth is preferable to foreign assistance. If the United States and other advanced economies can provide greater access to their markets, the economic benefits to exporting developing countries are quantitatively far larger than any conceivable assistance programs. The qualitative political benefits are profound as well. By getting away from the destructive donor-recipient relationship, the beneficiary countries can feel respected and empowered. If this enables our Latin American neighbors to take more responsibility for their own fates and for their relationships with us, then so much the better.

Inter-American Democratic Charter

When the Inter-American Democratic Charter was adopted by the General Assembly of the Organization of American States (OAS) in September 2001, it marked a milestone of democratic progress in Latin America and the Caribbean. Elected governments were in place everywhere except Cuba and it seemed as if nearly everyone could agree that that was the way it should be. To consolidate that state of affairs, the Charter provided that "in the event of an unconstitutional alteration of the constitutional regime that seriously impairs the democratic order in a member state," procedures would be implemented with measures up to and including suspension of the member state from participation in the OAS. In retrospect the Charter looks like a case of preparing to fight the last war. This time it was the diplomats, not the generals, doing the planning. They were formulating an ideal battle plan for countering the kind of military coups that plagued democracy around the hemisphere in the 1960s and 1970s. What we have seen in this decade, though, is democracy under threat from elected leaders, most notably in Venezuela but also elsewhere—in Morales's Bolivia, Correa's Ecuador, and Ortega's Nicaragua. In Venezuela, Chávez has systematically undermined institutions of government and civil society that threatened his aspirations for a monopoly on political power. His actions against political opponents, the press and labor unions, for example, recall the fabled Russian peasant who stole his neighbor's property by moving the fence between their fields every night one stone at a time. Chávez has not presented the international community with a bold affront like a coup or an invasion that has to be addressed, but in retrospect the result is obvious. Clearly, the United States is limited in what it can do to weaken Chávez's campaign by virtue of the legacy of past interventions, the regrettable U.S. reaction to the short-lived coup against Chávez earlier in this decade, and the fact that Chávez is a master at turning opposition from Washington to his advantage at home. Other Latin leaders should be able to help counter the assault on democracy in Venezuela. For the most part, they have ducked this responsibility. So a question arises: can the U.S. government find a way to persuade other governments to act to encourage or effect greater political competition in Venezuela? Their failure to oppose the excesses of Chávez serves to erode the legitimacy of their own democratic credentials.

Important Issues in Which Domestic Agencies Have the Lead

Drug-Related Violence in Mexico

One of the regional issues of greatest urgency to the United States is the matter of the disturbing level of narcotics-related violence in Mexico that not only challenges the Mexicans' ability to maintain order and stability but also risks spilling over into U.S. border cities and beyond. In their recent visits to Mexico, Secretary Clinton and the president have usefully acknowledged that the United States bears some responsibility, both because U.S. demand for drugs energizes the industry and because many of the powerful guns that figure in the drug wars were bought in the United States and smuggled south across the border. The traditional foreign affairs agencies have little authority to act against drug demand or to control guns, so the task falls to agencies that do not have international relations in their core missions. This creates a challenge to try to adjust the organizational structures and agency cultures in ways that will get the job done. The record of controlling guns in our own cities does not engender hope that we will do better on behalf of a foreign country. At the recent Miami Conference on Hemispheric Security, tentative suggestions were heard that the War on Drugs had failed. Plan Colombia had succeeded impressively in reducing murders and kidnappings but made little impact on the narcotics trade. The time may have come to take the business out of the violent hands of criminals by legalizing and regulating the drug industry. A few voices in Congress are making the same point.

Inter-Agency Coordination

The initiative of the U.S. Southern Command to bring representatives of civilian agencies into the organization is a well-intentioned effort to improve coordination of U.S. activities in the hemisphere.[5] It must be exercised with extreme caution, however, to keep from putting a military face on U.S. civilian activities in SOUTHCOM's area of focus. In Latin America, more than in any other region, this perception could be the kiss of death because U.S. military activities are widely suspect. A better approach to organizing civil-military cooperation would have the chief of plans and strategy of the Southern Command resident in Washington and assigned concurrently as a deputy assistant secretary of state in the Bureau of Western Hemisphere Affairs. He or she would replicate in reverse the assignment of a senior Foreign Service Officer as a deputy combatant commander and this would place a senior officer of SOUTHCOM at the center of policy making for Latin America and the Caribbean.

Fixing the U.S. Economy

Brazil's President Lula no doubt spoke for many of his fellow Latin American presidents when he remarked recently that the most helpful thing that the United States could do for Latin America would be to restore the American economy to health. This objective is one shared by all Americans so it should be easy to take his recommendation into account. Once again, it will be domestic agencies that lead, the Federal Reserve and the Treasury in particular. For people responsible for foreign relations, the main challenge will be to keep the international dimensions of the process in mind and thereby avoid beggar-thy-neighbor measures like protectionism. Additionally, they should seek ways to work with foreign governments in a manner that will reinforce the perception that our countries are acting in concert even if most of what must be done could be justified in terms of domestic policy alone.

What Do We Know About the Future?

Predictions are always hard, but it is relatively easy to see some tough sledding ahead for Latin American economies. The economic downturn came hard on the heels of impressive economic success in Latin America and the Caribbean. Increases in gross domestic product in the region had averaged about 6 percent *per annum* from 2003 to 2007, making it one of the most buoyant periods in recent history. Low interest rates and strong world growth had strengthened the prices of Latin America's exports of minerals, raw materials, and foodstuffs and impressive demand in China and India had driven historically large volumes of exports to those countries. Internally, governments had generally implemented wise economic policies and the global financial crisis revealed that for the most part their financial institutions and government accounts were in sounder shape than their counterparts in many member countries of the Organization for Economic Cooperation and Development. Latin American ratios of debt to GDP fell from an average of 52 percent in 2003 to 35 percent by the end of 2007.[6]

Now the crisis has hit and the effects are already being seen in Latin America. The World Bank predicts that world trade will shrink in 2009 for the first time since World War II. Figures for February 2009 show the exports of Argentina, Brazil, and Mexico down 24, 25, and 26 percent, respectively, from the same month a year earlier. At the same time, the flow of capital to the region is slowing, which increases the cost of borrowing by governments and companies. Both domestic and foreign investors will delay projects. Economic distress also accounts for fewer tourists arriving, even without the discouraging impact of swine flu in Mexico. Remittances home by Latin Americans working abroad had expanded to $69.2 billion by 2008, a level that dwarfed official assistance. They have now started to drop in reaction to recessions in the United States and Europe. Remittances to Mexico were down 8.7 percent during the first four months of 2009 and 18.7 percent in April, compared with the same periods in 2008.[7]

As these drivers are lost, overall economic growth rates are certain to slow or go negative, at least in per capita terms, which regrettably are the terms that matter to people. This fact will result, in turn, in increased unemployment and a reversal of hard-fought improvements in the distribution of income. Latin America is still the region of the world with the worst aggregate income distribution. There is no more effective way, in the short run, to influence the distribution of income for better or for worse than to change the rate of unemployment.

We cannot know with certainty how long or how deep the crisis will be, but one does not have to be Chicken Little to fear that economic distress could again fuel crime and produce the kind of elevated rates of death that were attributed to criminal violence during the economic crisis in Argentina early in this decade, which exceeded the level in war-torn Colombia.

Beyond the effects already being felt are the prospects of greater pressures for protectionism and tighter restrictions on immigration. The "buy American" provisions of the stimulus bill put fear into the hearts of U.S. trading partners. The Obama administration has tried to assure foreign leaders that it will not permit those provisions actually to bite and it may succeed. Clearly, though, the time is not propitious for reenergizing the Doha Round or negotiating any further free trade arrangements in Latin America. Since immigration to the United States is driven almost exclusively by economics, the recession is already having a profound effect on migration patterns. Census data released recently by the government of Mexico show that emigration declined 25 percent in the year that ended in August 2008 compared to the year before.[8] Like trade liberalization, though, immigration reform is harder to accomplish in the midst of economic difficulties, so any attempts at reform this year will be more easily understood in terms of gaining domestic political advantage than actually changing the rules of immigration in any significant way.

Conclusion

Politics in Latin America and the Caribbean could become ugly as the economic crisis causes populations to suffer and governments to lose their room for maneuver. This prospect could reinforce the tendency to blame others, and in this hemisphere, the United States has always been the favorite target for blame. One way for the United States to protect itself against such charges is to engage our Latin American neighbors in a dialogue about how to manage the crisis, not in the donor-recipient relationship that so often characterized the past, but as partners searching together for ways to mitigate and contain the costs of the economic crisis. If we can do that, the process could be part of the solution.

Notes

1. John H. Coatsworth, "United States Interventions: What For?" *ReVista,* Spring–Summer 2005, pp. 6–9.
2. *President Carlos Salinas . . . Talking with David Frost*, VHS (PBS, October 29, 1993).
3. Guillermo R. Delamer, Lyle J. Goldstein, Jorge Eduardo Malena, and Gabriela E. Porn, "Chinese Interests in Latin America," in *Latin American Security Challenge: A Collaborative Inquiry from North and South,* ed. Paul D. Taylor, Newport Paper 21 (Newport, RI: Naval War College Press, 2004).
4. Michael Shifter and Daniel Joyce, "No Longer Washington's Back Yard," *Current History,* February 2009.
5. James G. Stavridis, "Sailing to a New Port," *Naval War College Review* 60, no. 3 (Summer 2007), pp. 7–13.
6. Alejandro Izquierdo and Ernesto Talvi, *Policy Trade-Offs for Unprecedented Times: Confronting the Global Crisis in Latin America and the Caribbean* (Inter-American Development Bank, 2009).
7. Elisabeth Malkin, "Money Sent Home by Mexican Workers in U.S. Falls Sharply," *New York Times,* June 2, 2009.
8. Julia Preston, "Mexican Data Say Migration to the U.S. Has Plummeted," *New York Times,* May 15, 2009.

Panel II: Western Hemisphere

Summary of Discussion

Professor Laurence L. McCabe
Associate Professor of
National Security Affairs
Chair Latin America Regional Study Group
Naval War College

The Ruger Workshop's American Foreign Policy panel on the Western Hemisphere included insightful presentations by Mr. Peter Hakim of the Inter-American Dialogue, Dr. Shannon O'Neil of the Council on Foreign Relations, and Ambassador Paul Taylor on the faculty of the Naval War College. Peter Hakim offered a fresh look at new opportunities for improving relations with Latin America under President Obama's administration. He provided an intriguing list of ten challenges facing the United States that might be accurate indicators of the new administration's regional policies, to include policies on Cuba, Mexico, Brazil, Venezuela, and on regional immigration policies among others. Shannon O'Neil suggested future U.S.–Latin American relations will play out in four broad areas: public security, sustainable energy, economic advancement, and hemispheric migration. She believes it is critically important how the United States and the region address these four issue areas as these challenge areas will play a significant role in shaping the future security environment in the hemisphere. Finally, Ambassador Taylor offered an approach to translating the goodwill offered a new Obama administration into effective foreign policy. While noting the sometimes difficult history between the United States and Latin America, Ambassador Taylor proposes our relations with Latin America are at times impacted by events in other parts of the world. He also suggests there is an increasing number of events in Latin America that do not concern the United States. How the Obama administration crafts a regional policy in this increasingly complex security environment will define the future security environment. Each presentation provided innovative perspectives on political, economic, and security issues of importance to those responsible for foreign policy in the recently elected Obama administration. The presentations set the stage for a lively and animated discussion among the workshop attendees, particularly on the role of the U.S. military in the Western Hemisphere, the strategic intentions of hemispheric rising powers such as Brazil and Mexico, the ongoing efforts to reduce the production and distribution of illegal narcotics, the role of China in the hemisphere, and the often contentious and controversial relationship between the United States and our neighbors to the south.

The moderator opened the panel discussion by reminding the attendees of the controversial history between Latin America and the United States and how this historical baggage often has a disproportionate and dysfunctional impact on current foreign policy initiatives in the hemisphere. The moderator also reminded the attendees not to forget the English-speaking Caribbean and Canada in workshop discussions as they too play an important role in Western Hemisphere affairs.

The first discussion topic and line of questioning concerned the role of U.S. Southern Command (SOUTHCOM), the combatant command responsible for U.S. military activities in Latin America and the Caribbean. The question specifically addressed recent initiatives by SOUTHCOM to reorganize in a way that strengthens the interagency process by deemphasizing more traditional military organizational relationships. A participant's response acknowledged the unique vision of SOUTHCOM and its offer to be the face of the U.S. government in Latin America. Another respondent suggested the advantages of military leadership are many. The military can respond very quickly to any crises in the region in a way other agencies cannot. The military has flexibility to fund various operational responses and programs in the region—again, something other agencies lack. Conversely, several in the workshop indicated it might not be optimal to have the U.S. military be the lead agency in the region. It was the general consensus that a perception of the military taking the lead in policy formulation would generally be counterproductive to U.S. regional interests. A comment was made referencing the negative historical perspectives of the U.S. military in the region and the impact this might have on foreign policy implementation. Another participant suggested there is "too little for the military to do" in the region and that the United States would be better served if the money spent by the military was given to a different agency. Specifically mentioned was the recent activity in the region by the USNS *Comfort,* a hospital ship, and the recent reactivation of the U.S. Fourth Fleet in Mayport, Florida. It was suggested the USNS *Comfort* deployment was more a response to the "Cuban doctor" issue in the region and that the money would be better spent helping countries develop internal medical and health care capacity. It was also suggested the announcement of the establishment of the Fourth Fleet was confusing and threatening to the region—something that was avoidable. The general consensus of the group was that the U.S. government would be best served by a military with a more reserved, secondary role in regional policy implementation—not the agency leading policy formulation and implementation.

The next line of questioning and discussion concerned the United States' "war on drugs" and the questionable effectiveness of the expensive programs associated with it. A participant made the case that the global challenge of illegal narcotics is growing worse and that the talk of ramping down antidrug programs in the United States and the redirection of funds to softer programs is extremely troubling. The participant suggested the United States is sending the message that the war on drugs is "too hard" and that this is not the best answer. Another participant responded by pointing out the significant challenge of stopping the importation and use of drugs and that there is no good solution. The best approach might be to work to reduce the demand side while also working to reduce the level of violence associated with the drug trade. A member of the group then suggested a greater effort should be made to rehabilitate drug users instead of locking them away in prison. Another participant then mentioned the transnational criminal activity associated with the drug problem, in particular the small arms trade. The suggestion was made that the United States be more sensitive and responsive to the significant hardships the drug trade brings to small developing countries in the region.

The next comment raised the issue of relative peace in the hemisphere and that this should lead U.S. foreign policy away from military engagement in the region. The participant also raised the issue of Canada and asked the workshop participants to comment on the role of Canada in the region. The first response suggested we first consider the lack of conflict in the region and the relatively good relations currently existing between countries in the region. The participant reemphasized that because of the relative lack of interstate conflict the U.S. military should not be "out front" in the region. The point was made, however, that it is very important for the U.S. military to assist Latin American security forces to make the transition from an external focus to a domestic support role in accordance with their constitutions. A response pointed out the continuing and very important role of military organizations in many Latin American countries and that they are often the most capable agency available to assist the government in crisis situations. It was suggested that it is important to reshape the militaries in the region, not ignore or abandon them because of a lack of interstate conflict. One workshop participant suggested that retraining the military to support other government agencies is the preferred option for the U.S. government. This way, the United States might eventually be able to redirect funding to nonmilitary programs when other agencies are better able to manage foreign assistance. The participants continued with comments addressing the role of Canada in the region. Another then observed that Canada does not seem to have a strategy for the region despite several indications from the Canadian government that Canada was rebuilding its presence in the region. The suggestion was made that Canada should develop a more coherent approach to Latin America if for no other reason than to better focus program investments. Another participant commented that Canada is very active in the Caribbean region but does not seem to have a coherent engagement strategy.

The discussion shifted to the role of China in the region and to what extent the United States should be concerned about China's expanding role. It was suggested any "concern" of the United States' over China's expanding role is "terribly patronizing" as it presupposes Latin America is unable to manage a "diverse set of international relations." It was pointed out that Chile, Mexico, Peru, Brazil, and Argentina all have trade agreements with China and that trading with China is having a positive effect on the region's economies. Another respondent suggested that China is interested in long-term economic relations in the region, which can quickly lead to an unhelpful interest in the political stability of these countries. It was pointed out that China is very cautious about offending the United States in its relations with Latin American countries. Several agreed that Chinese activity in the region is not necessarily good or bad—it just is. No one is going to tell the Brazilians to stop trading with the Chinese. One participant also pointed out that the United States has a significant economic relationship with China. As such, it would appear inconsistent for the United States to complain about other countries in the region developing a similar relationship with China. It was suggested that the United States not waste time resisting or worrying about Chinese activity in the region, but rather begin to develop a strategy of managing the Chinese presence in a way that is in the interests of the United States. A contrary view was offered that suggested China's economic interests are self-serving and will only serve to limit real economic development in the region. It is for this reason that the United States should increase economic activity in the region with an emphasis on real, sustainable economic development. It was then noted that Brazil and Chile are excellent examples of Latin American countries that are developing real, sustainable, independent industries. Finally, the recent activity by Iran in the region was noted and it was suggested this phenomenon might be worth watching as the long-term interests of Iran are most likely different from those of China.

A participant next raised a question about U.S. Northern Command (NORTHCOM) and its role vis-à-vis Latin America and Canada. The point was made that while NORTHCOM is extremely important to homeland security, real progress is being made in the interagency community in the area of maritime cooperation. It was suggested that more attention should be given to the non-military-to-military activities going on in the hemisphere in the area of regional security.

The final issue discussed was Brazil and its role in the future of the region. A participant noted Brazil's long-standing desire to be not only a leader in South America but a leader on the world stage. It was suggested that the two best leaders in modern Latin America history were from Brazil. A workshop participant then stated Brazil most likely has a bright future as a global player; however, the real test will come when it has to take positions on issues of global security such as Iranian nuclear power, North Korea, human rights, etc. Brazil will have to make an important transition from a developing country often portrayed as a victim of the global system to a participant responsible for the global system. This will not be easy.

The discussion ended on a general note of agreement that the future of Latin America is generally bright; however, the regional powers will have to make significant political and economic transformations before they are taken seriously by global powers and ready to assume positions of leadership in the community of nation-states.

Panel III

Asia and the Pacific

Dr. Jonathan D. Pollack
Professor of Asian and Pacific Studies, Chair Asia-Pacific Study Group, Naval War College

Dr. Evan S. Medeiros
Senior Political Scientist, RAND Corporation

Dr. Emrys Chew
S. Rajaratnam School of International Studies, Singapore

Moderator:
Dr. John F. Garofano
Professor of Strategy and Policy, Jerome Levy Chair of Economic Geography and National Security, Naval War College

U.S. Asia-Pacific Strategy in the Obama Administration

Dr. Jonathan D. Pollack
Professor of Asian and Pacific Studies
Chair Asia-Pacific Study Group
Naval War College

Introduction

This essay will outline some prospective choices in the Obama administration's Asia-Pacific strategy. Every presidential administration aspires to a distinctive policy imprint, but it inherits the legacies of its predecessor and the cumulative consequences of past practices and policies. In addition, it must respond to events that it cannot fully anticipate or prepare for, and for which there is no established owner's manual. Absent a major crisis, there may also be a limited opportunity to refashion strategy before inherited policy or bureaucratic inertia takes hold. The Bush administration, for example, entered office intending to reassert American military primacy in Asia, in particular countering perceived challenges to U.S. strategic dominance, especially from China. In the aftermath of September 11, however, the administration sharply shifted course, focusing intensively on the war in Iraq and the uprooting of terrorist networks, thereby obligating American power and resources to very different priorities.[1] As a consequence, the salience and immediacy of the Asia-Pacific region for U.S. policy makers diminished.

Amid the Bush administration's preoccupations with ongoing military conflicts, however, a major regional transition was under way across the Asia-Pacific region, and it continues to shape U.S. policy choices.[2] The primary source of change involved internal economic and societal transitions within Asia, linked closely to an unprecedented expansion in international trade, financial flows, and the global telecommunications revolution. America's regional policies, though still reflecting longer-term balance of power calculations, were no longer characterized primarily in traditional national security terms. Senior U.S. officials, while mindful of China's military development, focused increasingly on the region's stability and enhanced prosperity, the prevention of interstate conflict, the growth of multilateral institutions, and the extraordinary expansion of trade and investment between the United States and the region's major economies. The most consequential changes occurred in U.S. relations with China and India. Contrary to expectations, threat-driven considerations were generally subordinated to the requirements of enhanced policy management, especially with Beijing. The Bush administration, for example, viewed China in largely bilateral terms, while also seeking to enmesh China in global institutions and continuing to hedge against longer-term strategic uncertainties. The potential for a major regional crisis, with the Taiwan Strait or the

Korean Peninsula (the latter case heightened by a renewed nuclear crisis), also persisted. But a larger transformation of Asia-Pacific politics, economics, and security underlies these shifts in U.S. strategy.

The United States has a clear need to refashion U.S. strategy in a manner that reflects and responds to the Asia-Pacific's dynamism and success.[3] While such policy shifts will not negate or ignore the military component of U.S. strategy, they need to be embedded in a larger conception of America's vital interests.[4] Toward this end, this essay focuses on the underlying assumptions influencing U.S. strategy, rather than detailed attention to any particular issue. The Asia-Pacific region is more economically vibrant, more politically self-confident, and more militarily robust than at any time in its modern history. Though many states face daunting political, economic, and societal challenges, and numerous states express unease about future power configurations, the region's successes far outweigh its failures and uncertainties. However, there is neither clarity nor closure on the contours of a reconfigured regional order, in regional expectations of the United States, or on long-term U.S. strategy. There is the additional question of whether the Asia-Pacific future is better conceptualized in subregional terms or on a more integrated basis across various subregions. The incentives for enhanced collaboration between the United States and the region are self-evident; the questions are how to achieve them, and what are the rules of the road that will govern future U.S. ties with Asia and the Pacific as a whole. At the same time, the Obama administration's focus on the war in Afghanistan and on potential instability in Pakistan could readily divert the attention of senior policy makers from this larger picture.[5]

Beneath a veneer of regional cooperation and multilateral institution building, the states of Asia and the Pacific continue to enhance their absolute and relative power; strategic trust and longer-term collaborative habits and practices remain scarce commodities, especially in Northeast Asia. There is also as yet no discernible power equilibrium among Asia's major powers. China, India, and Japan are all seeking to enhance their economic, political, and military weight and diversify their strategic options, even as all hope to build durable relationships with the United States in light of their respective strategic circumstances. Russia also seeks to reassert its claims to major-power status in the region. In addition, divergent national interests and the persistence of historically rooted conflicts continue to complicate the building of a coordinated, consensual strategy. Immediate as well as long-term political and security issues also intrude on strategic calculations. North Korea's nuclear weapons development, for example, is a major regional concern as well as a pressing issue for the future of the nonproliferation regime that cannot be addressed by the United States alone.[6]

China's renewed power ascendance constitutes the most consequential strategic development in the Asia-Pacific region.[7] Over the past decade, China has proved adept at keeping off the U.S. radar screen, even as it has advanced comprehensive modernization goals under the rubric of a 20-year "strategic opportunity." The United States and China's neighbors are seeking to incorporate a more powerful Chinese state within a shared (or at least complementary) vision of the regional future. At the same time, the states of the Asia-Pacific region are fashioning national-level responses to adapt to China's rise, and to how Chinese power and policy shape their separate interests.[8] Beijing is intent on achieving comprehensive modernization (including of its military power) while simultaneously broadening and deepening the possibilities for collaboration with nearly all external actors. Regional states, though desirous of closer working relations with China, have no incentive to be enveloped in a Chinese-defined regional order, even as leaders in Beijing insist they are not pursuing such an objective.

China's development is symptomatic of deeper trends at work across the Asia-Pacific region. The United States thus needs to recalibrate its future strategy in light of three interrelated factors: the increased maturation of regional political, economic, and military capabilities and institutions; American policy preoccupations in "out of area" conflicts; and an ever diversifying range of global issues that will require enhanced multilateral collaboration.[9]

Asia's Transformation and U.S. Policy Responses

The Obama administration seeks to define future U.S. leadership in terms of enhanced international engagement without a threat-dominant major power rationale, and Asia and the Pacific will be central to this equation. Even as many U.S. policies remain predominantly bilateral in design, a longer-term regional strategy must incorporate China as a full participant at the table, not as an unspoken subtext in the discussion. Marginalizing China's role in the regional future or basing U.S. strategy on threat-based concepts will not advance the building of a new regional order, but this process must be a two-way street. Long-term stability will require China to fully articulate its security perceptions and expectations, and to mesh Chinese power with larger collaborative ends. Regional powers must also build a deeper understanding of how competing national-level goals, interests, and capabilities intersect and interact. Absent a willingness of all states (but especially the United States and China) to communicate much more fully with one another, none will be reassured about the strategic intentions of others. If either the United States or China retreats into default options that implicitly challenge the legitimacy of the other's strategic objectives, the prospects for a collaborative regional order that can accommodate both countries and reassure states across the region will be far more problematic.

American military power will remain a central component in U.S. regional strategy. However, coercive capabilities, though important and potentially decisive in some circumstances, can no longer ensure that regional actors (including long-term U.S. allies) will remain supportive of the full spectrum of U.S. policy expectations. Unlike during the early decades of the Cold War, the Asia-Pacific region is not characterized principally by weakness, instability, or recurrent crisis. Despite the economic turbulence in much of Asia since the onset of the U.S. financial crisis in the fall of 2008, there have been remarkable economic and political advances across the region. Some regional strategic thinkers assert that this process portends the ultimate emergence of an indigenously defined political and security order, but few envision or advocate the development of a reconfigured regional order without a major American role.[10]

Notwithstanding these trends, various U.S. observers express worries about Asia's future, ranging from modest disquiet to acute pessimism. These concerns encompass heightened nationalism, failures in governance, the possibilities of economic and social disruption, and the potential for unregulated military modernization. But broad regional dynamics are highly beneficial to long-term American interests, even as the ultimate shapes of a new regional order and a reshaped global order remain uncertain. The United States is endeavoring to redefine its alliances and security arrangements while sustaining an ample U.S. forward military presence, premised on the increasing maturation of its regional partners.[11] It is also seeking to build more durable arrangements for military-to-military relations with China. Though terrorist threats persist in Southeast Asia, the region is not wracked by crisis, nor do any regional states perceive the necessity of an "either-or" choice between Washington and Beijing.

The big picture across the Asia-Pacific region lends continued optimism about the longer term. Nearly all regional actors seek enhanced well-being for their citizens, and the opportunities for autonomous political activity within numerous societies continue to grow. The expansion of the middle class in various Asian societies is without precedent in economic history. The levels of intraregional trade and integration across national boundaries are at all-time highs. The risks of war across the Taiwan Strait have diminished with the election of Ma Ying-jeou as president of Taiwan, and with heightened if still preliminary efforts by Beijing and Taipei to establish rules of the road for the longer term. Should cross-strait accommodation deepen, the "long pole in the tent" with the latent potential to generate an acute United States–China crisis will diminish. But North Korea and Burma remain East Asia's persistent strategic outliers, with North Korea's nuclear weapons capabilities and internal vulnerabilities posing a particular danger to regional security.

The challenge and opportunity confronting the United States and the emergent powers of the Asia-Pacific region is to vest regional states (especially the major powers) in an inclusive, future-oriented order. Though bilateral relations will remain a cornerstone of American strategy, they must be embedded in a larger concept of the region's long-term future. Absent a serious attempt to fashion such an order, individual states will focus on autonomous policy goals and will be less attentive to collective needs, and hedging strategies will increasingly dominate national-level calculations. Such an outcome would be highly adverse to long-term U.S. interests. America's deep military involvement in the Middle East and Southwest Asia will also continue to impinge on U.S. policy choices. The Obama administration therefore needs to ensure that preoccupations in the Islamic world do not severely curtail attention to the Asia-Pacific and other regions of lasting import to American interests.

The Role of U.S. Military Power

Many strategic observers across Asia continue to view America as an all-powerful, frequently overbearing military colossus astride the international system. But U.S. financial and budgetary turbulence (including the incurring of major new deficit spending to stimulate renewed domestic growth) will almost certainly constrain U.S. defense choices in the years to come. The administration's current estimate of the budget deficit for Fiscal Year 2009 is $1.84 trillion, or close to 13 percent of the U.S. gross national product, the highest level of deficit spending since World War II.[12] Staggering deficits are not indefinitely sustainable, and national defense expenditure will be an important factor in future U.S. policy debate. But the collective failure of the executive and legislative branches over the past decade to match expenditures with resources (for example, funding the Iraq and Afghanistan wars through supplemental authorizations not "counted" as part of the defense budget) ill served U.S. interests. Such budgetary sleight of hand is no longer credible. These habits have generated a mountain of unfunded future obligations (all obscured under budgetary artifices of one kind or another) that defy imagination.[13] The Obama administration has taken initial steps to introduce greater transparency into defense budgetary allocations, which will almost certainly inject added realism that many policy deliberations have heretofore lacked.

America undoubtedly retains global reach and global interests, manifested most fully by its military power. But what is this power intended to achieve? How relevant will it prove in light of the diffusion of power across the Asia-Pacific region? How does the United States adapt to and shape a world of multiple power centers amid an ever-increasing cacophony of voices on the long-term global future? In the 2006 *Quadrennial Defense Review*, Pentagon planners identified a group of pivotal national actors who were at a "strategic crossroads," with particular reference to China, India, and Russia.[14] But the United States cannot be exempted from this list. Though the United States has a clear incentive to link other major powers as fully as possible to existing international institutions and norms, this will entail far more than the acceptance by others of an international order designed and led by the United States.

U.S. Policy in the Longer Term

Can the United States enhance its regional role by accommodating to concepts and practices not invented and copyrighted by Washington? Can it forgo a measure of its freedom of action for the sake of shared interests and collective goals? What are the potential liabilities the United States might confront under these circumstances? At the same time, can regional actors subordinate their individual aspirations in light of shared interests in development and regional stability? Or will enduring or heightened national differences and rivalries mandate a continued American role as arbiter, balancer, and crisis manager? These constitute some of the central challenges facing the Obama administration in the Asia-Pacific region.

Defining the future goals of American power is therefore pivotal. Fareed Zakaria offers some possibilities in his well argued but inappropriately titled book *The Post-American World*.[15] Zakaria is not talking about a world in which the United States is irrelevant or inconsequential, but one where American power is no longer singular, relative to the capabilities of other national actors. He *is* discussing what he terms "the rise of the rest," or (more accurately) "the rise of some of the rest," and of how the United States should endeavor to incorporate rising powers (especially China and India, but particularly the former) within a redefined international order. In his view, the United States needs to legitimate various emergent major powers, and ensure them seats at the table, while retaining the capability to dissuade emerging powers from coercive actions that would disrupt the peace and undermine security and regional prosperity.

Sino-American relations are increasingly characterized by "mutually assured dependence" across a range of international issues, including energy supply and climate change, trade flows, global finance, and nuclear diplomacy on the Korean Peninsula.[16] The capability of the United States to exclude or marginalize a major power such as China is thus increasingly problematic. At the same time, a U.S. strategy focused predominantly on coercive capabilities and high-end threat possibilities skews American power to an overly narrow range of interests and policy goals. Indeed, it is far from certain that any Asia-Pacific state deems a "hard power"–dominant U.S. strategy as optimal or even sustainable, even though some prominent strategists continue to advocate such an approach.[17] The efforts of Secretary of Defense Gates to rebalance U.S. defense programs "to fight the wars we are in today and the scenarios we are most likely to face in the years ahead" speaks to this issue, as well.[18] Secretary Gates has also called attention to what he deems the militarization of American foreign policy. He has observed that the number of uniformed personnel on a single *Nimitz*-class aircraft carrier exceeds the entire active-duty membership of the U.S. Foreign Service. Few contest that the United States will retain capabilities to project its military power on a global basis, but this will be inadequate in ensuring a comprehensive American role. A larger conceptualization of U.S. regional interests cannot be defined principally by U.S. military-strategic requirements, even as any future approach must remain highly attentive to American security interests.

Since the end of the Cold War, successive American presidents have grappled with the consequences of the loss of a global adversary around whom American strategy was long organized, including the inherited "hub and spokes" alliance system in Asia and the Pacific. Each administration has also sought to confront the inertial tendencies inherent in the operation of large military bureaucracies. The approaches have varied, from George H. W. Bush's emphasis on adaptation and the preservation of American regional preponderance; to Bill Clinton's early efforts at an economically oriented strategy relying less on the traditional instruments of American power, but subsequently geared more toward a doctrine of counterproliferation and preventive defense; to George W. Bush's pursuit of strategic primacy and dissuasion, partially supplanted in his second term by an emphasis on selective diplomatic collaboration.[19] President Obama must now renew this quest, all in the context of an ascendant Asia.

Regional states largely fit into three broad categories: autonomous major powers, long-time allies pursuing an enhanced national identity while seeking assurance from the United States, and local actors prepared to collaborate with the United States and to facilitate complementary political and security goals (North Korea and Burma remain outside these categories.) In oversimplified terms, the United States can pursue (1) inclusion and integration,

(2) prevention and inhibition, or (3) preservation and hedging. These correspond to three impulses underlying these alternative possibilities: an opportunity- or incentive-based approach, a threat-driven approach; or an uncertainty-based approach. The tests of a strategy's viability and relevance, in turn, seem relatively straightforward. First, it should guide and shape the actions of policy makers, not simply serve as an empty slogan or bumper sticker. Second, it should be proportional to the totality of national interests. Third, it should be commensurate with the budgetary, manpower, and political resources required to achieve it. Fourth, it must enjoy a requisite level of public support. Fifth, it must be attentive to the potential effects either of failed or incomplete policy assumptions, or of the intended or unintended consequences of different courses of action. Sixth, it must remain mindful of how policy will influence the incentives, interests, and calculations of states whose behavior a given strategy seeks to influence. I will resist the temptation to engage in a "name that strategy" competition, but will instead turn briefly to a fundamental if often ignored premise of strategy—i.e., the presumed relationship between ends and means.

In a series of wide-ranging speeches delivered in the spring and summer of 2008, Secretary of Defense Gates excoriated those in the U.S. military afflicted by what he termed "next war–itis," lobbying for weapons systems that do little to address the military conflicts in which the United States remains deeply involved. In the *National Defense Strategy* released in mid-2008 (characterized at the time by Secretary Gates as a "blueprint" for the next administration) the document focuses explicitly on the "global struggle against a violent extremist ideology that seeks to overturn the international state system." Though the report also calls attention to threats posed by adversaries such as Iran and North Korea and the need to "consider the possibility of challenges by more powerful states" such as China and Russia, the tone of the document is measured and prudent. As the report notes, "China is one ascendant state with the potential for competing with the United States. For the foreseeable future, we will need to hedge against China's growing military modernization and the impact of its strategic choices upon international security. . . . Our interaction with China will be long-term and multi-dimensional and will involve peacetime engagement between defense establishments as much as fielded combat capabilities. The objective of this effort is to mitigate near term challenges while preserving and enhancing U.S. national advantages over time."[20]

This policy document underscores the increasing realism of Department of Defense (DoD) planning. A U.S. strategy skewed either to a unilateral assertion of American strategic interests or to building the foundations for a long-term military competition with Beijing will increase the likelihood of the outcome that the United States insists it does not want: a long-term adversarial relationship with Beijing that would compel regional states to choose between the two countries. DoD has remained mindful of the enhancement of Chinese military power over the past decade, and it has recalibrated U.S. strategic calculations accordingly. Leaders in Beijing continue to forswear either the intention or the capability to challenge the United States, though Chinese policy makers clearly seek to inhibit (or deter outright) any American actions that might put China's vital interests at risk. China is also actively deliberating the larger purposes and directions of its long-term military development that look beyond Beijing's traditional focus on the "Taiwan scenario." However, absent a fuller strategic understanding between Washington and Beijing, the latent elements of a strategic rivalry (if not outright confrontation) could readily take root in the defense processes of both countries. A deeper, ongoing strategic conversation between senior leaders and between the American and Chinese militaries must be an explicit priority for the Obama administration.

The United States also needs to weigh much more fully the opportunities and possibilities of Asia's ongoing transformation in conjunction with relevant regional states. Such an approach would entail at least six principal security goals:

- Preventing a strategic breakdown or major regional crisis (e.g., conflicts in Korea, in the Taiwan Strait, or between India and Pakistan);
- Enhancing communication related to potential contingencies that could involve multiple powers, with Korea as the preeminent example;
- Achieving sustainable alliance bargains that move beyond traditional approaches and would entail more meaningful responsibility sharing;
- Simultaneously achieving durable relationships with China and Japan, while facilitating and sustaining a longer-term strategic accommodation between Beijing and Tokyo;
- Undertaking a far more extensive set of exchanges between the U.S. and Chinese militaries, with particular emphasis on China assuming more of a "stakeholder" role in international security;
- Pursuing international arrangements where the United States has either been skittish or oppositional in the past (e.g., accession to the ASEAN Treaty of Amity and Cooperation, and ratification of the UN Convention on the Law of the Sea).

Evolutionary understandings or crisis prevention mechanisms could also help ensure what Hugh White characterizes as continuation of "consensual American primacy," while not denying regional powers the means to pursue their individual interests or acutely impinging on American freedom of action.[21] This will require the United States to adapt more fully to a region that bears modest resemblance to the past, but that would legitimate a longer-term U.S. leadership role.

There are at least three distinct projections of Asia's longer-term future and its relationship to American power. One school of thought sees the region's ongoing transformation largely supplanting an American hegemonial position, with multilateral institutions and cooperative security norms underpinning a new regional order. A second school of thought sees the region awash with actual or incipient rivalries that make longer-term stability impossible to achieve, thereby mandating open-ended U.S. security interventions to guarantee the peace. A third approach is premised on renewed American military primacy, with the United States configured principally to "high end" military contingencies. It is doubtful that any of these alternative approaches is viable as a standalone strategic concept. But lasting American engagement is both necessary and possible. Without a concurrent approach that vests the United States and regional states in a compatible vision of the longer-term future, neither America nor the region can expect to ensure mutual security and well being on which the vital interests of all will depend.

Notes

1. David E. Sanger, *The Inheritance: The World Obama Confronts and the Challenges to American Power* (New York: Harmony Books, 2009).
2. For a valuable overview of recent trends, consult David Shambaugh and Michael Yahuda, eds., *International Relations of Asia* (Lanham, MD: Rowman and Littlefield, 2008).
3. See Morton Abramowitz and Stephen Bosworth, *Chasing the Sun: Rethinking East Asian Policy* (New York: The Century Foundation Press, 2006).
4. Ralph A. Cossa et al., *The United States and the Asia-Pacific Region: Security Strategy for the Obama Administration* (Washington, DC: Center for a New American Security, February 2009).
5. David Kilcullen, *The Accidental Guerrilla: Fighting Small Wars in the Midst of a Big One* (Oxford and New York: Oxford University Press, 2009).
6. Jonathan D. Pollack, "North Korea: The Clenched Fist," *Washington Quarterly*, forthcoming.
7. C. Fred Bergsten et al., *China's Rise: Challenges and Opportunities* (Washington, DC: Peterson Institute for International Economics and Center for Strategic and International Studies, 2008).
8. See, in particular, Evan S. Medeiros et al., *Pacific Currents: The Responses of U.S. Allies and Security Partners in East Asia to China's Rise* (Santa Monica: RAND Corporation, MG-736-AF, 2008).
9. For the fullest statement of a future strategy with ample reference to Asia and the Pacific, see Anne-Marie Slaughter et al., *Strategic Leadership: Framework for a 21st Century National Security Strategy* (Washington, DC: Center for a New American Security, July 2008). Many of the contributors to this document are now serving in senior capacities in the Obama administration.
10. For relevant examples, consult Muthiah Alagappa, ed., *Asian Security Order: Instrumental and Normative Features* (Stanford: Stanford University Press, 2003).
11. See in particular Cossa et al., *The United States and the Asia-Pacific Region.*
12. Jackie Calmes, "Estimate of Budget Deficit Now Tops $1.84 Trillion," *New York Times*, May 11, 2009.
13. For a searing critique of these practices, consult Anthony H. Cordesman, "Cleansing the Poisoned Chalice? The Obama Administration and the Challenge of National Security Planning, Programming, and Budgeting" (Washington, DC: Center for Strategic and International Studies, April 21, 2009); see also Michele A. Flournoy and Shawn Brimley, "The Defense Inheritance: Challenges and Choices for the Next Pentagon Team," *The Washington Quarterly*, Autumn 2008, pp. 59–76; and John Lehman, "Wasteful Defense Spending Is a Clear and Present Danger," *Wall Street Journal*, July 18, 2009.
14. *Quadrennial Defense Review Report* (Washington, DC: Office of the Secretary of Defense, 2006), p. 29.
15. Fareed Zakaria, *The Post-American World* (New York: W.W. Norton and Company, 2008).
16. As developed in LtCol Matthew J. Bonnot et al., "Responding to China's Global Pursuit of Energy: U.S. Challenges and Strategic Options" (paper submitted to the faculty of the Naval War College, China Energy Advanced Research Project, June 1, 2007). For my views, see Jonathan D. Pollack, "Energy Insecurity with Chinese and American

Characteristics: Implications for Sino-American Relations," *Journal of Contemporary China* 17, no. 55 (May 2008), pp. 229–245.

17. See Dan Blumenthal and Aaron Friedberg, *An American Strategy for Asia* (Washington, DC: American Enterprise Institute, January 2009).
18. Secretary of Defense Robert M. Gates, "Defense Budget Recommendation Statement" (speech, Arlington, VA, April 6, 2009).
19. For an extended discussion, see Jonathan D. Pollack, "U.S. Strategies in Northeast Asia: A Revisionist Hegemon," in *Power and Security in Northeast Asia: Shifting Strategies*, ed. Byung-Kook Kim and Anthony Jones (Boulder and London: Lynne Rienner Publishers, 2007), pp. 55–95.
20. *National Defense Strategy* (Washington, DC: Office of the Secretary of Defense, June 2008), p. 3.
21. See Hugh White, "In Support of Accommodation: An Australian View of U.S. Policy toward China," in *Asia Eyes America: Regional Perspectives on U.S. Asia-Pacific Strategy in the Twenty-first Century*, ed. Jonathan D. Pollack (Newport, RI: Naval War College Press, 2007), pp. 153–167.

The New Security Drama in East Asia: The Responses of U.S. Allies and Security Partners to China's Rise

Dr. Evan S. Medeiros
Senior Political Scientist
RAND Corporation

In the theater of East Asia, a geopolitical drama is unfolding. The growing presence of China in regional economic and security affairs—generically referred to as the "rise of China"—is changing interstate relations. While the major powers in East Asia are the protagonists, there are no bit players in this drama. Think *King Lear,* not *Macbeth*. China's rise is affecting the perceptions, interests, and policies of all nations throughout East Asia. For the United States, the responses of its allies and security partners are uniquely consequential. These countries are the foundation of American presence in the region as well as the edifice of a regional security architecture that has produced decades of relative stability and prosperity.[1]

Much of the prevailing research about regional responses to the rise of China makes this drama sound like a slowly unfolding tragedy for the United States. Many argue that China is rapidly gaining regional influence at the expense of the United States. The use of superlatives abounds in the description of China's rise in East Asia, with the unproven implication that this uniformly redounds to Beijing's benefit and to American disadvantage. Joshua Kurlantzick notably argued that China's "charm offensive" is allowing it to displace the United States as the dominant power in East Asia.[2]

To understand and evaluate these evolving dynamics, the RAND Corporation conducted a year-long study of the responses of U.S. allies and security partners in East Asia.[3] The study sought to answer four questions: How have these nations responded to China? What forces are driving these reactions? How will the drivers change? What are the implications for American regional security interests? The study examined the responses to China of the five U.S. allies in the Asia-Pacific and of Singapore, a major security partner.[4] The RAND study analyzed the responses of these six nations in four areas: domestic politics and public opinion, economic policy, foreign policy, and defense policy. This structure allowed the study to explore a range of national responses as well as responses across each functional area (e.g., defense policy), generating conclusions about both country-specific and regionwide responses to China. This article highlights the most salient findings from this research.

Overall Regional Responses to China's Rise

In contrast to much of the current research, China's growing presence and interactions with U.S. allies and security partners are not fundamentally transforming the security order in the Asia-Pacific. China is having an influence on these relationships, but these changes are not as rapid or comprehensive as many presume.

First, the foundation of the U.S. alliances in Asia continues to endure. No allies or major security partners see China as a viable strategic alternative to the United States. The United States remains the security partner of choice, largely because it is the one nation seen as possessing the capability and resolve to balance China. Its allies and partners prefer that Washington do the "heavy lifting" of deterring China and, ultimately, preventing Chinese domination of regional affairs. U.S. allies are all intensely pursuing engagement strategies with China, driven principally by an economic logic. They want to benefit from China's large and growing economy, especially during the current global recession. But these goals exist alongside concerns about China's long-term intentions, particularly its military modernization plans. A recent project by the Center for Strategic and International Studies that uniquely polled elites throughout Asia confirmed this duality. The study found that China was ranked first as the "greatest threat to peace and stability in the next 10 years" and second as the "greatest force for peace and stability" in the next ten years.[5]

Second, China is affecting American relationships with its allies and security partners. On the one hand, China's rise makes some U.S. security commitments more relevant. These countries can interact with China more confidently because they know (and Chinese leaders see) that the U.S. commitments to them and to involvement in Asia continue. On the other hand, allies and partners are also positioning themselves to benefit from both the United States and China. This is a recalibration more than a transformation. None of these nations want to choose between the United States and China, and all reject having to make such a choice. Also, some of these nations use their interactions with China to generate leverage in dealings with the United States. Some of the smaller, middle powers in East Asia, like the Philippines and Thailand, have attempted such strategies. On balance, U.S. allies and security partners want continued American involvement in the region but sometimes only in certain ways, at certain times, and on particular issues.

Third, China is undoubtedly gaining influence with U.S. allies and partners in East Asia—in the defined sense of looming larger in their economic, diplomatic, and defense policies decisions. This is a natural and inevitable trend. The key question is how it is manifesting itself in these states' regional behaviors. Our research found that U.S. allies and partners in Asia have become more sensitive to some of China's preferences and interests, especially on China's self-identified "core interests" *(hexin liyi)*, which now include both Taiwan *and* Tibet.[6] There have been several instances in which specific nations have canceled visits and changed policies on these issues due to Chinese intervention. But this too is not terribly surprising. Sovereignty issues resonate with many postcolonial states in Asia and, more important, changes in Taiwan or Tibet policy are seldom costly for these states in the sense of undermining their material interests. Thus, these behaviors are not leading indicators of wholesale accommodation to China.

A related indicator of Chinese influence on these states is that Beijing has been effective at precluding the emergence of "anti-China" containment efforts, to the extent that there was ever a push for such an approach. China has been effective at accumulating "defensive influence," persuading nations to avoid taking actions China deems to be threatening. There is very little evidence that China has accumulated "offensive influence," in the sense of policies that could effectively degrade or dismantle U.S. alliances or security partnerships in the region. In the late 1990s, China tried and failed to offer an alternative regional security architecture, with the promotion of its "New Security Concept." Few nations were interested, or now are, in jumping onto this strategic bandwagon, even in the wake of the Asian financial crisis in 1997–98 and the disillusionment with American responses to it. More recent Chinese attempts to push U.S. allies have backfired, alienating regional states and enhancing their coordination with the United States. Prominent examples include Singapore in 2004 and South Korea in 2006.[7]

In assessing China's rise in East Asia, two additional considerations are noteworthy. First, China's growing presence and interactions in the region do not directly translate into influence—that is, using incentives and sanctions to alter other states' behavior. Many analysts too often mistake presence for influence. The fact that countries are trading more with China and negotiating with it in regional organizations does not, ipso facto, imply that China can change these states' policies, especially when policy changes require a state to compromise its material interests.

Second, the regional consensus favoring engagement with China has a tentative quality. There is creeping uncertainty about China's future: some nations fear a weak China, and some fear a strong China. Few are willing to bet their futures on Beijing's assurances about a "peaceful rise." China's large and growing economy (even during the current global recession) is not a geopolitical "tractor beam." While China's economy looms large for all nations, fears of China as a competitive threat have motivated much diversification in trade relations. There are nagging concerns among regional leaders about Chinese military modernization. As People's Liberation Army (PLA) capabilities improve, such as with the likely future deployment of China's first aircraft carrier, and as the PLA conducts more out-of-area operations, these nagging concerns could evolve into closer security coordination with the United States and its allies. The recent statement of concern about China's growing defense budget by South Korea's president during new security consultations with Australia is instructive in this regard.

A final regional response to China's rise is a nonevent—the lack of a regional rush, over the last decade, to increase military budgets and modernize conventional forces in response to concerns about China's military. The military budgets of Japan and South Korea have remained relatively flat in *real* terms, with gradual increases in South Korean defense spending. Southeast Asian militaries' budgets did not substantially increase either in the last decade; many just returned to the spending levels of the period prior to the Asian financial crisis (see figures 1 and 2). There are even some notable examples of a deep atrophy in external defense capabilities, such as in the Philippines. That said, Asia could be on the cusp of a limited change in this past trend. Australia's recently released defense white paper calls for a substantial increase in naval

Figure 1
Total Defense Budgets in Japan and South Korea, 1997–2007

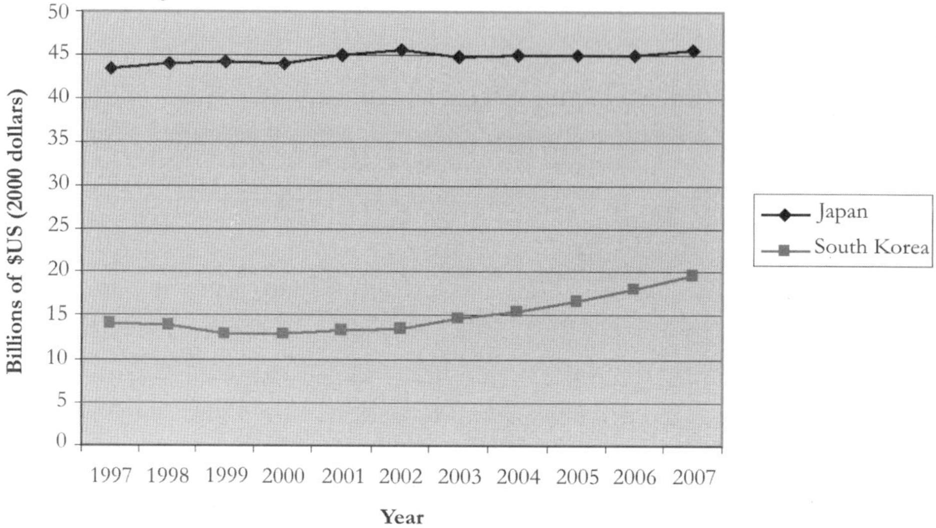

Source: The data for figures 1 and 2 are from Australia Defence Intelligence Organization, *Defence Economic Trends in the Asia-Pacific* (Canberra: Department of Defence, 2007), available at www.defence.gov.au/dio/documents/2007_DET.pdf.

Figure 2
Defense Budgets in Southeast Asia, 1997–2007

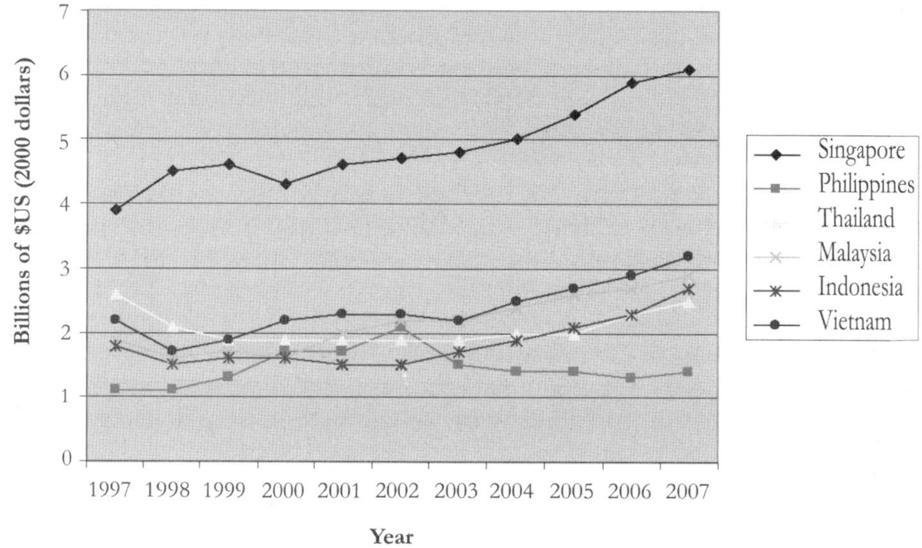

capabilities, especially submarines, in reaction to China's sustained naval expansion and the Chinese navy's growing presence in the South and East China seas. But most East Asian states are not likely to initiate major procurement programs in the next five years; many are suffering from the global economic crisis, allocating scarce government resources to much-needed economic stimulus programs.

Country-Specific Reactions to China

The particular responses of individual countries provide greater texture for understanding these trends.[8] Those of Japan, South Korea, the Philippines, Thailand, Singapore, and Australia are summarized below.

Japan

The rise of China in East Asia has clearly stirred Japan's competitive impulses, but its posture toward China remains characterized by considerable ambivalence and marked by growing anxiety. Many Japanese leaders are more willing than in the past to cite China explicitly as a potential military threat, and the two countries have engaged in heated disputes over territorial boundaries, historical issues, and regional leadership. These three sets of issues will drive competition between China and Japan in the coming years. Japan has edged closer to the United States and strengthened ties with other regional states, from India to Australia to Taiwan—moves that are increasingly justified by reference to China. Tokyo has also demonstrated a new willingness to use its military forces to, for example, patrol ocean areas disputed with Beijing.

At the same time, Japan's businessmen and economic planners remain convinced that their nation's economic well-being is tied to continued trade and investment with China. This remains the case during the current global recession; many in Japan viewed China as having pulled it out of the last recession, which began in the 1990s, and as being able to pull it out of the current one as well. A broad alliance of business, political, and media actors have supported the outreach to China since the prime ministership (2001–2006) of Junichiro Koizumi, and Beijing has reciprocated by taking a more conciliatory posture. Many strategists and politicians also foresee damage to Japan's position in Asia should a cold war develop between Tokyo and Beijing.

The long-term prognosis for Sino-Japanese relations is highly uncertain, and there are certainly grounds for concern about future instability. For the first time, both China and Japan are unified internally, possess substantial and growing economic and military capabilities, and are capable of influencing events beyond their borders. At the same time, the United States is pushing for Japan to assume a larger global role, especially in military terms. Domestically, the demise of the Socialist Party during the mid-1990s nudged the political center of domestic politics to the right. Japan's emergence from fifteen years of sluggish economic growth helped usher in the rise of nationalist sentiments that remain today. At the same time, a new breed of popular politicians has challenged the long-dominant bureaucracy for control of national policy, including foreign policy.

South Korea

The most basic—but not the most complete—answer to the question of what is driving South Korea's response to China is a generally benign view of China and the perceived economic benefits of stable relations with it. Given these conditions, there is considerable sensitivity toward China in South Korea today and reluctance either to challenge major Chinese interests or needlessly stimulate Chinese sensitivities. At the same time, growing concerns and anxieties about Chinese economic policy making and diplomacy show that the honeymoon in China–South Korean relations is decidedly over. The forces holding the relationship back, if not driving it in the opposite direction, include uncertainties about China's medium- to long-term intentions (especially regarding China's military modernization and its growing influence in North Korea), awareness of potential South Korean vulnerability to Chinese economic or other pressure, a widely shared awareness of the importance of the United States, and a continuing gap between South Korean aspirations and capabilities.

These cross-pressures suggest that, first, South Korea will continue to expand ties with China, with trade and investment leading the charge to the extent possible during a global recession. South Korea is likely to emphasize solving actual problems between the two countries, such as implementing confidence and security-building measures that could improve prospects for peace on the Korean Peninsula. By geography alone, sensitivity toward some Chinese interests will remain a characteristic of South Korean policies. Furthermore, the irritants in and constraints on the relationship will also continue, and an occasional spike in tensions is to be expected. As China continues to ensconce itself in North Korea, issues pertaining to the North could come to have as many negatives as positives for bilateral relations. Even short of this, a new strategic alignment between South Korea and China is not likely, in the absence of some major external event. South Korea will likely seek to maintain good relations with China on the basis of—rather than instead of—a continued close alliance with the United States. Another North Korean nuclear test, or clear Chinese unwillingness or inability to bring the North to resolve the nuclear issue peacefully, would reinforce this inclination.

This mixed picture suggests that barring unexpected developments, South Korea will stick with the United States, even at critical decision points that test the U.S.–South Korean alliance, as was the case with American Iraq policy. For Washington the real policy challenge is that China's rise may complicate its efforts to expand U.S.–South Korean security cooperation. Domestic politics in Seoul will strongly influence this. South Korean agreement to participate in American military operations based out of its homeland will be particularly difficult to obtain, although this will depend heavily on the context in Korean domestic politics, bilateral relations, and international relations. The key to the future of the relationship will be reconfiguring the alliance correctly.

The Philippines

The Philippines' response to China is strongly defined by the country's fundamental and myriad weaknesses. Chronic political instability, debilitating domestic insurgencies, and deteriorating external defense capabilities have left the Philippines unable to ensure stability within the main islands, let alone to protect its offshore territorial claims vis-à-vis China. These weaknesses have spurred Philippine efforts to reestablish close defense ties with the United States, mainly to cope with its own severe internal security challenges. Philippine leaders no longer view China as a major security threat, as they did in the mid-1990s. This ambivalence about China has been reflected in a severe atrophy of Philippine air and naval capabilities in the last five to ten years. However, distrust of China's ultimate intentions remains and is growing in some quarters, driven in part by domestic politics. Since 2007, China policy has emerged as a politically sensitive issue, constraining Manila's engagement with Beijing and lubricating interaction with Washington. Rebuilding of the Philippines' external defense capabilities remains a long-term goal, however.

The Philippine economy is less dependent on trade with China (and on international trade, more generally) than are the economies of some of its Southeast Asia neighbors (e.g., Singapore and Thailand). Like other Asian economies, however, China has become a major destination for Philippine exports, which motivates a perception that trade with China is important to the Philippines' future economic growth. This calculation could be changing as the China-centered processing trade rapidly declines due to the current global recession. A broad consensus in the Philippines over China's importance as an economic partner has, for the past five years, helped to strengthen bilateral ties. Yet the view that China is an important future economic partner is mixed with an incipient sense that China is also a competitive economic threat.

While there are forces driving the Philippines' response to China, it is important to stress that these forces are not "driving" Philippine policy anywhere in particular. The leadership is heavily focused on internal challenges, and the public is relatively inattentive to China and, for that matter, most other foreign-policy issues. To the extent that China has gained popular and elite attention, it has been linked to politically charged corruption scandals that fuel popular concerns about becoming too close to China.

Thailand

Thailand has a long tradition of "bending with the wind." In today's East Asia, that means accommodating—and seeking advantage from—both China and the United States. Among the six nations examined in the RAND study, Thailand was the most likely and willing to accommodate China. Thaksin Shinawatra, the former prime minister, modified this approach by trying to "blow the wind" as well as bend with it. He strengthened political and military, as well as economic, ties with China at the same time as he was taking bold new steps to buttress Bangkok's alliance with the United States. His successors, however, have returned to a more muted style of foreign policy—to the extent they have the time or resources to focus on foreign policy amid sustained political instability. The post-Thaksin governments have de-emphasized bold initiatives, particularly on the strategic and military fronts, and have refocused Bangkok's diplomacy on the Association of Southeast Asian Nations (ASEAN). Absent the reemergence of a political leader with a strong foreign policy vision and the political space to pursue it, Bangkok will continue to deepen gradually its economic, political, and, to a lesser extent, military relationships with Beijing, as well as with Washington.

While Thai foreign policy has seldom been all in one direction, several long-term trends suggest that relations with China have become more important to Thailand in the last decade. China's value as a trade and investment partner has grown substantially, but recent declines in trade with China could alter this calculation. Thailand has acquired some military hardware from China, and the two nations have conducted two joint military exercises. But these trends pale in comparison to the scope of Thailand's economic and security cooperation with the United States.

There are also limits to the Thai-Chinese relationship. Despite Thailand's past efforts to engage Burma (thereby removing a source of tension with Beijing), Burma's recent instability has once again made it an issue between Beijing and Bangkok. Thai leaders are intensely focused on establishing stability at home, a seemingly endless task since the 2006 coup. When they do focus on foreign policy, they state that they are committed to a balanced posture between China and the United States. Thai policy makers recognize the long-standing material and symbolic benefits of the U.S. alliance. Bangkok is also working to develop options with other countries. Economically, it has strengthened ties with India, Australia, New Zealand, and Japan. Politically and militarily, it cooperates with India, Singapore, Malaysia, and Indonesia, as well as with the United States and China.

China's regional behavior will be the largest variable in the evolution of Thai attitudes toward the rise of China; China has been heavy-handed with Thailand regarding its interaction with Taiwan and Tibet authorities. Events in Burma, the success or failure of ongoing negotiations with the United States and Japan for free trade agreements, and the future of political reform in Thailand are also important variables, albeit less widely appreciated ones.

Singapore

Singapore shows less ambivalence about the rise of China than do most Southeast Asian countries. The country's small size, geostrategic vulnerability, and continuing concerns about long-term Chinese intentions propel it toward a close, strategic relationship with the United States, despite its close ethnic links to China. Singaporean leaders see the United States as both the principal stabilizer in East Asia and the only realistic counterweight to potential Chinese assertiveness. Keeping the United States actively engaged and forward deployed in the region is a central Singaporean objective. China's rise, the spread of Islamic extremism, and heightened concerns about stability in neighboring countries have prompted Singapore to strengthen security cooperation further with the United States. At the same time, Singapore has expanded security links with the United Kingdom, Japan, Australia, and other nations with stakes in Asia's stability.

The benefits Singapore receives from increasing trade and investment with China, as well as from China's broader economic integration in the region, also drive bilateral relations. These policies are balanced, however, by Singapore's corresponding efforts to diversify its economic relationships to avoid excessive dependence on China. Singapore is doing so by negotiating a range of free trade agreements, in particular with Japan and the United States, as a means of countering China's intensive economic diplomacy; this also helps Singapore entrench the former countries economically in Southeast Asia.

Because of the relative clarity of Singapore's long-term vision, the future of Singapore's relationship with China has a greater level of certainty than that of any other Southeast Asian nation. As China becomes more powerful, Singaporean leaders will do everything they can to ensure a continued balance of power in the region, one in which China does not dominate economic or security affairs. This strategy will almost surely guarantee continued close diplomatic and security relations with the United States and other U.S. allies. However, in the absence of unprovoked Chinese aggression, Singapore will neither encourage nor support "containment" or an explicitly "anti-China" balancing coalition.

Australia

There are distinct cross-pressures in Australian-Chinese relations. First, rapidly growing merchandise trade (mainly in natural resources) and the perception among Australian policy makers that China is key to future prosperity have been the major drivers of bilateral relations. Second, few in Australia see conflict with China as likely or inevitable. Australia wants to avoid being drawn into a regional rivalry with China. Third, Australian policy makers possess a deep uncertainty, mixed with a growing concern, about China's role in Asian economic and security affairs. Recent Chinese investments in Australia's resource sector have prompted a debate about overreliance on China. Beijing's diplomatic activism, especially in the South Pacific, and its military modernization are generating worries among Australian policy makers and strategists.

How will these cross-pressures play out? Canberra will continue to expand its bilateral relations with Beijing, with economic ties at the fore, albeit more tentatively than in the past ten years. Concerns about Chinese investment in Australia and limited access of Australian businesses to key sectors of China's economy are now emerging. As China looms larger in Australia's foreign policy, Canberra will continue to be sensitive to, and will accommodate, some of Beijing's interests, such as its policies on Taiwan. Australia's concerns about China's diplomatic and military behaviors in Asia will persist. This in turn will limit the expansion of Chinese-Australian relations and enable greater alliance cooperation with both the United States and other regional powers. Australia's recent security-policy coordination with Japan and South Korea is notable in this regard.

Under the John Howard administration (1996–2007), Australia's concerns about China motivated a series of foreign and defense policies that expanded alliance cooperation and sought to ensure that the United States would remain highly influential in the Asia-Pacific region. The new Labor Party government, led by Kevin Rudd, has pursued a similar approach. Rudd chose to distinguish his foreign policy from that of his predecessor on global issues—such as Iraq policy, nuclear nonproliferation, and climate change—rather than on China policy. Kevin Rudd has made it clear that while China may be an increasingly important "partner" for Australia, the United States is a "strategic ally." He believes that a strong alliance bolsters Australia's position in Asia and that the alliance contributes to broader regional stability.

A new and more complex stage in Australia's relations with China (and the United States) began this year with the publication in May 2009 of a new defense white paper, which is Rudd's first and the nation's first since 2000.[9] This important document cited China's improving power-projection capabilities and uncertainty about both American defense capabilities and the U.S. role in Asia to justify a significant increase in defense procurement. The white paper called for acquiring up to twelve conventional submarines, additional amphibious lift, and land-attack cruise missiles (among other items).[10] Unsurprisingly, Beijing reacted negatively to this assessment, assuming that this procurement was directed at countering Chinese military capabilities. Washington continues to digest the explicit and implicit messages from one of its most stalwart allies in the Asia-Pacific. American strategists should be concerned that some in Australia view U.S. defense strategy and Secretary of Defense Robert Gates's proposed cuts as indicating an eventual inability to maintain robust power projection into the western Pacific.[11]

Assessing the Impact of the Global Financial Crisis

The global financial crisis and resulting recession raise numerous questions about economic and security relationships in the Asia-Pacific, including about China's relative influence over U.S. allies and partners. Many commentators have speculated that the current crisis is a strategic tipping point akin to the era just after World War II in which the United States eclipsed Britain as the global economic hegemon. I would recommend much caution in accepting such dire assessments.

First, it is far too early to make such grandiose conclusions about the effects of the crisis on the global balance of power. It remains uncertain how severe and lasting the crisis will be, especially among East Asian economies. Key questions remain unanswered: Who will be hurt the most? Who will recover the fastest, and how? Which states or institutions will help East Asian states recover? It is likely that both the United States and China will play roles—individually, jointly, and in concert with international organizations.

Second, it is uncertain that this crisis increases China's economic clout while diminishing that of the United States. China's economy was challenged by the crisis in ways that highlight existing questions about the sustainability of its current growth model, which emphasizes exports and investment over consumption. The steep declines in aggregate external demand from the United States and European Union (EU) triggered rapid and dramatic declines in China's exports and imports beginning in fall 2008. This in turn led to a reduction in exports as a driver of growth, leaving consumption and investment to carry much of the load. This is the first time in the last thirty years that China has experienced a sustained and deep decline in total trade. The economic effects—both direct and indirect—of this on employment and trade-related investment are highly uncertain. This could prove to be a constraint on Beijing's ability to sustain a moderate level of growth while stimulating greater domestic demand.

Beijing is addressing its predicament through a four-trillion-RMB stimulus package, which seeks to increase internal demand (e.g., consumption and investment) to replace the loss of external demand (e.g., exports). The stimulus package is facilitated by a wave of spending financed by central and local government on infrastructure and real estate projects. (Total bank lending in the first quarter of 2009 was more than in all of 2008!) To boost internal demand, Beijing is making added efforts to stimulate domestic *consumption,* especially in the rural areas, as part of its economic restructuring and, ultimately, the rebalancing of the Chinese economy. Analyses by major international investment banks indicate that China's initial stimulus is working, which has led many of them to revise upward their estimates of Chinese growth in gross domestic product from around 6.5 percent to 7.0–7.5 percent for 2009. In other words, China will almost certainly recover from the crisis faster than the United States and other major Western economies.

But China's approach may not be as beneficial to its economy as initial indicators suggest; time will tell. China's direct and indirect stimulus spending is probably not sustainable for more than two or three years, given the scope of deficit spending and related bank loans. Government-directed bank lending has been so intensive in 2009 alone that many now worry that China is fueling a new wave of bad debts, which would gut the last round of successful bank reform, initiated in the late 1990s. A key determinant of China's success will be its ability to stimulate domestic consumption as a driver of growth and not simply rely on government-funded investment in order to transition from a short-term policy response to global recession to a long-term strategy for sustainable growth.[12]

Lastly, it remains decidedly unclear that China's projected quick recovery will aid struggling East Asian economies. China's stimulus package may not position it to emerge as a new engine of regional prosperity. In other words, China will not necessarily be East Asia's economic savior. Due to the declines in Chinese imports and exports (as a result of recessions in the United States and EU), the regional network of processing trade in East Asia seems to be unraveling. Asian economies that are both trade dependent and heavily involved in processing trade with China—namely, Singapore, Taiwan, Thailand, the Philippines, and some South Korean sectors—will not enjoy the benefits of stimulus-driven growth in China. These economies are suffering the most right now, and unlike after the Asian financial crisis, they cannot simply export their way to renewed growth. By contrast, regional economies whose trade with China is in capital goods and commodities, such as Japan, Australia, Indonesia, and other sectors in South Korea, will benefit from China's stimulus package. This situation could, over time, result in an adjustment in some regional perceptions of the perils of overreliance on trade with China, leading to diversification in trading partners and bilateral relations.

China does possess an important economic tool that it could use to be viewed once again as the fulcrum of regional growth: outward direct investment. China has the world's largest foreign-exchange reserves, and as a result of reforms initiated in the late 1990s, its major banks and some corporations are de-leveraged and quite profitable—at least for now. Thus, the government has substantial financial resources it could use to invest in East Asia and globally. As the United States and Japan found in past decades, investment in countries can, over time, produce political influence by employing local people and creating a political constituency in favor of the investing nation. There are incipient indicators that China is ramping up its overseas investments—taking advantage of cheap prices and needy companies. For example, China has accelerated its acquisition of ownership stakes in resource-producing companies in Australia, Russia, Kazakhstan, and Brazil. Chinese outward direct investment will be an important variable to watch in assessing its mechanisms and channels for translating economic capabilities into political influence.

Which Way Will They Go?

The preceding analysis suggests several preliminary conclusions about this evolving geopolitical drama in East Asia. These conclusions represent neither a climax nor a denouement but a developing plotline.

First, the United States remains well positioned to achieve its long-standing regional objectives, however the Barack Obama administration chooses to characterize them. The United States does not face a crisis of confidence, and the foundations of its influence endure. It is still early days in Asia's response to China; most countries are still coming to terms with what it means for China to be a more influential actor. This has prompted an abundance of reactions, including many contradictory ones. Accordingly, there is still abundant geopolitical space for Washington to expand and improve its security partnerships in the region. If the United States is to do so, its Asia policy needs persistent attention. Although the George W. Bush administration's Asia policy left the region in fine condition, renovation of regional relationships is needed. In the face of China's rise (as well as the growing prominence of India and Japan), the United States needs to improve the legitimacy of its role and the credibility of its commitments in the Asia-Pacific. That effort will require an adaptation to the changing constellation of the equities of U.S. allies and security partners. None want to provoke China or be drawn into a containment effort; none want China to dominate the region; none want the United States to leave or even substantially draw down its presence; and all want China to play a major role in managing regional challenges. American policy needs to reflect these changing regional realities.

A second major finding of RAND's work on regional reactions to China was that there was no strong correlation between high levels of economic integration with China and accommodation of it. Japan, Singapore, and Australia all have large, growing, and highly complementary trade and investment relations with China. Their trade with China represents a larger share of their total world trade than that of other East Asian nations, and the business communities in these countries have been, on balance, bullish about China.

However, this is not reflected in their foreign and security policy making in any direct manner. Policy makers in all three nations harbor deep uncertainty about China's future and have growing concerns about its emergence as a regional security threat. The governments in all three countries have responded in part by enhancing their alliance links with the United States, each other, and others in Asia. Australia has begun to improve its regional power-projection capabilities in particular ways. A distinct diplomatic priority in all three nations is ensuring that the United States remains active and influential in East Asia, so that China does not dominate. For Tokyo and Canberra, Chinese defense modernization is increasingly a factor in their military procurement and planning, a set of assumptions that is poised to become more prominent as the PLA deploys additional power-projection capabilities and increasingly operates outside China's littoral.

A third important finding is that domestic politics matters a lot in determining nations' responses to China's rise. For most East Asian states, China's rise generates a variety of contradictory reactions, some drawing them toward China and others making them wary. What determines which way they go? A key independent variable is domestic politics. The political conditions in East Asian nations and, especially, the views of political leaders mediate the extent to which diplomatic and economic interactions with China result in accommodation of China, alienation from the United States, or both. The changes in South Korean responses to China following the 2008 election of Lee Myung-bak offer a prominent example. Although relations with China had not fundamentally changed by early 2008, President Lee reoriented South Korea more toward the United States and created a permissive environment for questioning Korea's growing reliance on China. Lee has now positioned South Korea as yet another medium-sized regional power raising concerns about Chinese military modernization. Ultimately, the perspectives and preferences of these nations' top leaders will have a defining influence on how they respond to the myriad of challenges posed by China as well as by U.S. policy in East Asia.

Notes

1. To be sure, a growing number of voices have called into question the permanence of the U.S.-based security architecture in East Asia. These perspectives call American presence in East Asia a historical aberration following World War II, and they view its current presence as outdated, given the alleged acceleration in regionalism. See Kishore Mahbubani, "America's Place in the Asian Century," *Current History* (May 2008), pp. 195–99.
2. Joshua Kurlantzick, *Charm Offensive: How China's Soft Power Is Transforming the World* (New Haven, Conn.: Yale Univ. Press, 2007). Interestingly, Kurlantzick seems to have stepped back from some of his book's main conclusions; see his "So Far, It Just Isn't Looking Like Asia's Century," *Washington Post,* 7 September 2008.
3. See Evan S. Medeiros, Keith Crane, Eric Heginbotham, Norman D. Levin, Julia F. Lowell, Angel Rabasa, and Somi Seong, *Pacific Currents: The Responses of U.S. Allies and Security Partners in East Asia to China's Rise* (Santa Monica, Calif.: RAND, 2008).
4. The United States has concluded mutual defense treaties with five countries in the Asia-Pacific: Australia, Japan, the Republic of Korea (South Korea), the Philippines, and Thailand.
5. Bates Gill, Michael Green, Kiyoto Tsuji, and William Watts, *Strategic Views on Asian Regionalism: Survey Results and Analysis* (Washington, D.C.: Center for Strategic and International Studies, February 2009), pp. 4–7.
6. In his first major speech in the United States, State Councillor Dai Bingguo characterized the issue of Tibet as a core national interest. He stated, "Taiwan and Tibet-related issues concern China's core interests. The Chinese people have an unshakable determination to defend our core interests." See Dai Bingguo, "Address at the Dinner Marking the 30th

Anniversary of the Establishment of China-US Diplomatic Relations Hosted by the Brookings Institution" (Brookings Institution, Washington, D.C., 11 December 2008), available at www.fmprc.gov.cn/eng/.

7. In 2004, China publicly and strongly criticized Prime Minister–elect Lee Hsien Long for a visit to Taiwan following his election but before his inauguration. China's loud and harsh rhetoric raised alarm bells in Singapore about China's growing assertiveness. In response, Singapore subsequently took several steps to broaden its defense and diplomatic ties to the United States. It also took measures to constrain China's diplomatic influence in East Asia, including leading the charge in 2004–2005 to eliminate China's control over the location and agenda of, and participants in, the East Asia Summit. Singapore did not radically alter its Taiwan policy in response to Chinese pressure; Singapore continues to train its military forces at facilities in Taiwan.

   A similar course of events transpired in South Korea in 2006, when then–Chinese ambassador to South Korea Ning Fukui publicly warned South Korean policy makers to restrict the geographic scope of operations for U.S. forces based in Korea to the defense of the peninsula and not other regional contingencies, such as a Taiwan conflict. This statement piqued latent but growing sensitivities in Seoul about China's meddling in South Korean foreign policy and China's growing assertiveness. This event contributed to a change in the national conversation in South Korea about the challenges and threats posed by a rising China. The 2007 change in government created a more permissive political environment for the reflection of these concerns in national policy making.

8. This section draws from the summary of Medeiros et al., *Pacific Currents,* pp. xviii–xxiii.

9. Australia's Department of Defence then published biennial "updates" to the 2000 white paper—in 2003, 2005, and 2007. For copies see merln.ndu.edu/whitepapers.html.

10. Australian Department of Defence, *Defending Australia in the Asia Pacific Century: Force 2030* (Canberra: Department of Defence, May 2009).

11. For an example of this view see Andrew Shearer, "Australia Bulks Up," *Wall Street Journal Asia,* 6 May 2009.

12. Patrick Chovanec, "Undoing Chinese Bank Reform," *Wall Street Journal Asia,* 7 May 2009.

U.S. Foreign Policy and Southeast Asia: From Manifest Destiny to Shared Destiny

Dr. Emrys Chew
S. Rajaratnam School of International Studies, Singapore

From postcolonial state to global superpower, America's relations with Southeast Asia—as with the rest of the world—have been driven by a peculiar sense of "manifest destiny." Founded upon such transcendent values as "life, liberty, and the pursuit of happiness," the United States as champion of those values in the world has, time and again, rightly or wrongly, made a case for American exceptionalism if not interventionism. In its quest for security and prosperity, and in little over two centuries of its existence, the United States attained a measure of global authority surpassing George Washington's loftiest aspirations. Not since Rome (and Britain even) had any Western nation achieved such supremacy.

Yet America's global transformation into a new "empire of liberty," with all its inherent ambiguities of power, did not thereby deliver the freedom from fear that Washington had envisioned: from Pearl Harbor to Ground Zero, from Vietnam to Afghanistan.

Just as the United States has sought to refashion nations abroad in its image—from past ages of Western imperialism, world wars and decolonization, through to the Cold War and the "war on terror"—the diverse nations that constitute Southeast Asia have played their part, too, in shaping the imperatives and dynamics of U.S. foreign policy. These cross-cultural interactions, perceptions, and reactions reveal both the extent and the limits of American power in the region. This historical study examines the distinctive phases and emphases of U.S. foreign policy in Southeast Asia, as well as evolving Southeast Asian perspectives on U.S. foreign policy.

U.S. Foreign Policy in Southeast Asia

Since its birth as an independent nation with an independent foreign policy, the United States has cast itself uniquely as champion of a new world order built upon universal values of self-determination and human rights. Throughout the history of U.S. foreign policy, however, such notions of American exceptionalism have been manifested unevenly in terms of both the power of America's example and the example of America's power.[1] Global pressures in war and peace, and the rise of American world power and influence, have tended to make more explicit what was always implicit in the ideas, institutions, and instruments of U.S. foreign policy. Woven into U.S. foreign policy tradition are almost contradictory, alternating strands of unilateralism and universalism, liberal as well as fundamentally conservative values, where ideals and national interests intertwine but have not always complemented one another; the Republican administrations have tended to place greater emphasis on military-strategic interests while Democratic administrations have emphasized human rights issues. The regions of the world—including Southeast Asia—have felt the influence and impact of such foreign policy imperatives, at times as assertive and expansive under a Democratic president as under a Republican.

There have been three main phases of U.S. foreign policy in Southeast Asia. First, a period of early adventurism and expansionism (1776–1946), culminating in America's colonial policy in the Philippines during the era of Western imperialism, followed by the end of that colonial experiment through America's promotion of national self-determination in the era of world wars and decolonization. Second, a period of anticommunism and ambiguity (1946–1989), where America's containment policy during the Cold War was marked by a certain ambivalence in its support of authoritarian regimes while proclaiming liberal-democratic values in the bid to counter the communists. This period came to be dominated by the imperatives of the "domino" theory and "quagmire" thesis, as Southeast Asia became a critical frontier and the United States was increasingly bogged down by military-strategic commitments in Vietnam. Finally, a period of unparalleled authority mingled with uncertainty (1989–2009), in which America's post–Cold War global hegemony was challenged in such a manner as to require post-9/11 counterterrorism strategies dealing with the Islamic extremist threat. Southeast Asia, as home to the largest concentration of Muslims in the world, became a crucial frontier once again in America's military-strategic calculations.

Back in the nineteenth century, U.S. envoys had negotiated commercial treaties with Siam (Thailand) and Cochin China (southern Vietnam) as early as the 1830s. It was, however, in the 1890s that the United States first took on substantive military-strategic commitments in Southeast Asia, when Alfred Thayer Mahan's classic *The Influence of Seapower upon History* (1890) supplied a persuasive rationale for a new battleship navy and a more ambitious U.S. foreign policy across the Pacific. At the onset of the Spanish-American War, battleships of the U.S. Navy sank the Spanish fleet in Manila Bay (1898), a spectacular victory that galvanized U.S. Admiral George Dewey and Presidents William McKinley and Theodore Roosevelt. In McKinley's words: "There was nothing left for us to do but to take them all, and to educate the Filipinos, and uplift and civilize them, and by God's grace do the very best we could by them, as our fellow-men for whom Christ died."[2]

America's "benevolent assimilation" and "Americanization" of the Philippines, with self-evident "Orientalist" assumptions of cultural and moral superiority, moved well within the mainstream of Western imperialism.[3] But when the Democrats won the presidency in 1912, the Wilson administration introduced a program of "Filipinization," giving Filipinos more seats on the governing executive council and larger roles in the bureaucracy. Congress passed the Jones Act (1916), committing the United States to granting independence as soon as the Filipinos could establish a "stable government"; even though the pledge was vaguely worded, it still was unprecedented in that no imperial power to date had yet promised independence or even autonomy to its colonies. By 1935, America had granted self-government to the "Commonwealth of the Philippines." Admittedly, with the need to defend its interests against Japanese aggression in the Pacific War, the United States could only make good on its promise of independence to the Philippines on July 4, 1946, while the retention of military bases and close economic ties would confer an almost neocolonial status for decades.[4]

From the celebratory discourse surrounding America's "liberation" of the Philippines through to controversial debates about America's defense of "liberty" in the Vietnam War, there was nevertheless a persistent faith in the ability of superior American political, economic, and social models to cross and transform cultures. Even the most enlightened of American presidents accepted the need for international trusteeships to prepare indigenous peoples for self-government; and hence adopted a patronizing, ultimately dismissive view of indigenous societies, which in turn echoed a fundamental belief in racialized cultural hierarchies that shaped the broader Euro-American encounter with non-white peoples at home and abroad. For all their anticolonial sympathies and internationalism, Presidents Wilson and Franklin Roosevelt were still paternalistic: the devoutly predestinarian Wilson assumed the superiority of Western civilization, the continued dominance of the West, and the role of American exceptionalism in regenerating the Philippines; the patrician Roosevelt saw the Vietnamese as children, a "small and passive people" incapable of governing themselves and thus needing external assistance from the West.[5]

Subsequently, just as one main phase of U.S. foreign policy in the region was ending, another was beginning. The outbreak of the communist-inspired Hukbalahap rebellion in the Philippines (1946) was followed by the eruption of communist insurgencies in Burma, Malaya, and Indonesia (1948). The Democratic Republic of Vietnam (DRV) became a fully fledged communist state by 1950–1951, and the Vietminh began to launch full-scale military assaults on the French across the Tonkin Delta region. By the early 1950s, the United States saw Southeast Asia as a crucial front line in the global Cold War that America had to win for the preservation of the "free world." But there were troubling inconsistencies in the way America managed its relations with Western colonial allies (as they contemplated decolonization) and Southeast Asian nationalist groups (as they pursued self-determination), for which there would be long-term consequences. Meanwhile, America decided to contain the spread of communism through the establishment of a U.S.-led Southeast Asia Treaty Organization (SEATO) in 1954–1955, which included only two Southeast Asian nations perceived to be reliably anticommunist at the time: the Philippines and Thailand.

In Indonesia, the Truman administration urged the Dutch colonial regime to promise independence for a nationalist group led by Achmed Sukarno. The Americans were prepared to support Sukarno (who had declared Indonesia's independence in August 1945) because he was noncommunist, and thereby exerted pressure on the Netherlands to recognize Indonesian independence in December 1949. Sukarno did not join SEATO, however, and instead hosted the first Afro-Asian conference of supposedly "non-aligned" nations at Bandung (1955). The response of the Eisenhower administration was to subvert Sukarno's regime by funneling arms and cash subsidies via the CIA to insurgents in an abortive rebellion on the Javanese outer islands (1957–1958).[6]

In Indochina, by contrast, the United States supported the French colonial regime against the Vietnamese independence movement led by Ho Chi Minh. Although the DRV had been proclaimed by Ho Chi Minh in September 1945, using words drawn from the American Declaration of Independence, it was not recognized by the French, because of their renewed colonial interests, or by the Americans, because of their aversion to Ho's communist credentials. The DRV's independence was recognized only after its Vietminh victory over the French at Dien Bien Phu and the Geneva Conference (1954). By that time, President Eisenhower had applied the domino theory to Vietnam: "You have a row of dominoes set up, you knock over the first one, and what will happen to the last one is the certainty that it will go over very quickly. So you could have the beginning of a disintegration that would have the most profound influences."[7] Even as President Kennedy reiterated that the United States would "pay any price, bear any burden, meet any hardship, support any friend, oppose any foe, in order to assure the survival and success of liberty,"[8] America found itself supporting an authoritarian South Vietnamese regime under Ngo Dinh Diem until the Kennedy administration allowed army generals to dispose of him (1963). Following the 1964 Gulf of Tonkin Incident and the Johnson administration's decision to "Americanize" the war effort with the deployment of U.S. combat units, the Vietnam War escalated into a "quagmire" from which America was unable to disengage until 1975.[9]

Vietnam left such deep scars in the American psyche that it led to a corresponding loss of U.S. foreign policy interest in Southeast Asia for the rest of the century. Whereas the Clinton administration eventually assigned Southeast Asia to an important position in America's post–Cold War vision of a Pacific community, the emphasis on human rights presented a stumbling block. Only the terrorist attacks of September 11, 2001—an "Occidentalist" assault by al Qaeda operatives on the very basis of American "hyperpower"[10]—had the ability to truly revive America's strategic focus on the region. Home to over 200 million Muslims, Southeast Asia came to be viewed by the George W. Bush administration as a potential breeding ground with "safe havens" for Islamic militants. The uncovering of regional terrorist networks as well as some terrorist attacks, including several targeted at American interests in Southeast Asia, seemed to confirm this view.[11] Southeast Asia was transformed into a key frontier in America's latest global struggle—the "war on terror"—even as President Bush affirmed that "the survival of liberty in our land increasingly depends on the success of liberty in other lands."[12]

Southeast Asian Perspectives on U.S. Foreign Policy

Southeast Asia is "exceptional" in its own right, however. It is a porous, fragmented geographic region of tremendous variety and fluidity, consisting of both "mainland" and "maritime" components. Encompassing the world's largest archipelago and major sea-lanes connecting the Indian and Pacific oceans, this region has been the historic setting for waves of cross-cultural interaction, involving cooperation and collaboration as well as competition and conflict. In both space and time, the lands and peoples that constitute "Southeast Asia" have found themselves repeatedly positioned betwixt larger forces—from both East and West—including China, India, the European colonial powers, and the United States.[13]

Inasmuch as they have evolved distinctive histories and identities within the region, the countries that are now member states of the Association of Southeast Asian Nations (ASEAN) have developed a range of perspectives on U.S. foreign policy.[14] Generally, these perspectives have been shaped by internal and intraregional factors (such as the evolution of indigenous societies and their relations with one another) as well as external and extraregional factors (such as good or bad experiences of colonial authority and relations with the outside world).

Southeast Asia's long-term interactions with archaic Indo-Islamic and Sino-Confucian civilizations, on the one hand, and modern Western civilization, on the other, have created a rich potential for cross-cultural tension. Such creative tension has resulted either in cross-cultural clashes or in cross-cultural fertilization between deeply embedded "Asian values" and newly imported "Western values." Broadly, cultural perspectives within the region stem from (and tend to lead to) conservative worldviews that value deference to authority, social hierarchy and religious harmony, the greater good of the community over the individual, and family loyalty in addition to personal virtue. Juxtaposed against such indigenous values would be "Western" liberal and atomistic views of society that emphasize the autonomy of individuals, normally under the universalizing banner of "liberty," "democracy," or "human rights," which might in turn lead to moral license, permissiveness if not decadence.[15] For better or for worse, the lands and peoples of Southeast Asia have endeavored to negotiate their middle way through the entanglements of East-West cultural relativism. Just as America developed its own brand of "manifest destiny," a variegated set of hybrid cultural values (including democratic principles operating in a largely authoritarian matrix) has gradually taken root across many of Southeast Asia's multiethnic societies, all of which believe they are masters of their own destinies and yet part of a wider regional consensus embodied by ASEAN.

By no means unproblematic, the evolution of this "values debate" between the cultures of East and West has been made more complex in the region by the whole Western colonial discourse—especially the political legacies of the "civilizing mission" and "the white man's burden." Between the sixteenth and the nineteenth centuries, indigenous political systems in Southeast Asia were subjected to the global projection of increasingly competitive, aggressive forms of European imperialism that were in turn legitimated by Western notions of transcendent law and unitary sovereignty. What followed was an irreversible transition from the traditional politics of the *mandala* to the norms of a "Westphalian" system: the finely balanced, layered concept of sovereignty shared by precolonial states located between India and China, which had also opened up various autonomous spaces for the inhabitants of Southeast Asia's port cities, was progressively displaced by the Western idea of indivisible, monolithic sovereignty imported under colonial conditions from Europe.[16]

Southeast Asia was divided somewhat arbitrarily into various Western colonial spheres and colonial states, whose borders would harden into the boundaries of future nation-states. For a number of them, however, the experience of divide-and-rule under Western colonial regimes proved so traumatic that it may have altogether delegitimized the concept of empire, even a more benevolent American hegemony. With the exception of Thailand, never formally colonized, most of the nations in the region had to earn their freedom by winning a hard-fought struggle for independence—a struggle that turned especially violent in Burma, Indonesia, and Indochina. Even as the postcolonial order that emerged after 1945 inherited the legacy of a system of sovereign states with fixed maritime and territorial boundaries, one of the principal reactions to the colonial past has been the instinctive nationalist tendency to prevent or preempt any recurrence of extraregional domination, particularly over hard-won issues of national sovereignty and jurisdiction. The vitality and, in some cases, volatility of Southeast Asia's postcolonial discourse on "nation-building" has prevented either the unwelcome assertion or the uncritical acceptance of any form of latter-day *Pax Americana* in place of former Western colonial regimes. Still, in view of their internal dynamics and their individual experiences of external power, there are nations that would be predisposed toward maintaining closer strategic relations with the United States (such as the Philippines and Singapore) in their pursuit of autonomy, just as there are also nations that would prefer a more cautious, measured approach (such as Indonesia, Malaysia, Thailand, Vietnam, and Cambodia) in their guarding of sovereignty.[17]

What may be agreed upon is that the distinctive phases and emphases of U.S. foreign policy have had, at times, a decidedly polarizing effect across Southeast Asia. In some instances, scholars and their indigenous sources have characterized the extreme reactions to U.S. foreign policy—various shades of "anti-Americanism" directed against what is perceived to be America's economic, military, or cultural imperialism—as part of a more ambivalent "love-hate" relationship.[18] This shifting kaleidoscope would encompass both elite and popular perceptions of U.S. foreign policy in the region as well as around the world: from America's policies as colonial power, through to its policies as leader of the "free world" and then as latter-day crusader against "evil-doers." Underpinning all indigenous perspectives of America's changing roles (including the rhetoric and doctrines of American presidents) would be the most basic of questions: America, our friend or foe, our benefactor or burden?

During the colonial period, the Philippine Insurrection (1899–1902) against American rule demonstrated from the start how strongly Filipinos wanted independence, even from their American "liberators."[19] Yet America's subsequent promise of independence to the Philippines, and its clear determination to follow through on this promise, won Americans many admirers across the region. Thereafter, the United States was not perceived as a "real colonialist"; the anticolonial attitudes of Presidents Woodrow Wilson and Franklin Roosevelt—seen as defenders of freedom and democracy for all oppressed peoples despite the "Orientalist" prejudices of their New World paternalism—further endeared them to nationalist leaders in Southeast Asia, including Ho Chi Minh.[20] At the onset of the Pacific War, Roosevelt had written: "Our course in dealing with the Philippines situation . . . offers, I think, a perfect example of how a nation should treat a colony or a dependency."[21] "Don't think for a minute that Americans would be dying in the Pacific tonight," the president told his son, "if it hadn't been for the short-sighted greed of the French and the British and the Dutch."[22] In his inaugural speech of July 4, 1946, newly elected Philippine president Manuel Roxas expressed heartfelt gratitude, remarkable for the leader of a nation emerging from a half-century of colonial rule:

> The world cannot but have faith in America. For our part, we cannot but place our trust in the good intentions of a nation which has been our friend and protector for 48 years. To do otherwise would be to forswear all faith in democracy, in our future, and in ourselves.
>
> As we pursue our career as a nation, as we churn through treacherous waters, it is well to have a landfall, that we may know our bearing and chart our course. Our safest course, and I firmly believe it is true for the rest of the world as well, is in the glistening wake of America whose sure advance with mighty prow breaks for small craft the waves we fear.[23]

During the course of the Cold War, America as the capitalist superpower was seen to play a more ambivalent, polarizing role in Southeast Asian politics. America's anticommunist containment policy and military-strategic support was regarded as vital to the independence and survival of some nations, especially against perceived Soviet and Chinese threats. Hence, whilst keeping the "dominoes" from falling and retaining access to key military bases, the United States ended up supporting authoritarian regimes such as that of Ngo Dinh Diem in South Vietnam and Ferdinand Marcos in the Philippines—examples of "the tail wagging the dog" in this region. Singapore was also criticized for its close strategic relations with the United States, as its former prime minister Lee Kuan Yew has acknowledged: "In the 1960s and 1970s, Singapore was berated in the Chinese media as a lackey of the American imperialists."[24] Yet America's largely benign presence in the region would yield beneficial results for the destinies of many, according to Lee's successor Goh Chok Tong: "The U.S. involvement in Vietnam bought precious time for the ASEAN countries to put their house in order and to lay the foundation for the grouping to develop into a cohesive organization. ASEAN economies began to take off, spurred by U.S. investments and a friendly American market."[25]

Others developed less sanguine views on America's Cold War involvement. Indonesian leaders, diplomats, and scholars still resent America's interference in Indonesia's domestic politics: subverting Sukarno when he went down the Bandung path of "non-alignment" and challenged the U.S.-led Western alliance in *Konfrontasi* with Malaysia and Singapore (1963–1966); and then backing former general Suharto and his authoritarian, even corrupt, "New Order" regime (1967–1998). According to these elite perceptions, America's legacy as provider of economic and military assistance as well as guarantor of stability in the region was a distinctly mixed blessing that left a bittersweet aftertaste.[26] This ambiguity has, of course, stemmed from essential differences in the basic threat perceptions of small or medium powers in the regional context and a superpower in the global context: whereas America has tended to view the dangers to its national interests and to the Southeast Asian states primarily in military and security terms, the indigenous elite in Southeast Asia are more inclined to perceive threats in economic and internal terms. In addition, while enjoying the security guaranteed by the superpower, the smaller powers have no desire to be pawns in an American "great game"—contest or crusade—against another global power.[27]

In the post–Cold War milieu, U.S. foreign policy has again proved controversial and polarizing, though for somewhat different reasons. Initially, there was a momentary reduction and rearrangement of America's military-strategic commitments in Southeast Asia: most notably, the closure of the U.S. bases in the Philippines (1991) due to resurgent Filipino nationalism, and the provision of alternative military facilities in Singapore, out of characteristic pragmatism and continuing perceptions of America as an essentially benign hegemon. Singapore's then–foreign minister S. Jayakumar observed: "[T]he United States remains an indispensable factor of any new configuration for peace, security and economic growth in the Asia-Pacific. Only the United States has the strategic weight, economic strength and political clout to hold the ring in the Asia-Pacific."[28] Conversely, the new era of American "hyperpower" also saw more assertive championing of human rights issues and American ideas of good governance that impinged on the asserted sovereignty of Southeast Asian nations such as Singapore, Malaysia, and Indonesia, leading to a clash of political cultures between them.[29] Indonesia's then–foreign minister Ali Alatas sought to deflect charges of human rights violations with a call for "understanding of the traditions and social values of developing nations, many of which were endowed with ancient and sophisticated cultures."[30]

More damaging and divisive has been the fallout from a much-heralded "clash of civilizations" between neoconservative America and a supposed monolith called militant Islam. As the Bush administration assembled a "coalition of the willing" to fight al Qaeda and its affiliates around the world—expanding the theatre of operations from Afghanistan to Iraq and beyond—Southeast Asian nations proved broadly cooperative, though the extent of their cooperation would be constrained by domestic factors. Given its long history of collaboration with the United States, the Philippines committed troops and logistics teams to Iraq (as far as popular support would allow) in return for American defense assistance to enhance the "counterterrorism" capabilities of the Philippine armed forces and police; Singapore made available naval bases that have a geostrategic reach transcending Southeast Asia, further deploying naval and air support in the Persian Gulf for the reconstruction of Iraq.[31] In the predominantly Muslim nations of Indonesia and Malaysia, however, official support would be more qualified. Adding to the cumulative history of suspicion and resentment was more recent anti-Americanism directed against the U.S.-led invasion of Iraq plus associated cases of prisoner abuse—all taken as damning evidence of the unilateralism and anti-Islamism of U.S. foreign policy. Nonetheless, these countries have been cooperative in terms of information-sharing and pursuing the active elements of putative terrorist organizations (such as Jemaah Islamiyah).[32]

There are, of course, other realities and priorities to consider in the post–Cold War international order. These include the rise of China (although India is rising, too), the pace and intensity of globalization, and the importance of regionalism as well as multilateralism in view of these challenges. Since the end of the Cold War, with China's re-emergence as a regional player of growing stature in the Asia-Pacific, ASEAN countries have attempted a balancing act between the United States and China: facilitating the retention of U.S. involvement and forward deployment in the region, while simultaneously engaging China in political and military-strategic discourse. Lee Kuan Yew has underscored the impact of China's regional ascendancy on ASEAN's strategic relations with the United States: "Regional perceptions of the value of American access to Singapore facilities underwent a sea change after China published maps in 1992 that included the Spratlys as part of China. Three ASEAN countries (Malaysia, Brunei, and the Philippines) had also claimed these islands. That November Ali Alatas said that Indonesia had no difficulty in seeing the merits of U.S. access to Singapore's military facilities."[33]

Analysts have thus subdivided the region into three categories. First, nations engaging with China but still placing greater emphasis and faith in their long-term strategic relations with America: the Philippines and Singapore. Second, nations charting a middle course between America and China, mainly due to geographical distance from China and unease over pursuing closer strategic relations with America: Indonesia and Malaysia. Third, nations whose security strategies are dominated by their proximity to China: Burma, Thailand, Vietnam, Laos, and Cambodia.[34]

Given the vast economic opportunities, and the need to safeguard the flows as well as fruits of trade and investment in an age of global interdependence, ASEAN countries see the way forward in terms of both regionalism and multilateralism. In ASEAN's view, an expanded, reinforced regional architecture that engages and enmeshes both China and the United States can only be a positive, constructive development. ASEAN's aspiration is to embed them in a cooperative mechanism, thereby reducing potential for misunderstanding and enhancing prospects of stability. But while the United States is a member of the Asia-Pacific Economic Cooperation Forum (APEC) and the ASEAN Regional Forum (ARF), it is not part of the East Asia Summit, which is strategically important because—apart from ASEAN—it includes Asia's three major powers: China, India, and Japan. Despite compelling statistical evidence indicating that ASEAN has become a more important trade and investment partner for the United States than Latin America, Russia, the Middle East, and Africa, there is an underlying sense that the United States is reluctant to nurture relationships with nascent institutions that may not yield immediate results, just as it is unwilling to accord its Asian interlocutors an equal measure of respect. Conversely, there are lingering doubts over America's fitness to lead the "free world": in the Bush administration's singular obsession with the "war on terror," the United States appeared to lose its way in the world on other issues—from climate change to nuclear nonproliferation—even as other nations increased their power and influence in Southeast Asia and elsewhere.[35]

Ultimately, however, even though partnership and multilateralism are themes that resonate far and wide, there is no escaping the fact that "past is prologue." At the operational level, the U.S. Navy's key role in organizing what became a massive, multinational humanitarian relief effort following the Boxing Day tsunami of 2004—off the coast of Indonesia's Aceh province—certainly improved America's image in Indonesia and across the region. Still, there remains significant unease arising from perceptions of U.S.-dominated Western "media imperialism" facilitating a potential "fifth column" within indigenous society, or nongovernmental organizations serving as possible "Trojan horses."[36] In the wake of Cyclone Nargis, which struck coastal Burma in May 2008, neither the U.S. Navy nor nongovernmental organizations were permitted by Burma's military regime to intervene for fear that under the cover of humanitarian relief the United States had a political agenda that included regime change.[37] Echoing anticolonial sentiment from the days of European naval dominance, there continues to be underlying suspicion that extraregional powers such as the United States would use the threat posed by natural disasters, trafficking in weapons (conventional or nuclear), drugs, and humans, as well as piracy and terrorism, to justify their longer-term naval presence in the region.

Framing a Pacific Future

Whatever their differences of perspective, the nations of Southeast Asia would all prefer a greater measure of clarity and consistency in U.S. foreign policy: less prescriptive, more sensitive. While proclaiming the virtues of liberty and democracy to Burma's military rulers, or other parts of Southeast Asia with more volatile and authoritarian political traditions, America would do well to remember its patchy historical record of supporting right-wing dictatorships in this region and elsewhere.[38] While championing its notions of good governance and human rights, America could display deeper cross-cultural sensitivity and patience when it comes to the apparent lack of progress, promoting more constructive diplomatic and developmental approaches over military solutions or economic sanctions. After all, to what extent has America itself practiced what it has often preached to others?

The stress on liberty and democracy abroad—as visible indicators of modernity and civilized norms—raises questions about America's own long-term evolution at home, when modern America has periodically exhibited strong premodern features. In the so-called "land of the free," formerly the home of the Amerindian brave, slavery remained lawful in the United States until 1863; and even then, with its reservations and segregated communities, twentieth-century America remained "a caste society whose marker was color, used to exclude a large social fragment from civil and political rights until the 1960s or later."[39] In conducting its "war on terror" at the start of the twenty-first century, the United States again proved inconsistent with its own principles in dealing with terrorist suspects and political detainees in Guantánamo and abroad. With the application of torture being all that was liberal about the procedures, how was that culturally or morally superior to detention without trial under the internal security laws of Malaysia and Singapore? Vindicating the dreams of the founding fathers at long last, it has taken the almost ironic election of an exceptional man of color to the White House to restore some measure of hope in the promise of America for the rest of the world: inaugurating a new era of internationalism—both responsible and responsive—in which the United States pledges to listen more than dictate; dismantling Guantánamo while engaging with others—especially the Muslim world—on the basis of "mutual interest and mutual respect."[40]

The nations of Southeast Asia would also prefer a greater degree of commitment and compromise in U.S. foreign policy: less unilateralist, more multilateralist. Singapore ambassador Tommy Koh has observed, "Since the end of the Vietnam War, U.S. attention to Southeast Asia has been episodic rather than consistent, focusing more on security and defense issues. U.S. attention has been less engaged in the dynamics of the region—including economic growth and the development and strengthening of a Southeast Asian regional architecture that is high on the agenda of not only ASEAN, but many Asian nations. Since the September 11th terrorist attacks, policymakers in Washington have tended to look at Southeast Asia primarily through the unidimensional lens of terrorism."[41] The United States would do well to commit itself to achieving its foreign policy agenda in partnership with multilateral institutions in the region, paying more attention to the regional agenda rather than resorting to "coalitions of the willing" whenever American foreign policy aims appear to be thwarted. Such an approach would add substance to the new post of U.S. ambassador to ASEAN, created in 2007 in a rare display of bi-partisanship by Congress with backing from the Bush administration.

Exactly how America's historic sense of "manifest destiny" adapts to the needs and demands of competing regional agendas in an increasingly "globalized" age remains a work in progress. At least there is now acknowledgment of a "shared destiny" rather than just "manifest destiny": in the words of America's forty-fourth president, "our stories are singular, but our destiny is shared, and a new dawn of American leadership is at hand."[42] Secretary of State Hillary Clinton, speaking on behalf of the Obama administration, has already declared that the United States "is not ceding the Pacific to anyone" in measured response to the rise of China and the Australian government's defense white paper, which in May 2009 raised the possibility of American dominance fading in the Asia-Pacific region in the decades ahead. "We have longstanding bilateral relationships with nations like Australia and others," Clinton affirmed, "and we have a very active multilateral agenda that we intend to reinvigorate, such as our membership with ASEAN and other fora within the Pacific region."[43]

If the status quo vis-à-vis Southeast Asia is maintained, residual anti-Americanism in parts of the region will likely still be outweighed by America's continuing importance as economic partner, security guarantor, and cultural exemplar. But the current shift in U.S. foreign policy is helpful to the cause. Although Secretary Clinton's visit to Indonesia in February 2009 drew sharp protests from hundreds who demonstrated against Clinton's pro-Israel sympathies and America's occupation of Iraq, officials welcomed her pledge of support for Indonesian president Susilo Bambang Yudhoyono's suggestion of a "Comprehensive Partnership." Indonesian foreign minister Hassan Wirajuda added that Indonesia could be an effective bridge to help America reconnect with the Muslim world.[44] Clinton's trip was further intended to prepare the way for a future visit to ASEAN countries by President Obama. A promising new dawn, no doubt, but in the wider formulation and articulation of foreign policy on all sides, only time will tell whether the intertwined destinies of the United States, Southeast Asia, and other regional players lead on to a brighter, more pacific future.

Notes

1. Whereas America was founded upon the ideal that all human beings are created equal and endowed with "certain unalienable rights," it was also Thomas Jefferson, the principal author of America's Declaration of Independence (1776) and third president of the United States, who recognized that the exercise of power in the real world can corrupt such an ideal: "Not in our day, but at no distant one, we may shake a rod over the heads of all, which may make the stoutest of them tremble. But I hope our wisdom will grow with our power, and teach us that the less we use our power, the greater it will be." See Thomas Jefferson Randolph, ed., *Memoirs, Correspondence, and Private Papers of Thomas Jefferson* (London: Colburn and Bentley, 1829), vol. 4, p. 272.
2. General James Rusling, "Interview with President William McKinley," *The Christian Advocate*, January 22, 1903, p. 17.
3. See Edward Said's seminal works, *Orientalism* (New York: Random House, 1978) and *Culture and Imperialism* (New York: Random House, 1993).
4. George Herring, *From Colony to Superpower: U.S. Foreign Relations since 1776* (Oxford; New York: Oxford University Press, 2008), pp. 321–329, 366–367, 411–412; Stanley Karnow, *In Our Image: America's Empire in the Philippines* (New York: Ballantine, 1989), p. 247; Robin Jeffrey, ed., *Asia: The Winning of Independence* (London: Macmillan, 1982), p. 263.
5. Lloyd E. Ambrosius, *Woodrow Wilson and the American Diplomatic Tradition: The Treaty Fight in Perspective* (Cambridge: Cambridge University Press, 1990), pp. 9–13. Also see Mark Philip Bradley, *Imagining Vietnam and America: The Making of a Postcolonial Vietnam, 1919–1950* (Chapel Hill, North Carolina: University of North Carolina Press, 2000).
6. Michael Leifer, *Dictionary of the Modern Politics of South-East Asia,* 3rd ed. (London; New York: Routledge, 2001), pp. 17, 63, 241. Also see Audrey Kahin and George Kahin, *Subversion as Foreign Policy: The Secret Eisenhower and Dulles Debacle in Indonesia* (New York: The New Press, 1995).
7. "Domino Theory Principle," Press Conference of April 7, 1954, in *Public Papers of the Presidents: Dwight D. Eisenhower* (Washington, DC: U.S. Government, 1954), pp. 381–390.
8. "John F. Kennedy Inaugural Address," January 20, 1961, available at http://www.bartleby.com/124/pres56.html.
9. Herring, *From Colony to Superpower*, pp. 634–635, 661–662, 726–729; Leifer, *Dictionary of the Modern Politics of South-East Asia,* pp. 105, 118–119, 129. Also see George C. Herring, *America's Longest War: The United States and Vietnam,* 4th ed. (New York: 2002).
10. See Ian Buruma and Avishai Margalit, *Occidentalism: The West in the Eyes of Its Enemies* (New York: Penguin, 2004).
11. Kumar Ramakrishna and See Seng Tan, eds., *After Bali: The Threat of Terrorism in Southeast Asia* (Singapore: Institute of Defence and Strategic Studies, 2003).
12. "George W. Bush Second Inaugural Address," January 20, 2005, available at http://www.bartleby.com/124/pres67.html.
13. The term "Southeast Asia" is part of the diplomatic and academic discourse that evolved mostly in the second half of the twentieth century: from its usage in Britain's wartime South East Asia Command, followed by the American-led Southeast Asia Treaty

Organization (SEATO) and then the independent Association of Southeast Asian Nations (ASEAN), through to its application in late colonial and postcolonial scholarship. See Paul Kratoska, Remco Raben, and Henk Nordholt, *Locating Southeast Asia: Geographies of Knowledge and Politics of Space* (Singapore: Singapore University Press, 2005); Donald E. Weatherbee (with Ralf Emmers, Mari Pengestu, and Leonard C. Sebastian), *International Relations in Southeast Asia: The Struggle for Autonomy* (Lanham, Maryland: Rowman & Littlefield, 2005), pp. 1–21.

14. ASEAN was formed on August 8, 1967, by Indonesia, Malaysia, the Philippines, Singapore, and Thailand. Since then, membership has expanded to include Brunei, Burma, Cambodia, Laos, and Vietnam. The only nation-state in Southeast Asia that is not yet a member is East Timor.
15. Michael D. Barr, *Cultural Politics and Asian Values: The Tepid War* (London; New York: Routledge, 2002), pp. 3–11, 177–187.
16. O. W. Wolters, *History, Culture, and Region in Southeast Asian Perspectives,* rev. ed. (Singapore: Institute of Southeast Asian Studies, 1999), pp. 27–40, 126–154; Norman G. Owen, ed., *The Emergence of Modern Southeast Asia* (Singapore: Singapore University Press, 2005), pp. 52–82. Southeast Asia's precolonial polities—ranging from localized units to centralized kingdoms—often knew how to share power and divide sovereignty, thus accommodating ethnic and religious differences. Most characteristic of the early political history of this region is what is known as the *mandala* system (Sanskrit, *manda* = core, *la* = container), whereby clusters of small settlements (vassals) coalesced around strong rulers (overlords) in a loose geopolitical or economic alliance.
17. Although Thailand is one of only two Southeast Asian nations retaining a formal alliance with the United States—the other is the Philippines—Thai foreign policy is traditionally described as "bending with the wind" in order to avoid any form of colonial domination. See Michael K. Connors, "Thailand and the United States: Beyond Hegemony?" in *Bush and Asia: America's Evolving Relations with East Asia,* ed. Mark Beeson (London; New York: Routledge, 2007), pp. 128–144.
18. See, for example, Daniel Novotny, "Indonesian Foreign Policy: A Quest for the Balance of Threats" (PhD thesis, University of New South Wales, 2007), pp. 103–177; Melani Budianta, "Beyond the Stained Glass Window: Indonesian Perceptions of the United States and the War on Terror," in *What They Think of US: International Perceptions of the United States since 9/11,* ed. David Farber (Princeton, New Jersey: Princeton University Press, 2008), pp. 27–48.
19. Herring, *From Colony to Superpower*, pp. 326–327.
20. Marilyn B. Young, *The Vietnam Wars 1945–1990* (New York: HarperCollins, 1991), pp. 2–10.
21. Franklin D. Roosevelt to William Philips, November 19, 1942, cited in William Roger Louis, *Imperialism at Bay: The United States and the Decolonization of the British Empire* (New York: Oxford University Press, 1987), p. 180.
22. Elliott Roosevelt, *As He Saw It* (New York: Duell, Sloan & Pearce, 1946), p. 116.
23. Manuel Roxas, quoted in Jeffrey, ed., *Asia,* p. 26.
24. Lee Kuan Yew, "The Fundamentals of Singapore's Foreign Policy: Then and Now" (S. Rajaratnam Lecture, April 9, 2009).
25. Goh Chok Tong, "ASEAN-U.S. Relations" (keynote speech, ASEAN-U.S. Partnership Conference, September 7, 2000).
26. Novotny, "Indonesian Foreign Policy," pp. 112–121.
27. Robert C. Horn, "Southeast Asian Perceptions of U.S. Foreign Policy," *Asian Survey* 25, no. 6 (June 1985), pp. 683–684.
28. S. Jayakumar, "The Southeast Asian Drama: Evolution and Future Challenges" (lecture, Georgetown University, April 22, 1996).
29. Michael Leifer, *Singapore's Foreign Policy: Coping with Vulnerability* (London: 2000), pp. 99–108; Karminder Singh Dhillon, *Malaysian Foreign Policy in the Mahathir Era*

(1981–2003): Dilemmas of Development (Singapore: Singapore University Press, 2009), pp. 117–118.; Novotny, "Indonesian Foreign Policy," pp. 129–131.

30. Ali Alatas, quoted in *Straits Times,* June 16, 1993.
31. Ben Reid, "Bush and the Philippines after September 11: Hegemony, Mutual Opportunism and Democratic Retreat," in *Bush and Asia,* ed. Beeson, pp. 145–161; Emrys Chew, "A Merlion at the Edge of an Afrasian Sea: Singapore's Strategic Involvement in the Indian Ocean," *RSIS Working Paper Series,* 164 (July 2008), p. 22.
32. David Bourchier, "The United States, Bush and Indonesia: Bitter Memories, New Eggshells"; and Helen Nesadurai, "Malaysia and the United States: Rejecting Dominance, Embracing Engagement," both in *Bush and Asia,* ed. Beeson, pp. 162–178 and 179–195, respectively.
33. Lee Kuan Yew, *From Third World to First: The Singapore Story, 1965–2000* (Singapore: Times, 2000), p. 538.
34. Evelyn Goh, ed., *Betwixt and Between: Southeast Asian Strategic Relations with the U.S. and China* (Singapore: Institute of Defence and Strategic Studies, 2005).
35. See the Asia Foundation monograph, *America's Role in Asia: Asian and American Views* (San Francisco: The Asia Foundation, 2008), pp. 38–48. Tommy Koh makes a strong case for the rising economic importance of Southeast Asia vis-à-vis U.S. foreign policy calculations: "The U.S.-ASEAN economic relationship is substantial, growing, and mutually beneficial. U.S. investment in ASEAN is about US$100 billion, exceeding U.S. investments in China, Hong Kong, and Taiwan combined. U.S. investment in Southeast Asia earns the highest rate of return in the world at approximately 20 percent. The United States is ASEAN's second-largest trading partner and largest foreign direct investor. ASEAN is America's fifth-largest trading partner and third-largest export market. Few Americans know that Southeast Asia imports twice as many American goods as China does. Two-way trade has grown 40 percent since 2001 and amounts to US$170 billion. The United States has concluded a free trade agreement (FTA) with Singapore and has attempted to negotiate FTAs with Malaysia and Thailand, while also concluding bilateral trade and investment framework agreements (TIFAs) with other ASEAN countries. . . . Energy passing through the Strait of Malacca is three times more than what passes through the Suez Canal and 15 times more than what is transported through the Panama Canal. This is the energy lifeline for China, Japan, and South Korea, as more than 80 percent of its oil and natural gas comes from or passes through Southeast Asia. In September 2007, the three coastal states (Indonesia, Malaysia, Singapore), the United States, and other user states met in Singapore, under the auspices of the United Nations' International Maritime Organization (IMO); and created a cooperative mechanism to further ensure safe, secure, and efficient shipping in the Straits of Malacca and Singapore."
36. Novotny, "Indonesian Foreign Policy," pp. 163–170.
37. Tommy Koh, "The United States and Southeast Asia," in *America's Role in Asia,* p. 43.
38. In demanding international isolation of Burma for its harsh military rule, America betrayed profound lack of cultural understanding and historical perspective. "The most striking aspect of the Burma debate today is its absence of nuance and its singularly ahistorical nature," observes Thant Myint-U, grandson of former United Nations Secretary-General U Thant. "Dictatorship and the prospects for democracy are seen within the prism of the past ten or twenty years, as if three Anglo-Burmese wars, a century of colonial rule, an immensely destructive Japanese invasion and occupation, and five decades of civil war, foreign intervention and Communist insurgency had never happened. A country the size and population of the German Empire on the eve of the First World War is viewed through a single-dimensional lens, and then there is a surprise over the predictions unfulfilled and strategies that never seem to bear fruit." See Thant Myint-U, *The River of Lost Footsteps: Histories of Burma* (London: Faber and Faber, 2007), especially pp. 31–41.
39. John Darwin, *After Tamerlane: The Rise and Fall of Global Empires, 1400–2000* (London: 2007), p. 26.

40. "President Barack Obama's Inaugural Address," January 20, 2009, available at http://www.whitehouse.gov/blog/inaugural-address/.
41. Koh, "The United States and Southeast Asia," pp. 37–38. The collective wisdom of the monograph's "Asian" authors suggests that America should sign the Treaty of Amity and Cooperation (TAC), which would allow America to be an effective dialogue partner with members of the East Asia Summit. It would also demonstrate America's confidence in regional organizations such as ASEAN, ARF, and APEC (see pp. 7, 48–54, 233–41).
42. "Barack Obama's Victory Speech," November 5, 2008, available at http://www.america.gov/st/elections08-english/2008/November/20081105101958abretnuh0.580044.html.
43. Geoff Elliott, "Hillary Clinton Firmly Commits the US to Asia-Pacific Security," *The Australian*, May 21, 2009. Secretary Clinton affirmed that the United States, which has a large naval presence at its Pacific base in Hawaii, "will be engaged—we are a trans-Pacific power and a trans-Atlantic power." Hence, the United States would focus on "deepening and broadening our engagement—we don't think it is a zero sum game; the fact that a country like China is becoming more successful or Indonesia is now a very successful democracy—we see that as to the good for the entire Pacific region."
44. Erwida Maulia, "Obama Asks Indonesia to 'Join Hands,'" *Jakarta Post,* March 14, 2009; Retno Marsudi, "Indonesia, the US: A New Partnership," *Jakarta Post,* April 27, 2009. In Jakarta, Clinton also took the opportunity to announce a major review of U.S. policy toward Burma. See the UPI news article, available at http://www.upi.com/Top_News/2009/02/18/Clinton-says-Myanmar-policy-under-review/UPI-10091234960636/.

Panel III: Asia and the Pacific

Summary of Discussion

Dr. John F. Garofano
Professor of Strategy and Policy
Jerome Levy Chair of Economic Geography and National Security
Naval War College

Dr. Jonathan Pollack noted the centrality of Asian security issues in U.S. debates today, in no small part because of the remarkable success the region has experienced in terms of political and economic development. There is peace, significant prosperity, and hope for successful political transitions. Yet there are many unanswered questions. These include the prospects for continued Chinese economic growth and the future stability of its political system.

Beijing has many incentives to cooperate with Washington. Washington has many incentives for cooperation as well, but it is possible, as one commentator put it, that the United States may be suffering from "enemy deprivation syndrome." There is a danger we may "shoe-horn" China into templates such as that of the Cold War with the Soviet Union. China does not fit well into such preexisting concepts, however. It has established real, working relationships with its neighbors, and is accepted more or less as a normal actor. The real challenge will be to establish a set of relationships that do not presume antagonism and are geared toward the long-term prospects for the region. These relationships must also be responsive to the political transitions under way.

Any future framework or set of relationships must include attention to concepts of responsibility sharing. Is the United States amenable to responsibility sharing in the region? The ground has begun to shift, accelerated by our economic difficulties, toward a region that does not have the United States as a singularly dominant power. Do we accept this? How do we conceive of our role? We need to acknowledge this transition if we are to move beyond existing frameworks and conceptions and toward a more sustainable long-term approach to keeping the United States in the region in a constructive way.

Dr. Evan Medeiros examined how states in the region are actually responding to China's rise. Using an interdisciplinary approach, the RAND Corporation has been looking for some years at how the region is responding to the rise of China, the drivers of these responses, and the implications for security. The most current update to this research looks at the actions of the six closest U.S. allies and friends—Japan, South Korea, Thailand, the Philippines, Australia, and Singapore. The goal is to examine how China's growing economic and military power has or has not changed the overall balance of power, and to examine trends. The research shows that none of these countries views China as a viable strategic partner; the United States is seen as the only power with the capability and intent to balance against China. These six states harbor many concerns about China's long-term interests and actions. Much of what we see in the way of deeper involvement with China is a function of these states, especially the smaller ones, viewing the United States as committed to remaining in East Asia and to balancing if necessary. To some extent, these states leverage the U.S. commitment for their own purposes by extracting concessions from Washington or their neighbors.

There is a regional consensus on engaging with China as opposed to containing or isolating it. Beijing has successfully reassured its neighbors that balancing is not necessary because China's intentions are benign over the long term. China's successful strategic communications have not, however, translated into any kind of serious interference in America's alliance relationships. Furthermore, there is no evidence of incipient Chinese hegemony or of bandwagoning by Asian states. Nor is there evidence of internal balancing, or arms racing, out of fear of Chinese growth. Internal balancing is somewhat "elastic" and could change with, say, the deployment of a Chinese aircraft carrier or with aggressive out-of-area activities involving the People's Liberation Army. In terms of the domestic constituencies of the states studied, there is little evidence that public opinion has any impact on policy, even if that opinion is somewhat more pro-China than in the past.

Overall, Beijing has simply had great difficulty in translating its growing economic might and trade presence into political leverage. Dr. Medeiros referred to his own recent study, presented at the Naval War College's Levy Chair lecture series two months prior, finding that the expanded scope of China's trade and investment in the region has not led to greater influence. The Global Financial Crisis (GFC) bears on this as well. China's strategies for recovery are unclear. It has adopted a large stimulus package which appears to be working, but it is too early to tell whether recovery will be sustained within China or what its impact will be on the region. All of this means the United States remains well positioned to achieve its strategic goals in East Asia. There remains time, too, to decide how it should respond to the rise of China. We do need to account for the equities of our allies and friends. There will be costs and risks associated with taking a hard line of balancing militarily or economically against China.

Dr. Emrys Chew of Singapore began his talk by referring to his own early education in Singapore, at a school founded by American missionaries toward the end of the nineteenth century. Not only is the U.S. presence in Singapore historic, but so is the influence of other nations, as is nicely captured by the emblem of this school, which has an image with the wings of an American eagle, the head of a British lion, and the body of a Chinese dragon. With that notion, Dr. Chew discussed the changes in U.S. policy toward the region following September 11, 2001. Southeast Asian states have developed a range of perspectives on U.S. foreign policy. Generally these policies are shaped by both internal and external factors, not least of which include their many hard-fought independence struggles during the last century. Colonialism and great-power politics of the past have left a lingering suspicion of any scent of neocolonial or heavy-handed power politics.

Toward the United States, Dr. Chew stated that the subregion has had a love-hate relationship somewhat reflective of historic U.S. policies. The United

States first liberated the Philippines, Dr. Chew noted, but then fought a war to determine its political future. There is deep respect for the United States—"even Uncle Ho admired Uncle Sam because of FDR"—but there are also ambivalent feelings. During the Cold War, America as a capitalist, anticommunist superpower prevented dominoes from falling but supported authoritarian regimes. There may be some lingering resentment in Indonesia, for example, due to U.S. involvement there in the 1960s and support for President Suharto. Dr. Chew noted an apparent contradiction between what America preaches and what it has done in the past, from dealing with Native Americans to detentions without trial and allegations of torture.

Thus, elites in Southeast Asia have a mixed view of the U.S. presence. While U.S. elites focus on international, power-related issues when thinking about the subregion, Southeast Asian elites focus on internal, economic, and political concerns. Yet Dr. Chew noted that by and large the region's elites, especially in Manila and Singapore, have supported the war on terror. Such support is more qualified in the Muslim-majority states. Overall, alignment with Washington's policies has varied within the subregion. The Philippines and Singapore have been most closely aligned with the United States, Indonesia and Malaysia steer a middle path, and the rest are more closely associated with China.

Whatever their differences in perspective, leaders around the region would like more clarity and consistency in U.S. policy, and would like it to be less prescriptive and exhibit more cross-cultural sensitivity. Since there is regional agreement that the way forward is regionalism and multilateralism, they would like to see more emphasis on diplomacy and development rather than on military solutions or economic sanctions. There is a sense that manifest destiny has given way to shared destiny, and rhetoric matters. The United States would do well by attempting to achieve its foreign policy agenda with multilateral partners in East Asia. Indeed, U.S. policy seems to be shifting in this direction. Secretary Clinton's visit to Indonesia was welcome, but even more welcome was reference to a renewed interest in the Peace Corps and related efforts. Analysts predict that a visit by President Obama will do much to restore the U.S. image in the region. Dr. Chew concluded with a description of President Obama's favorite foods from his days as a youth in Indonesia.

Discussion focused primarily on great-power politics and Northeast Asia, with China and North Korea as leading topics. One questioner asked how the United States should engage with China, and how specifically it should manage a transition to a multipolar system. Respondents argued that China should not be singularized, for it feeds Beijing's anxiety. We have to recognize explicitly that power is more dispersed in Asia than it used to be. Still, there is an aspect of unreality to policy and academic debate about China, which doesn't seem to recognize that China will grow militarily, as all states have grown.

The question was raised as to what kind of new regional framework would aid in dealing with North Korea. Responses were not overly optimistic or specific. North Korea was described as an old man who knows his time is near. The regime is not likely to disappear tomorrow but upheaval, if it takes place, will likely arise internally and from the younger generation of party members. In general, the threats to North Korea are not external but internal. It was suggested that in Washington there is an enormous amount of "hand-wringing" involved over U.S. policy past and present. The reality is the North's immediate neighbors have profound unease about what could go wrong and will go to great lengths to prevent the unraveling of North Korea. On the other hand there is no military option for the North. The best approach may be risk mitigation, and being candid about how we got here.

Russia received relatively little attention in the papers and several participants thought its role in the region may not be appreciated. Respondents stated that Russia is a marginal actor in East Asia. Moscow may be becoming a slightly more important player in the energy field, but only selectively. To be a greater force there must be a rebuilding of institutions within Russia. The Foreign Ministry, for example, simply has little capability right now. It was also claimed that East Asian countries don't cue policies off Russia. Both Russia and China value relations with the United States more than with each other, too, and between those two powers there is limited cooperation, except for coordination on Iran and North Korea. Otherwise there is divergence even in areas of traditional common interest, such as defense and arms trade. Russia also is fed up with China copying its weapons and platform designs, and wondering if it really wants to rearm a potential adversary.

Disagreement was registered. In particular, the United States should be sensitive to the way China is advantaged by its relationship with Russia. China gains border security. It gets 95 percent of its imported weapons from Russia. Beijing gains critical support at international forums for issues such as Tibet and East Turkistan, among other hot-button issues. Russia helps China balance the United States in inner Asia. All of these things accrue to China from its relationship with Russia.

Australian defense policy was discussed in the context of Australia's recent white paper. One participant asked whether the white paper was perhaps driven by either the China threat or by the vision of U.S. decline. The white paper was described as something intended to shape forces out to 2030 and beyond. It was noted by several that Canberra is putting its money where its mouth is. Further, recent meetings with Washington and a joint communiqué have highlighted the extensive and deep areas of cooperation. Further, the Australian military clearly has linked itself to the United States in tangible ways that are difficult to break.

It was suggested that the credibility of U.S. extended deterrence may be weakening, as evidenced by Japanese discussion of a nuclear option. Most believed this was a recurring symptom of Tokyo's wanting reassurance from Washington, but nothing more. On the other hand, it was noted that there is perhaps no more important relationship for Asian security over the long term than that between Tokyo and Beijing, and that, as one individual put it, "we ignore Japan at our peril" given the sheer size of its economy and military. But its "nuclear reviews" are mostly hedges to send signals to allies and neighbors.

There was extensive discussion of specific U.S. policy options. It was argued that the United States will have to look at its options differently, and will be seen differently, due to the obvious alterations in power caused by the rise of

China. After September 11, 2001, the United States made basic security policy decisions that both downgraded the role of Asia in our own eyes and had lasting impact on the region's perceptions of the United States. China and some other countries welcomed the fact that Asia was no longer a top priority. We now have to find a way to reposition and find a new center of gravity. One example of why this is so important is the Pakistan-China relationship, and the importance of Pakistan for U.S. policy toward South Asia and terrorism. Another respondent argued that this question has indeed been asked, but China hasn't been interested in helping us on Pakistan.

It appears the Obama administration will pursue more trilateral dialogues. One official has suggested reinvigorating United States–China–Japan trilateral discussions, to complement United States–Australia–Japan talks, and Canberra is already reaching out to Seoul possibly to begin one including Tokyo. Trilateral meetings have helped at crucial points in the recent past, and further progress can be made, and without denigrating the United States–Japan alliance or others. There was little disagreement on this issue of trilateral discussions.

Several commentators claimed Washington should make a decision to attend all major summit meetings in the region, as many in the region have clamored for and whose lack many have complained about during the Bush years. It was argued in opposition, however, that there should be clear and practical reasons for attending summits and meetings that have in the past been used to "bash" Washington for its policies. The United States is often used as a "punching bag" at these forums, so there should be good reason to attend them. One participant said, however, that the United States was tough and "could take it" and the benefits would outweigh the rhetoric and atmospherics.

Signing ASEAN's Treaty of Amity and Cooperation was also considered "low-lying fruit," something Washington could do with little risk. This would provide membership in the East Asia Summit, which brings all major actors to a table to discuss security issues. Washington should, in any case, choose which regional institutions will have priority in its vision and diplomacy. The Bush administration chose APEC even though it is huge, unwieldy, underinstitutionalized, and has no real security agenda.

It was agreed that the GFC has presented a tremendous challenge for U.S. policy in the region. There are arguments, gaining ground as this goes to press, that China will emerge more quickly from the crisis than will the United States. This could have long-term repercussions on images of the United States. It is also possible, however, that China will recover quickly but fail to restructure its economy as it must, toward domestic consumption, while the United States may be more adept at making the kinds of structural changes necessary for long-term economic health. This also would have major repercussions on regional perceptions of these two major powers vying for influence. In any case, there was no dissension expressed to the view that no significant player in the region wants the United States to depart or become less engaged; the United States is simply too important economically, militarily, and politically.

Panel IV

South Asia

Dr. Daniel Markey
Senior Fellow for India, Pakistan, and South Asia, Council on Foreign Relations

Ambassador (Ret.) Teresita C. Schaffer
Director, South Asia Program, Center for Strategic and International Studies

Moderator:
Dr. Timothy D. Hoyt
Professor of Strategy and Policy, Co-chair Indian Ocean Regional Study Group, Naval War College

A U.S. Strategy for Pakistan: Future Directions

Dr. Daniel Markey
Senior Fellow for India, Pakistan, and South Asia
Council on Foreign Relations

Since his election, President Obama has moved quickly to shift the focus of U.S. foreign and defense policy away from Iraq and toward Afghanistan and Pakistan. This shift reflects a long-held view that the terrorists responsible for 9/11 remain firmly entrenched in South Asia and that the war in Iraq represented a distraction from this central threat to American security. Having conducted a major interagency strategic review, endorsed significant troop increases, and requested an expansion of U.S. military and civilian assistance, it is clear that the Obama administration has initially committed itself to a far more aggressive effort in Pakistan and Afghanistan. That said, these initial steps have not locked Washington into a clear or precisely defined set of policies; important pieces of the new U.S. approach still remain in flux, open to interpretation and change.

Looking to the future, the United States will be best served by recognizing (1) that Pakistan, not Afghanistan, poses the paramount challenge to American security; (2) that building a strong partnership with Pakistan while working to transform perceptions of the strategic environment in South Asia holds the greatest potential for sustainable U.S. security; and (3) that even under the best of circumstances, success in the region will take a long time and may prove extremely costly.

Defining the Policy Challenge: Pakistan More than Afghanistan

Pakistan and Afghanistan comprise two facets of the same security environment, a fact acknowledged by the Obama administration in its adoption of the moniker "AfPak" when discussing the region. To reflect the hierarchy of strategic threat the two countries pose to U.S. national security, though, the terminology should in fact be "PakAf."

Pakistan is a country of 176 million, as compared to Afghanistan's 33 million. It is the country in which, by nearly all accounts, Taliban and al Qaeda leadership find sanctuary and whose security apparatus has long supported domestic Islamist militant groups as an asymmetric means to achieve strategic equilibrium with India, its neighbor and historical antagonist over the disputed territory of Kashmir. It is a state with a history of alternating authoritarian military and largely ineffectual and corrupt civilian rule. Hollowed-out state institutions, inadequate civilian control over the military, and an unsettled ideological debate about its own strategic interests leave Pakistan unable, and perhaps unwilling, to fulfill Washington's expectations of it as a partner against Islamist militancy.

By all accounts, Pakistan's internal instability has reached historically unprecedented levels. Extremist militants with the professed goal of imposing *sharia* law throughout Pakistan spread to within sixty miles of the capital in late April before being confronted and driven back by Pakistani security forces. This low point in Pakistani history reflected the inability or unwillingness of the Pakistani state to confront definitively these challenges to its sovereignty. And while few observers fear an immediate threat to Pakistan's nuclear arsenal, the geographic proximity of al Qaeda to such weapons of mass destruction drastically raises the strategic stakes for Islamabad, Washington, and the world.

Beyond the daunting immediate threats Pakistan already poses to U.S. security interests is an even more ominous long-term proposition: the next generation of globally linked extremists and terrorists is likely to hail from Pakistan. As of 2006, 59.3 percent of Pakistan's population was under the age of twenty-four. Because the fertility rate in Pakistan remains high, the "youth bulge" will persist into the coming decades; the percentage of Pakistanis under the age of twenty-four is still projected to be 51.4 percent as of 2030.[1] Combined with the anti-Americanism that already pervades much of Pakistani society, the deficient state of Pakistan's under-resourced public education system and the paucity of economic opportunities could potentially stir up a cauldron of disaffected young people ripe for indoctrination by Pakistan's expanding radical Islamist movement.

The purpose of highlighting the present and future threats found in Pakistan is not to diminish the challenges the United States faces across the border in Afghanistan, which remain real and daunting in their own right. It is merely to underscore that, while the threats to U.S. national security exist on both sides of the Pakistan-Afghanistan border and need to be addressed in tandem, the repercussions of failure in Pakistan are greater because of its size and nuclear status. Moreover, Washington's ability to influence change in Pakistan is limited by a clear red line imposed by Islamabad that prohibits significant numbers of U.S. military personnel from operating in Pakistan. For these reasons, the Obama administration should prioritize Pakistan as the chief national security challenge in South Asia.

U.S. Strategic Alternatives

Fortunately, statements from top U.S. officials suggest that the White House recognizes Pakistan's paramount importance to U.S. national security, and the unprecedented level of violence within Pakistan throughout early 2009 has helped this proposition gain currency throughout Washington's foreign policy community. Yet even if this supposition is adopted as fact, the critical and challenging next step remains: identify the best strategy to confront urgent threats while simultaneously addressing the underlying, longer-term challenges.

U.S. strategic interests in Pakistan—whether short or long term—will most likely be met by engaging with and bolstering those individuals, groups, and institutions within Pakistan who are actual or potential allies in the fight against violent Islamist extremism. There is little doubt that working with Pakistani partners will not be easy, in part because Washington and Islamabad do not perceive their strategic interests identically and in part because of a lingering bilateral mistrust reinforced by a lengthy history of "disenchanted alliance."[2] And in the end, a strategy of "partnership" may fail—it will no doubt be frustrating and costly—but it is worth attempting if only because the alternatives are even less likely to achieve success. In this context, it is worth examining the shortcomings of two credible alternative strategies: containment and coercion.

Poor Alternatives: Neither Containment nor Coercion

If U.S. efforts to win effective cooperation from Pakistani partners fail and the security environment in the region continues on its present trajectory, a strategy of containment—walling off the threat posed by terrorists and extremists—may be necessary. But containment is unlikely to be particularly effective or inexpensive over the long term.

Containment made good sense during the Cold War, when the United States faced a strong Soviet state under unified command. Today, the threat from Pakistan comes from strong subnational actors, not the state itself. Given Washington's inability even to pinpoint the location of top al Qaeda leaders along the Pakistan-Afghanistan border, it is clear that containing subnational terrorist groups will be extremely difficult. The complexity of the challenge would be multiplied if the United States is forced to operate without Pakistani cooperation. High-tech U.S. surveillance cannot substitute for human intelligence when it comes to unraveling complicated networks of individuals and groups.

Moreover, a strategy of containment against terrorists operating in Pakistan prematurely forfeits the prospect of cultivating an effective partnership with the vast majority of 176 million Pakistanis and their government. Once the United States embarks on a strategy of containment, it may be nearly impossible to reengage Pakistan with the goal of establishing a lasting partnership. Containing Pakistan may also increase the likelihood that an adversarial regime takes power in Islamabad, since Pakistan's political leaders would have little incentive to seek cooperation with Washington. U.S. tactics of containment would undoubtedly prove unpopular, offering a weak government in Islamabad plenty of reasons to play the anti-American card. Nor is containment flexible enough to address the longer-term threat posed by Pakistan in a more collaborative manner, even if it is successful in mitigating the security threat over the near term; it would need to be a strategy of indefinite duration.

Finally, once implemented, if a containment strategy begins to show signs of failure—if, for instance, there are serious indications that Pakistan's nuclear arsenal has become more vulnerable to seizure by terrorists—the United States will face a dilemma similar to the Bush administration's perception of the situation in prewar Iraq. Yet any U.S. invasion and occupation of Pakistan would almost certainly make Iraq look like child's play. In short, containment could very well lead the United States down a path to even more difficult and far more costly options.

The second strategic alternative centers on coercion, specifically the use of assistance coupled with the threat of withdrawal of support, or even sanctions, should Pakistan fail to meet U.S. conditions. Various types of coercion strategies regularly come up for consideration in Washington. For instance, the Pakistan Enduring Assistance and Cooperation Enhancement (PEACE) Act of

2009 legislation put forth by the House Foreign Affairs Committee makes a clear attempt to tie U.S. military assistance to Pakistan's compliance on several issues, including nuclear proliferation and counterterrorism.[3] The United States should not threaten sanctions unless it is willing to follow through and impose them. But because of Washington's dependence on Pakistan to achieve its security interests in the region, the threat of sanction is simply not credible. Unfortunately, some members of the U.S. Congress appear not to have come to terms with this credibility gap inherent in a coercive approach to assistance. Nor does stipulating that sanctions can be waived by the president for national security reasons (as is the case in the PEACE Act of 2009) resolve the matter. Instead, such waivers simply transform congressional conditions into toothless annoyances.

The problem with the coercive approach is that it will only compel Pakistani action if the United States has points of leverage—that is, if Pakistanis perceive the need to cooperate with the United States more than vice versa. Today this is simply not the case. Washington has few good instruments to address threats based in Pakistan without the cooperation of Pakistani security forces. Pakistan also remains an essential conduit for U.S. and NATO military supplies en route to Afghanistan, a logistics chain that cannot be easily replaced. But many Pakistanis perceive their security environment much differently. In particular, they do not see U.S. intervention in the region as a stabilizing force, and they continue to see India as Pakistan's primary threat. While few Pakistanis—especially those in the military—would prefer an outright break with the United States, many do believe that their security threats are manageable without Washington's assistance. Pakistan has done without America in the past; it might try to walk that path again in the future. Indeed, there are indications that Pakistan's military and intelligence services are already hedging their bets in anticipation of U.S. abandonment.[4]

Historical precedent is instructive in debunking the utility of coercion toward Pakistan. In 1990, Washington cut off aid to Pakistan as a condition of the Pressler Amendment that took effect when President George H. W. Bush failed to certify that Pakistan did not possess a nuclear weapon. The Pressler Amendment was adopted in 1985 and Pakistan's continued development toward a nuclear capability after that date demonstrates that the threat of American sanctions did not change Pakistani behavior that it deemed in its national interest. More troubling, cutting U.S. assistance severed important opportunities for building relations with Pakistan, including the International Military Education and Training (IMET) funding. Ending IMET was meant to penalize Pakistan but has proved detrimental to American interests as well. Over the course of the 1990s, Pakistani army officers lacked the opportunity to interact professionally with their American counterparts. Without this firsthand exposure to the ways and norms of the U.S. military and American society, a generation of the Pakistani officer corps has been left to its own devices to formulate opinions—typically negative ones—about the United States. In contrast, the previous generations faced no such restrictions. Pakistan's chief of army staff, General Ashfaq Parvez Kayani, for one, graduated from both U.S. Army Infantry School at Fort Benning and the Command and Staff General College at Fort

Leavenworth. At the very least, his prior experiences now offer him an understanding and appreciation for the U.S. military that his junior officers lack.

The revocation of U.S. aid in the 1990s also left a deep and enduring scar in relations between Pakistan and the United States. Pakistanis viewed the invocation of the Pressler Amendment in 1990 as more than a coincidence, coming shortly after the Soviet Union withdrew its forces from Afghanistan and Pakistan's tactical utility as a conduit of U.S. aid to the anti-Soviet mujahideen ceased. This bilateral trust deficit remains a central roadblock to deeper Pakistani cooperation with the United States. Overcoming it will be essential to shifting Pakistani opinions about the durability of partnership with Washington. Imposing a coercive policy again would simply reinforce a destructive pattern of U.S. engagement with Pakistan.

A Two-Pillar Strategy

A better long-term strategy to advance U.S. goals in Pakistan should be based on the twin pillars of induced bilateral partnership and the reshaping of Pakistani perceptions of the regional strategic environment. Washington must seek to induce, rather than coerce, allies and partners within Pakistan's civilian political leadership, military, and wider public. U.S. policy makers must understand that their actions play into an ongoing debate within Pakistani society, in which allies can be won or lost and adversaries can be undermined or empowered. American security and development assistance should be used to strengthen these elements of Pakistani society to better advance the areas where their interests coincide with those of the United States.

Too often lost in the discussion of Pakistan is the fact that the Taliban and other extremist groups do not represent the goals and aspirations of the people, the vast majority of whom would prefer to live in a peaceful, prosperous, and moderate country. Washington should therefore pursue policies that will be perceived in Pakistan as supporting these goals. Along these lines, the United States will find that helping to build a stronger Pakistani civilian administration, capable of delivering law and order, is in both U.S. and Pakistani interests. Similarly, enhancing the educational and economic prospects for millions of young Pakistanis will contribute to U.S. security over the medium to long run. Training and equipping effective counterinsurgency forces within the Pakistani military is also mutually beneficial. In each of these areas, the United States should patiently but persistently seek the most efficient means to deliver its assistance, as waste and corruption will undermine trust on both sides.

But assistance in these areas should not be tied to a rigid set of conditions. Instead, to demonstrate its long-term commitment to partners within Pakistan, Washington should clarify that if resources are misused or fail to produce desired results they should be reprogrammed rather than curtailed. On the civilian side, one way to improve transparency and accountability for the expenditure of U.S. assistance would be to create a multilateral trust fund, possibly administered by the World Bank, which could work with the United States, other donors, and Pakistan's government and nongovernmental groups to identify, formulate, and implement effective assistance projects. Such a trust fund would also permit Washington to leverage its investments by encouraging contributions from other donors, such as the EU and the Japanese.

As the second pillar of its effort, the United States should work to reshape the strategic environment as understood by Pakistanis. U.S. assistance, military operations, and diplomacy should all be employed to create new incentives that will convince fence-sitters within Pakistan's political and military leadership of the benefits of working with the United States and the costs inherent to opposing American efforts in the region. In particular, Washington should work to create conditions that diminish political and military uncertainty along Pakistan's borders. To be clear, this does not mean that the United States should somehow aim to resolve Indo-Pak and Pak-Afghan disputes in Pakistan's favor. Instead, Washington should make a clear commitment to regional stability, demonstrating that militancy will be defeated decisively in Afghanistan and lending under-the-radar support to a normalization of relations between India and Pakistan.

Transforming the regional security environment is necessary because from a Pakistani perspective, India and Afghanistan represent the essential, linked threats that have long inspired Islamabad's patronage and support to militant and extremist organizations as a means of asymmetrical power projection. The persistent ambivalence in some Pakistani circles about how to deal with these militant groups is tied to an underlying fear of Indian encirclement. The United States can address pieces of this dynamic by devoting greater, sustained resources to the fight in Afghanistan, thus eliminating the incentives for Pakistan to hedge its bets and support Taliban and other anti-Kabul factions. In other words, by demonstrating a convincing commitment to victory over the Taliban in Afghanistan, the United States will show Islamabad it has nothing to gain by active or passive assistance to the Afghan Taliban and related groups because they have no political future in Kabul.

At the same time, the United States should approach both New Delhi and Islamabad to encourage a return to dialogue and normalization of Indo-Pak relations. Washington should clarify to its Pakistani partners that the United States is committed to strategic partnership with a rising democratic India, that this relationship is not intended to threaten Islamabad, but that India's rising prominence in global politics cannot be held hostage to Pakistani fears. In short, the time has come for Pakistan to reconcile itself to a new strategic reality vis-à-vis India.

Recognizing that the United States will never have the leverage required to impose a resolution to the blood-drenched dispute between India and Pakistan, Washington should still make an effort to mitigate causes of insecurity. The Mumbai terrorist attack of November 2009 and the subsequent uptick in tensions between Islamabad and New Delhi should have impressed the new Obama administration with the urgent need to keep a lid on cross-border tensions. Washington should encourage both sides to return to their "composite dialogue" that was a casualty of the Mumbai attacks. That dialogue would also benefit from including discussions of Afghanistan, since Pakistanis bitterly complain about extensive—and threatening—Indian intelligence operations in Afghanistan. While the United States should not seek to adjudicate this dispute, it might play a helpful role in the sharing and verification of intelligence as a means of building confidence on both sides.

Overall, this twin-pillar strategy for Pakistan focuses on long-term goals. U.S. efforts may not yield rapid progress even with $1.5 billion per year in civilian aid to Pakistan and sixty-thousand or more U.S. troops on the ground in Afghanistan. They will require patience, especially in a three-to-five-year timeframe. But, in contrast to the lack of sustainability of containment and impracticality of coercion without leverage, this strategy attempts to bridge the near-term and longer-term security challenges posed by Pakistan. In addition to meeting the urgent challenges of today, a strategy of inducement will help Washington confront the looming challenge that Pakistan will pose over the next generation.

Implementing a Twin-Pillar Strategy: First Steps

The broad contours of a twin-pillar approach to Pakistan are outlined above, but in order to move from strategy to implementation, there are three specific areas where the White House can and should take quick action.

First, with respect to meeting the urgent security challenges posed by al Qaeda and other terrorist groups based within Pakistan, the United States will need to continue forceful intelligence and military operations in Afghanistan and along the Pakistani border, some of which may be unpopular among Pakistanis. But in conducting these operations, the U.S. military should work in ways that will do the least possible to jeopardize prospects for longer-term partnership. In particular, it should be understood that accelerated or geographically expanded use of Predator-type drones on Pakistani territory under present political conditions would be counterproductive. Drone strikes have served as a useful tactical disruption for a small number of targets, but they are unpopular in the current Pakistani political climate and raise costs for the U.S.-Pakistani partnership. In addition, drones are now—by some accounts—losing their tactical utility, as top terrorist leaders relocate outside the Federally Administered Tribal Areas (FATA) and into Pakistan's settled areas. This migration could contribute to instability and terrorism in places like Karachi or southern Punjab.

Second, in order to build capable Pakistani partners and to confront urgent security challenges, Washington must make rapid and extensive investments in critical institutions in Pakistan's security sphere, especially the police, paramilitaries, and army. To be most flexible and effective, these investments may require new funding mechanisms that circumvent the normal bureaucracy and red tape that often impose lengthy delays. On the civilian side, it will be impossible to formulate and implement smart assistance programs unless the State Department and USAID expand the scale of their operations inside Pakistan. But in order to enable the movement of civilian officers within Pakistan's difficult security environment, new facilities, procedures, and personnel will also be required.

Finally, when it comes to conducting its diplomatic efforts, U.S. policy makers must understand that statements that undermine confidence among Pakistanis in the stability of their state or the U.S. commitment to partnership are harmful to American interests. It is important for the Obama administration to voice U.S. concerns about the threats it perceives in Pakistan, including the extension of the Taliban's writ in the North-West Frontier Province. But it is also important that the administration's warnings should not, in themselves, contribute to a brain drain or capital flight among the educated Pakistani elite that would further undermine stability in Pakistan.

Urgent Action Needed

Of course, these steps represent only the very beginning of a much longer, exceedingly complicated, and costly process. This process should begin quickly: the challenges posed by Pakistan are vast and will not be met through half measures or passivity. Further delays could prove fatal: the many forces undermining Pakistan's stability are now ascendant if they are not yet dominant. The United States can reverse this momentum, not by seeking to contain the threat from a distance or by leveling coercive threats, but by cultivating and empowering a wide range of Pakistani partners.

Notes

The author wishes to thank Daniel Simons for his assistance with this paper.

1. Moeed Yusuf, "Prospects of Youth Radicalization in Pakistan: Implications for U.S. Policy," Brookings Analysis Paper 14, October 2008, p. 2.
2. Dennis Kux, *The United States and Pakistan 1947–2000: Disenchanted Allies* (Washington, DC: Woodrow Wilson Center Press, 2001).
3. *Pakistan Enduring Assistance and Cooperation Enhancement Act of 2009,* HR 1886, 111th Cong., 1st sess.
4. Pakistan's continued relationships with certain militant organizations along the Afghan border (especially the Haqqani network) and in Punjab (Lashkar-e-Taiba) appear to be a part of this hedging strategy.

India and the United States: Making the Partnership Strategic

Ambassador (Ret.) Teresita C. Schaffer
Director, South Asia Program
Center for Strategic and International Studies

The United States and India have transformed their relationship in the past fifteen years from a prickly set of conversations with relatively little substance to a serious bilateral engagement based on a growing array of common strategic and economic interests. The cliché once used to described it was "the world's oldest and largest democracies," or, in the pithier title of Dennis Kux's classic book, "estranged democracies." The new buzzword is "strategic partners." I will argue that the partnership is serious; that it is not yet strategic; and that in order to become strategic, it needs to be reinvented. Not to be outdone by my friend Dennis, I also invite you to buy my book, which will spell this argument out in much greater depth, and which should be available from the Center for Strategic and International Studies (CSIS) in early June.[1]

Understanding the U.S.-Indian relationship and its larger import starts with a look at both countries' strategic outlooks. Next, we will examine what the two countries have done to build up a vibrant bilateral relationship. Finally, we will look at the weak areas in our engagement—the failure to identify where we have a common view of the world, and to work that into our conversations—and highlight the way we could reinvent this partnership to maximize its strategic benefit for both.

Strategic Convergence

The revolution in U.S.-Indian relations starts with the new Indian foreign policy that resulted from the end of the Cold War. Four features of this policy were particularly important.

First, it was based on a new calculus of India's power. India's governments and its strategic thinkers came to regard economic power as a central element in their national power and national success. India's per capita economic growth roughly doubled between the 1960s and the years since 2000. Sustaining that growth was the only hope of having India move into the same global power "club" occupied by China.

Second, and partly as a result, the United States became India's most important external friend. Russia was still a major military supplier, but it had neither the economic weight nor the ability to help India move into the global governance circles that its leaders now sought.

Third, relations with China and with East Asia were much more important than they had been. Both economic and military relations expanded dramatically.

Fourth, Indians of all political stripes remained strongly committed to what they call "strategic autonomy"—the concept that India must neither allow anyone else to dictate its foreign policy, nor permit such an impression to be created. In practice, this means that despite its increasingly close and productive ties with Washington, New Delhi is still very sensitive about taking the U.S. side in multilateral settings.

Looking more specifically at defense, Indian national security policy is based on preeminence in South Asia, countering any major threat that intrudes into this space, and deterring major threats from beyond.[2] India's immediate neighborhood—the inner ring of its security environment—is a dangerous place. Senior security managers often make the point that their greatest national security challenge stems from internal insurgencies connected to problems in their neighbors' territories. Pride of place goes to Pakistan, not just to Kashmir but to a pattern of Pakistani support for other threats to India's security. Other threats have come at India from Bangladesh, Nepal, and Sri Lanka, or from linkages of domestic groups to threats in these countries.

The outer circle of India's strategic space extends from the Middle East to Malacca. Here, India's immediate goal is to protect its lifelines for trade, investment, and energy, both from large strategic threats and from such dangers as terrorism and piracy. This strategic environment is centered on maritime and littoral security, and naval and air power are key. India's large procurement budgets for the navy and air force aim to build up its power projection capability to give India presence and capacity throughout this extended region.[3] Even India's most hawkish security analysts do not foresee circumstances in which India would use it for offensive purposes beyond the subcontinent.[4]

India sees the Indian Ocean as a single strategic environment, with its navy as the dominant littoral force. India does not want to see its primacy challenged. In contrast to earlier years, India's security managers now see the U.S. presence in the Indian Ocean as benign.

The long-term challenge to India's east stems from China, still regarded as its main strategic rival. At the same time, with dramatic improvements in their political and economic ties, India hopes to build a peaceful and profitable relationship. In the longer term, Indians worry about possible Chinese plans for an eventual military presence in the Indian Ocean.

The area to India's west is more troublesome. Two-thirds of India's oil imports come from the Middle East; this percentage will rise in the next two decades.[5] China's assistance to Pakistan in building a new port in Gwadar, on the Arabian Sea coast near the Iranian border, and its ties with Burma are looked on with great suspicion in India. India's large diaspora of workers in the Persian Gulf region is both a source of remittances and a vulnerability.

India has little expectation that the United States will help it deal with Pakistan. Rather, it hopes that the new security relationship with the United States will help it deal with security challenges outside the inner perimeter of South Asia, both politically and militarily. It wants the political stature that comes from being taken seriously by the United States. It wants access to the full range of U.S. technologies, civilian and military, including the opportunity to produce top-of-the-line military equipment originating in the United States.

Indian leaders recognize that U.S. strategic goals in the Indian Ocean largely dovetail with India's. As a result, for the first time since India became independent, it is apparently comfortable making parallel security arrangements in a region where India has always considered itself preeminent. The major short-term contingencies India faces, in the words of one observer, are "Kargils and tsunamis"—short but intense land border engagements and more diffuse humanitarian disasters in which India's military assets become a tool for building a more peaceful and cooperative expanded neighborhood.

In the years since the Cold War ended, U.S. foreign policy thinking too has shifted. The world of two blocs is gone. In its place, the George W. Bush administration expected to put a policy organized around the central idea of the terrorist threat. In practice, however, we now have not one but several lead areas and concepts: the wars in Iraq and Afghanistan; a power structure in which emerging countries like China and India will play a greater role; a gradual shift in the emphasis of U.S. policy away from Europe toward Asia; the imperative of restoring global economic growth; and the big problems, especially important for the Obama administration, that can only be addressed globally, starting with proliferation of nuclear weapons, energy, and climate change. India figures critically in all these issues.

Looking more specifically at security interests, there are strong parallels between those of the United States and India's. Start with Indian Ocean security—the heart of India's "outer perimeter." The U.S. Maritime Strategy released in 2007 by the U.S. Navy, Marine Corps, and Coast Guard specifies that "credible combat power will be continuously postured in the Western Pacific and the Arabian Gulf/Indian Ocean." In other words, the two principal centers of U.S. sea power are both in India's area of strategic interest. For both the United States and India, as noted above, China's expanding economic and military power represents both an opportunity and a challenge.

There are both similarities and contrasts when it comes to U.S. and Indian interests in the Middle East. India has not been a shaper of events there, and has had grave misgivings about U.S. policy in the region. Neither country wants to see a nuclear-armed Iran. However, we disagree on what to do about it. India gets some 10 percent of its oil from Iran, and is wary of public confrontations. Despite these differences, shoring up U.S. political and military relations to the immediate east of the Gulf is one of the arguments for strengthening the U.S.-Indian relationship.

The U.S. relationship with Pakistan has always been a prickly subject in Indian-U.S. relations. In fact, India's policy leadership shares the prevailing U.S. view that everyone would benefit from a relatively stable Pakistan. However, the U.S. desire to avoid roiling Pakistani sensibilities has inhibited cooperation with India even on issues that are of great interest to both (such as antiterrorism). The Pakistan factor has also affected our ability to work with India in Afghanistan. We have very similar interests in that Afghan stability and having an Afghan state that is capable of withstanding subversion or blandishments from the outside is important to both. However, in deference to Pakistan's sensitivities, the United States has kept India at arm's length on Afghan security issues.

The Success Story: Building the Bilateral Infrastructure

In the first two decades of serious U.S.-Indian engagement, most of the big accomplishments related to our bilateral relations, not to interaction on the global scene. The first achievement was a dramatic expansion of economic ties. This was a nearly inevitable result of India's sharply higher economic growth and greater integration with the global economy. India's exports to the United States more than doubled between 2001 and 2008, from $9 billion to over $20 billion. India's imports from the United States nearly quadrupled from 2001 to 2007, from $3 billion to nearly $12 billion, and nearly doubled again, to $21 billion, in 2008. If one includes India's exports of software and other Information Technology (IT) services, the export figures would be twice as high. Investment rose by even greater percentages. The private economic relationship has now expanded to the point where it is one of the major drivers of U.S.-Indian relations, and will remain so more or less regardless of what the governments do.

The bigger surprise came in the security area. Military-to-military cooperation is almost unrecognizable now, compared to the 1990s. The common interests the two countries face, especially in the Indian Ocean and in Southeast Asia, have made a sharp increase in joint exercises natural and mutually beneficial. Nonetheless, the strategic dialogue has lagged behind this activity. The two sides have somewhat different expectations from their defense relationship. U.S. defense officials look on defense trade as the way to expand a security relationship. Using similar equipment facilitates interoperability, and creates the platform for harmonizing the way two military services think about their operational space. For India, the key is access to U.S. technology, and the big challenge is overcoming the concern that the United States is an unreliable supplier, with the U.S. Congress reserving the right to change the terms of military sales after the contract has been signed.

The most dramatic accomplishment of the past decade was the Indian-U.S. agreement on civil nuclear cooperation. The U.S. intention in proposing it was to remove India from its nuclear isolation, thus satisfying a thirty-year Indian ambition, to liberalize trade in high-technology goods, and, from a political and strategic perspective, to lay the groundwork for strategic cooperation on an Asia-wide and global scale. The agreement itself was intensely controversial in both countries. In the United States, it represented a major shift in U.S. nonproliferation policy. The agreement eventually passed the Congress with a large bipartisan majority. On the Indian side, however, the parliamentary maneuvers and draftsmanship that made this passage possible served to remind Indians that the United States had attached some strings to its offer of nuclear cooperation. This combined with India's complicated coalition politics to put the Indian-U.S. relationship at the center of a political storm in India.

But when all is said and done, these three areas—economic ties, defense relations, and nuclear and high-tech trade—represent extraordinary achievements that transform the political landscape between India and the United States. At the same time, the two governments developed a rich array of consultative mechanisms, which had the very useful effect of familiarizing each side with the other's quite different bureaucratic and political structures.

The "Global Gap"

This initial phase of building the U.S.-Indian partnership did relatively little, however, to move the partnership on to a regional or global stage. This raises the question of what kind of partnership the United States and India would like to have—and can sustain.

The two countries have different historical reference points. For the United States, most of our existing partnerships, both the formal multilateral ones like NATO and the less formal ones like the relationships with Egypt or Israel, started with shared global or regional security goals. In virtually every case, the United States was by far the stronger partner. This creates an expectation in the United States that international partners will look at the world in much the same way we do, and will readily follow the U.S. lead, perhaps not always but certainly as a rule.

India, on the other hand, has had little experience of lasting partnerships in its foreign policy. The watchwords have been "strategic autonomy" and "nonalignment" rather than "collective security." This leads Indian politicians to look with some degree of suspicion on the notion that India will normally accommodate itself to a powerful foreign friend.

This creates something of a disconnect between the United States and India, despite the increasing harmony of interests, and despite the increasingly productive bilateral relationship. Another disconnect arises out of their different priorities. For Indians, the principal benefit they expect from ties with the United States lies in the bilateral realm. For the United States, the prize it hopes for eventually is a closer harmony in the regional and global policies of the two countries. The United States looks ahead to a time when a rising China and a rising India will be part of an Asian balance of power in which U.S. interests are also reflected and protected, for example. It also hopes that a more powerful India will be pragmatic enough to help resolve major global issues like global warming, nonproliferation, and financial reform.

In fact, the United States and India have had difficulty working together multilaterally. In the United Nations, we work well on peacekeeping issues, but India's voting record in the General Assembly puts it on the same side as the United States only 14 percent of the time, below the already dismal 18 percent average of other countries' voting concurrence with the United States. In the Doha Round of multilateral trade negotiations, India emerged as the spokesman for the countries who opposed the U.S. proposals for agricultural trade, and was thus a leading contributor to the breakdown of the talks in the summer of 2008. India's approach to the negotiations reflected above all its belief that India had little to gain from a successful Doha Round. Nonetheless, India stood between the U.S. trade negotiators and their hopes for the future.

Can we remedy the "vision gap"? Yes, provided we have patience and realistic expectations. Limiting the relationship to the bilateral would be possible, but would fail to take advantage of the complementary interests that are likely to keep pushing India and the United States together in spite of periodic disappointments and controversies. The partnership is ripe to "go global," and it's time to figure out how. The first sign of an early potential success lies in the discussions on financial reform, which have avoided the prickliness of so many other multilateral encounters. The two countries need to build on their bilateral success, and start adding global issues to their dialogue. Three rules should govern: candor, inclusiveness, and no surprises. Of the big multilateral issues that are close to President Obama's heart, India and the United States are surprisingly close on financial reform, and have serious disagreements on nonproliferation and on climate change. But the issues cannot be addressed without India, and the process of bringing India into the world's governing discussions is likely over time to change the way India deals with these issues.

The end result, if both sides invest the necessary time and skill, will not be an alliance nor a seamless agreement. Rather, they can craft an expanding agenda of selective cooperation, and a relationship that can drive the increasingly important Asian balance of power. This should be our goal.

Notes

1. *The United States and India in the 21st Century: Reinventing Partnership*, in press, CSIS.
2. See presentation by MG (Ret.) Dipankar Banerjee in *Report on the National Workshop on Changing Contours of Indo-U.S. Relations,* ed. K. P. Vijayalakshmi, Arvind Kumar, Sonika Gupta, and S. Chandrashekar (Bangalore: International Strategic and Security Studies Programme, National Institute of Advanced Studies, Indian Institute of Science, 2006), p. 137.
3. Between 2004 and 2006, the air force received 36–44 percent of procurement funds and the navy 23–24 percent; the army, vastly larger than either, was held to 26–29 percent of procurement funds. See International Institute for Strategic Studies, *Military Balance* 106, no. 1, June 2006, pp. 224–227. See also Ministry of Defense, "Defense Expenditures," http://mod.nic.in/aboutus/welcome.html; Ministry of Finance, *Economic Survey of India 2008*, Table A-6, available at http://indiabudget.nic.in/es2007-08/chapt2008/tab13b.pdf.
4. See, for example, Brahma Chellaney, *Asian Juggernaut: The Rise of China, India and Japan* (New Delhi: HarperCollins, 2006), which argues for a realist policy and for preventing Chinese domination of Asia, but makes no argument for any kind of offensive goals.
5. Planning Commission, *Report of the Expert Committee on Integrated Energy Policy,* (Government of India, August 2006), p. 59, available at http://planningcommission.nic.in/reports/genrep/rep_intengy.pdf.

Panel IV: South Asia

Summary of Discussion

Dr. Timothy D. Hoyt
Professor of Strategy and Policy
Co-chair Indian Ocean Regional Study Group
Naval War College

In "A U.S. Strategy for Pakistan: Future Directions," Dr. Daniel Markey of the Council on Foreign Relations notes that the United States really needs more of a Pakistan strategy than an Afghanistan strategy. Pakistan has five times the population of Afghanistan, an arsenal of nuclear weapons, troubled civil-military relations, and is the sanctuary for Taliban and al Qaeda activists. He recommends a policy of identifying and partnering with individuals, groups, and institutions in Pakistan who are (or could be) allies against Islamist extremism. Dr. Markey argues that U.S. goals in Pakistan should be based on two pillars: induced bilateral partnership and reshaping Pakistani perceptions of their regional strategic environment.

Ambassador Teresita Schaffer's "India and the United States: Making the Partnership Strategic," notes how the Indo-U.S. relationship has changed from one of frustration (in the Cold War) to one of great promise. She notes, however, that many refer to the relationship as a "strategic partnership"—a title that it does not yet deserve. Key drivers in the evolution of the relationship include changes in the international environment, the expansion of bilateral economic interaction, and substantial increases in security cooperation. Expanding military-to-military contacts, growing defense trade, and the new agreement on civil nuclear cooperation are all significant changes in the relationship, and work to increase its momentum. Ambassador Schaffer notes, however, that neither the United States nor India has much experience working with strong partners, and that their history of multilateral cooperation is modest at best. Patience, realistic expectations, and careful diplomacy will create an expanding agenda of selective cooperation—an attainable goal, and a significant change in what was once a difficult relationship.

During the discussion period, some of the conversation focused on the difficulties of partnering with Pakistan. Pakistani institutions—particularly the police—were seen as deeply dysfunctional or corrupt. Other comments noted the ineffectiveness of Pakistan's political system, and its inability to maintain a handle on the more extreme elements in its society (including the Taliban).

A participant emphasized that partnership with Pakistan is a long-term process, but one which involves identifying and cultivating potential partners. There are members of the police force who see the need to work against the Taliban, and who could become much more effective with our support. Similarly, the participant noted that large parts of the country are not nearly as troubled or as threatened by the Taliban as the north and west.

In addition, the United States must address both short-term and long-term interests through the partnership program. The short-term need, which is urgent, is supporting the Pakistan army and security forces against an increasingly aggressive Taliban force. [Note: the workshop occurred in May, and by the time of this edit (July) the army had demonstrated substantial tactical successes against the Taliban in the North West Frontier Province.] In the long-term, however, Pakistan faces threats to its civil society and political system from Islamist radicals that the army cannot resolve. Indeed, the army's involvement in politics may accelerate that threat—which suggests a need for a broader approach to partnership targeting other institutions and potential allies.

Other comments raised the possibility that the most critical problem in Pakistan is the erosion of authority of the state and its unwillingness to respond to oppression and increasing aggressiveness on the part of the Taliban. A follow-on comment asked if the "Cold War" strategy—trying to identify and strengthen non-hardline factions—could be used, both to identify and approach potential partners. A third comment noted that the United States tends to exaggerate the degree to which we can influence Pakistan's internal decisions. A respondent agreed, noting that in many cases we are working with people who are either not wholeheartedly committed or very effective. The key is to identify people and institutions who might want to do the kinds of things we want and to empower them.

One potential partner for the United States is the United Kingdom, whom one commentator identified as particularly alarmed about the presence of alienated Pakistanis in the United Kingdom and their connections with previous terrorist attacks. A respondent noted that the United Kingdom's strategy toward Pakistan is very complementary to our own, and that we work in close collaboration. The respondent also noted the problem of population mobility in a globalized world. Terrorists have the ability to move fairly freely unless governments take security measures, but those security measures—such as terrorist watch lists—often inconvenience innocents and raise resentments against the United States. In a later response, the respondent noted that China may be a cooperative partner as well, given its close relationship to Pakistan and overlapping interests with the United States.

Another issue raised was the failure of Pakistan to develop a land reform program. Pakistan's agricultural production remains in the hands of a small number of families, who oppose not only land reform but also both economic and education reform. According to some reports, the Taliban have been able to promise land redistribution and use it as an effective tool in some areas of the country. A discussant explained that there is deep inequality in Pakistan, and that the leading landowners have been described as "feudal." On the other hand, the United States must continue to work with these elites, because they have enormous political and economic influence in Pakistan. The United States has recently agreed to raise economic aid in Pakistan to $1.5 billion per year, and this may provide a means to spur economic redistribution. The discussant noted that economic development holds the promise of making the value of land less central to the country's economy, which in turn may marginalize land reform as a rallying cry for radicals. One potential long-term threat to Pakistan's stability may be economic dissatisfaction and underemployment of a growing young population. Careful planning can minimize the revolutionary potential of economic inequities.

A discussant warned of a trend the discussant perceives in Congress. Many Americans want to narrow the objective in the region to eliminating al Qaeda. This is a seductive political solution, because it promises quick results to the problem. This approach, however, ignores the much more serious long-term threats in the region, and particularly within Pakistan. It does reflect, as one commentor pointed out, the history of U.S. engagement in the region. In the past, we have rarely been willing to engage in either Afghanistan or Pakistan for an extended period. Why would the people in either country feel like things are different now? Other questions revolved around the U.S. relationship with India, and whether the Indians really believe we will stay in the region. Does India begin to look at Iran as a more desirable partner than the United States, simply because it promises routes of commerce to Central Asia?

A participant noted that the U.S.-Indian economic relationship is important, but by itself may not be enough to sustain the current momentum in bilateral relations. The participant also noted, however, that the United States and India share many significant interests, including an unwillingness to allow Asia to be dominated by a single power. India is shifting its attention to Southeast Asia—the "Look East" policy—to a much greater extent than ever before. In addition, the United States wants Indian support on a range of global issues, which will require a closer and more positive bilateral relationship. The Indians generally do not trust the United States when it comes to Pakistan, based on years of experience. Tehran has been stirring up trouble in Afghanistan, which actually makes things more difficult for India. The United States and India, and even Iran, have quite similar interests in Afghanistan.

The participant also noted that India would not necessarily be alarmed by a long-term approach to Pakistan that relied on partnerships and economic development. The participant pointed out that what the Indians really dislike is the United States coming to India and asking India to consider concessions. The concept of a long-term change in Pakistan's strategic environment, as recommended by Dr. Markey, is that it does not require good U.S.-Indian relations or essentially false promises on Kashmir.

A discussant pointed out that the United States is setting very short-term time lines, in part because both U.S. and Afghan leaders must think about domestic politics. The United States must show progress in some meaningful manner, which requires an adjustment of goals. Buying time is essential—because real solutions in both Afghanistan and Pakistan will take longer than a few years. The discussant also suggested that one way to provide new aid to Pakistan is to put it in a trust fund, managed by an international nongovernmental organization or financial institution. As groups in Pakistan develop projects, they could come to the trust fund and apply for grants, which then could be monitored and evaluated. This would reduce the opportunities for corruption, and make the process more transparent and effective. It would also avoid the usual problem in Pakistan—which is that aid is put into the general budget and then disappears. But continued economic assistance is vital for Pakistan's struggling economy, and the United States and international community must find ways not only to provide it, but to enhance its effectiveness.

Panel V

The Greater Middle East

Ambassador (Ret.) David C. Litt
Executive Director, Center for Stabilization and Economic Reconstruction, Institute for Defense Business

Dr. Marc Lynch
Associate Professor of Political Science and International Affairs, The Elliot School of International Affairs, George Washington University

Dr. Heidi E. Lane
Associate Professor of Strategy and Policy, Greater Middle East Area of Study Coordinator, Naval War College

Moderator:
Dr. Hayat Alvi-Aziz
Associate Professor of National Security Affairs, Naval War College

New Directions for U.S. Foreign Policy in the Greater Middle East

Ambassador (Ret.) David C. Litt
Executive Director, Center for Stabilization and Economic Reconstruction
Institute for Defense Business

American foreign policy in the Greater Middle East[1] has remained relatively consistent from one year to the next since the end of World War II, regardless of which party occupies the White House: support for the State of Israel, search for regional stability, promotion of economic development and good governance (even if erratically), and rejection of terrorism. Periodically, however, a tectonic shift shakes the ground—once every ten years or so (the Suez crisis, the Six-Day War, the fall of the shah, etc.). Each event and its immediate "aftershocks" upset the existing order, redefine the environment, and spawn significant readjustments in the way the U.S. and other actors view and behave toward each other. With the decade following September 11, 2001, coming to a close, the new U.S. administration has a radically different take on how to implement U.S. foreign policy, even if the content so far remains more or less consistent with that of the previous administration. This is an ideal opportunity to prepare now for the next seismic event, which could well emerge soon from within the Greater Middle East if the pattern holds. Perhaps more than any other geographic region, the Middle East has the potential for directly affecting America's security, and that of our closest allies, in a variety of ways—from violent extremism, to energy vulnerability, to state-on-state warfare. Significantly, all of these security threats have proved to be interdependent and mutually reinforcing.

This paper will argue for both continuity and significant change: continuity in addressing regional conflict (albeit more effectively) and change in the attention and resources the U.S. government devotes to diplomacy, especially public diplomacy, and socioeconomic development. The administration's initial steps have been in the right direction and are encouraging. We must rebuild America's image and reputation in the region and the world. We must shift the center of gravity of the instruments of national power away from nearly exclusive reliance on the military and toward enhancing our civilian prowess in foreign affairs, especially diplomacy and development. The president cannot, however, make these changes alone. The Congress has a major role, as do our friends and allies around the world. We have no time to lose; America's ability to influence events in the region has suffered tremendous blows, many of them self-inflicted. This trend is not irreversible, but it will require time, money, hard work, and some risk.

These efforts should begin now; none can be postponed for long. The first step, however, is restoring America's credibility in the region. To paraphrase Edward R. Murrow: to be successful, we must be influential; to be influential we must be persuasive; to be persuasive we must be credible.[2] The reputation of the United States in the Middle East has always—at least until recently—run along two seemingly divergent tracks: respect for our ideals and condemnation of our policies. Perceptions of American behavior, especially in the last decade, brought increasing criticism that we had abandoned our ideals in pursuit of achieving our goals through military power. Excessive use of American military power in a region already prone to accept narratives of victimization, occupation, and colonialism, combined with the images of abuse and abhorrent behavior by U.S. forces, provided our adversaries with volumes of evidence of "America's real intentions," especially toward Arabs and Muslims, to present before the court of public opinion. American tactics in Iraq provided the most damning evidence: images and stories of U.S. prison guards' shameful conduct at Abu Ghraib, the aggressive behavior of private security personnel in Baghdad,[3] violent and sometimes erroneous invasions of privacy and the home by U.S. forces across the country,[4] all contributed to creating an image of America as a nation that had forsaken its values.

Rebuilding America's credibility of course requires first and foremost changes in policies and attitudes on the part of the government, but intensive American public diplomacy throughout the region must accompany those changes. Public diplomacy begins with the president, who has already launched the campaign in the broadcast and internet media calling on the Muslim world to give the United States another look.[5] Other senior officials, and not just the secretary of state, must likewise inform regional audiences on a regular and sustained basis that a new approach is under way. The messages must be clear and consistent. The U.S. government respects Islam and the people and cultures of the Middle East. We will listen more, and we will not view "success" only in terms that we define. When we say "security," we often mean freedom from some external threat; other peoples in different circumstances define "security" in terms of economic and social well-being. "Democracy" does not just mean elections; "democratic institutions" are not just political parties. Conflict-ridden societies often seek "justice" as a primary objective. We should speak in terms of popular participation, transparency, and accountability, principles that most societies view as admirable.

At the same time, our nation has interests that our government must protect; we will be particularly firm in our goals of securing the safety of Americans and our allies. We will be—and be perceived as—a wise and generous nation that uses all instruments of its national power judiciously and to good effect, including the appropriate exercise of military power. We must persuade citizens of the region that this does not mean that our strategies can trod upon the liberties and sovereignty of others.

Moreover, we should eschew the reputation of a go-it-alone Goliath. Only collaborative efforts of many nations can resolve the tribulations of the Greater Middle East. Naturally this involves nation-states and international organizations external to the region with significant influence and resources—the EU, Russia, Canada, and the United Nations agencies to name a few. This is also true, however, for trying to solicit the moral, political, and often financial support from nations within the region to achieve common goals. We have earned an unfortunate reputation for coming to countries, especially those in the Persian Gulf, with a tin cup in our hand without showing to their leaderships the common courtesies, personal engagement, and diplomatic respect that is coin of the realm in the region. We talk of partnerships and strategic relationships without putting substance behind it. To most Arab leaders, personal relationships count for a lot, and our rushed behavior, diplomacy-on-the-fly, is a transparent form of disrespect.[6] Arab leaders came to view America's version of how to garner respect as intimidation through military power. That works sometimes, but in this region we often confused respect with fear, which mutated too easily and too often into contempt. Many Middle Easterners saw the U.S. government as arrogant, narrow-minded, insensitive, and exclusionary. Taking the time to build strategic bilateral relationships in the Middle East properly will go a long way to garnering effective collaboration in resolving regional issues, not to mention global ones.

No improvement in our image or our relationships will gloss over ineffectual policies. The administration's next immediate challenges will involve the simmering conflicts to which the Middle East plays host, principally (but not exclusively) Iraq, Iran, and the Israeli-Palestinian conflict. As President Obama has acknowledged on several occasions, Iraq is far from resolved, Iran's policies and pretensions in the region are problematic to say the least, and Israel's continuing confrontations with Palestinian and Lebanese radicals show no sign of abatement. The administration has demonstrated its bona fides in addressing these crises right out of the starting gate, but it must enhance the government's capability of dealing with them.

Iraq's reconciliation process is underdeveloped at best. Former Sunni insurgents had agreed to stop fighting the coalition and the Shia-led government, and to become part of the system. Those agreements, however, are beginning to unravel as the government apparently is not meeting the expectations of some for reintegration of Sunni fighters into the economy, and especially into the security forces. Mistrust lies barely beneath the surface; good will is ephemeral. Ethnic and sectarian disputes abound, awaiting political resolution, including the potentially seismic social, political, and economic confrontation over Kirkuk and parts of Nineveh province. True reconciliation must be deeply rooted, and take place above the level of governmental or political party leadership, involving the nation's ethnic and sectarian leaderships on a societal level. Frankly, the U.S. government and military have only a very limited role in advancing reconciliation, but it is a valuable one. Part of it should be to urge prompt resolution of critical issues, such as sharing the nation's hydrocarbon patrimony, which are sometimes central to the broader ethnic and sectarian disputes. Our nation, together with our allies, should promote transparency and accountability in the struggle for political and economic power in Iraq, especially regarding the Arab-Kurd-Turcoman fault lines. In addition, we should do what we can to advocate nonpartisanship in the Iraqi security establishment as it seeks to guarantee a secure environment for working out thorny political and economic disputes peacefully, if not amicably. Evidence in the public domain suggests that sectarian fissures continue to rend the security forces.

Broadly speaking, American diplomacy and efforts to support good governance, transparency, and accountability in Iraq must not diminish. The State Department and USAID run important programs to build capacity in the Iraqi government at both the national and local levels. At the same time, we should leverage the existence of improved physical security wherever possible, to attract help from other nations, including Arab nations, and international organizations who could lend Iraqis a hand in finding nonviolent solutions to their problems.

The perennial struggle to build a just, comprehensive, and durable peace between Israel and all of its Arab neighbors continues—and that is the good news. The bad news is that sustainable progress seems even more elusive today. That is not for want of a serious effort on the part of the new administration. The appointment of former senator George Mitchell as special envoy for the Middle East went a long way to enhancing the government's credibility. It demonstrated the non-partisanship, seriousness, and patience with which the administration will seek to find and help implement a solution. A consensus exists in support of the administration's having reached out to Syria in new and credible ways. Some believe that a viable agreement between Israel and Syria could occur before the conclusion of an Israeli-Palestinian agreement and the creation of a viable and independent state of Palestine, but neither will take place any time soon. We must dampen expectations on that score.

Finally, America will wrestle with the conundrum of Iran. In June 2009 Iran's flawed and disputed election briefly sent violent tremors through Iran's variegated society. The episode revealed strong undercurrents of political moderation and progressiveness in Iran, a yearning for better relations with the United States and the rest of the world, and, unfortunately, the regime's heavy-handed mechanisms of repression to prevent these from happening. Even before all of this took place, the U.S. administration had, with a few deft strokes, altered the political landscape for trying to shape America's relations with Iran. The Nowruz message to the Iranian people and government, indications of change in policy toward meeting Iranian diplomats, and the extension to Iran of the Obama administration's signature willingness to listen, had given the Iranian government some pause, and generated some initial cautious but open responses to Washington. These were meaningful, but as yet superficial moves.

In the wake of the election fiasco, Washington's public posture has been measured but clear. The United States has yet to announce, however, any new strategic policy toward Iran, preferring to watch and wait. This is appropriate. Nevertheless, it is not likely that the overall goals—especially with respect to independent Iranian control over the nuclear fuel cycle and its pursuit of a nuclear weapons program—will change in any substantial way. Now, even more than before the election, the path to persuading Iran diplomatically to make the best policy choices with regard to a peaceful nuclear energy program lies in unambiguous international unanimity over those choices. Having bared its repressive fangs, the Iranian regime has chosen to distance itself further from an international community heretofore willing to give it a chance. We must be prepared to brandish nonviolent but coercive tools to steer Tehran in the right direction, should it continue to ignore the prescriptions of the international community. There is little we can or should do to affect the internal electoral processes in Iran. However, we can leverage the contradictions that the Iranian government has exposed through its manipulating and repressing a population clearly concerned about further international isolation because of the regime's misguided policies. The U.S. administration's outreach to Europe, China, and Russia will be critical on that score. This will entail deft diplomacy, since Moscow and Beijing will not be swayed solely by the Iranian regime's repressive actions. Furthermore, Iran's behavior makes it harder to acknowledge the possibility of helpful Iranian roles in the region, with respect to Afghanistan for example, as any good negotiating strategy might include. Trying to work constructively with Iran would have other serious challenges. Iranian policies are corrosive and antithetical not just to the United States and Europe, but to some Sunni Arab interests as well—most notably in the Gulf, among the Palestinians, and in Lebanon.

Perhaps it is fortuitous that, in some ways, the aftermath of the Iranian election might be a Pyrrhic victory for the regime. Iran will find it harder to prevent international solidarity opposing its nuclear policies. Iran's resort to repression and intimidation was an act of self-strangulation and Iran's continuing flouting of international will with respect to nuclear energy could merely draw the noose tighter.[9]

All in all, these efforts will require time, work and resources from actors beyond the president, Secretary Clinton, Senator Mitchell, and White House senior Middle East advisor Dennis Ross. American diplomacy and development efforts must be strong and effective, especially to build capacity in governance, security and economics, and to promote broad participation, transparency, and accountability in the region as the foundations of a lasting peace. Building bridges all across the Middle East and especially into Israel and its immediate neighbors will involve all U.S. embassies in the region and their diplomatic counterparts, particularly the EU states, Turkey, Russia, and the nations of the Gulf.

So much for "continuity." Now for the "change" part. America is not prepared to undertake these missions given its current foreign affairs infrastructure, especially in terms of human resources, despite the administration's assertions of a new emphasis on diplomacy and development. We have cut the corps of U.S. Foreign Service diplomats and development officers by an order of magnitude over the past two decades, much of that under pressure from Congress which disparaged the utility of civilian foreign policy functions—diplomacy, public diplomacy and development. When the sea change occurred in our perceptions of our own vulnerabilities in September 2001, and the U.S. scrambled to adjust to the new foreign policy reality, we failed to institute the necessary permanent changes that properly structured policy would require. Former secretary of state Colin Powell's Diplomatic Readiness Initiative (DRI)[10] temporarily succeeded in obtaining congressional funding for an increase of over 1000 new State Department Foreign Service officers, but that much-needed program ended in 2004. The ramp-up in staffing for Iraq and Afghanistan rapidly consumed much of the surplus that the DRI had created, and generated additional requirements for particular skill sets, most notably in public diplomacy and economic development. To fill the gap, the U.S. government resorted to using the military to perform civilian functions, and outsourced the rest to the growing ranks of contractors from the private sector, nongovernmental organizations, and academia.

While this ongoing human resource shortfall affects U.S. policy implementation globally, it is most egregiously true of our ability to carry out the tasks enumerated above in the Greater Middle East. Nowhere do the interlinking of economics, security, and long-term stability apply as much as in this perilous region; and nowhere are these three more fragile. This piece of real estate was the cradle of Western civilization, the bringer of laws, the great monotheistic religions, written language, commerce, the arts, mathematics, and applied science, but today it lags seriously behind the rest of the developed world, and much of the developing world, in all of the above.

Despite recent attitudes within the U.S. government to the contrary, the United States cannot and should not pretend to be the purveyor of all of the above to the region. First, many citizens of the Middle East have rejected the blemishes and failings of our own society—the self-indulgence, the prurience, the wanton violence. They have challenged our claims of superiority as hypocrisy and objected to our interventions as attacks on their sovereignty and independence. Our reputation further suffered as the face America showed to the Islamic world was a military one, decked out in full "battle rattle." Many Americans ignored the absence in many parts of the Arab Middle East of such seventeenth-century Western values as individual rights, the triumph of reason, and natural law that were foundational in the development of republican institutions in North America and Western Europe. Many Middle Eastern societies value instead consensus, tradition, justice, and trust built on personal relationships. Most Americans also were unaware of the legacy of broken promises and unfulfilled national aspirations of peoples of the region at the hands of the European powers following the demise of the Ottoman Empire. So we must tread lightly and with eyes wide open in efforts to help the Greater Middle East regain what it lost over several centuries.

This is not to argue that, in reaction to previous failed efforts to "spread democracy" across the Arab world, the United States should now become aloof, or turn a deaf ear to the desperate socioeconomic needs of the Middle East. To the contrary, we still have a lot to offer. We should recognize that the United States remains very alluring in certain aspects: its economic agility and resilience (particularly in today's global economic doldrums), its appealing managerial and administrative capabilities (despite headline-grabbing critiques of Wall Street's practices and executive compensation), our innovative educational systems and their relevance to developing career opportunities for youth, and our promotion of and respect for the arts and culture.

Fortunately, none of this is new to those foreign policy practitioners who remember America's diplomatic and development policies and practices of days gone by. The heyday of our ability to share what we had and offer to assist others in need occurred in the 1960s and 1970s. Granted, those laudable capabilities existed also side by side with robust intelligence and military activities that many in the Middle East viewed as unsavory or harmful. But America's promotion of the exchange of ideas, free enterprise, and praise for the rule of law and popular participation in government, brought publics throughout the Middle East flocking to U.S. cultural centers, libraries, and commercial offices.[11] The U.S. cultural centers in Damascus and Abu Dhabi attracted significant numbers of young Syrian and Emirati citizens, especially for English language training, exchange programs, and information on application to U.S. universities. Similarly, the Commercial and Agricultural Offices of the U.S. Missions in those countries were popular with local and U.S. businesses interested in expanding trade relations. The subsequent three decades, however, saw the transformation of our military and paramilitary power into a Mr. Hyde–like monster in the eyes of many in the region. And during the exact same period, we smashed the counterpart Dr. Jekyll into pieces, as diplomatic, development, and public diplomacy resources shrank considerably.

As we turn this page in our history, we have an opportunity to resource foreign policy properly, by ramping up and retooling the civilian side of our capabilities, while resetting and reorienting our military power into its appropriate channels. This is not just a desirable change, it is imperative if the administration is to be able to implement its new-found focus on diplomacy and development. It is not sufficient for Congress to budget money for development programs; someone must manage them effectively and responsibly. USAID's cadre of Foreign Service officers is so miniscule, that the agency no longer has the organic expertise in health, education, and agriculture that it once enjoyed.[12] It now relies on the private sector to fill that need without adequate policy and programmatic supervision. Additionally, it is not sufficient to call for an escalation in our diplomatic efforts to prevent, mitigate, and manage conflict throughout the Middle East. We need skilled and experienced diplomats to design, build, and implement those policies. Most of them must have professional-level skills in the languages of the region, especially Arabic and Farsi. These cadres of Foreign Service officers do not grow overnight; in fact, we can borrow the U.S. Special Operations Forces' (SOF) mottoes—the SOF Truths[13]—and apply them directly to Foreign Service personnel as well:

- Humans are more important than hardware.
- Quality is more important than quantity.
- Diplomats cannot be mass produced.
- Competent diplomats cannot be created after the emergency arises.

Last year the American Academy of Diplomacy and the Henry L. Stimson Center collaborated on a report laying out a credible and realizable "strategic and targeted set of staffing and related funding increases in the International Affairs (Function 150) portion of the federal budget."[14] Without such increases, the U.S. government will be unable to execute its foreign policy effectively. Even with immediate passage of the needed authorizations and appropriations, much time and effort will be required to retool and redevelop the experience and expertise that we will need in the State Department and USAID. The administration has in fact requested increased funding for approximately two thousand new personnel in the two agencies over the next few years. We must recruit, vet, hire, and train new officers, as well as establish new human resource policies to develop the expertise in the requisite fields and skills that we will need over the next generation or so.

Beyond hiring new personnel, the United States should adopt a comprehensive, interagency approach to managing conflict, mitigating the risk of violence, and collaborating with other international actors to improve the quality of life and governance in the Middle East. Several proposals are on the drawing board, including the Project for National Security Reform, and, if adopted, could help satisfy some of these requirements. It is critical, however, to identify the skills and capabilities appropriate to address the issues, and assure that our foreign policy apparatus has them, whether the issue is semi-arid agricultural development, critical infrastructure protection, educational reform, alternative energy technology transfer, safeguarding intellectual property rights, land tenure, or access to water. Of course, knowledge of regional languages is vital to establishing meaningful relationships. This is not to argue that our governmental foreign affairs personnel should necessarily be engineers, agronomists, or water treatment specialists who are called on to implement projects. Rather, we should acquire a sufficiently robust organic expertise within our departments and agencies in order to oversee properly the efforts of contractors on the ground, as we have done in decades past. We also must become somewhat more risk-tolerant in sending civilian personnel into the field. Our approach to date has been to eliminate all risk—which normally equates to staying within the confines of our fortress-like buildings in difficult environments or unstable areas. We should consider better ways to manage risk to civilians so that we can do our jobs and achieve our objectives.

As we address some of these human resource issues, a preferred outcome would be an interagency determination of what current and future skills are needed, and where they should reside, rather than depend on stove-piped agencies to make their own determinations and seek duplicative congressional funding, as is often the case today. Management and human resource planning in particular should take place across agencies involved in foreign policy development and implementation. Moreover, if we successfully recruit and hire personnel, our newly bulked-up cadres of skilled international affairs officers across many agencies must then receive the education and training to perform effectively in the *comprehensive* approach to foreign policy issues. These educational opportunities should take place together with private sector actors—both for-profit and not-for-profit—upon whom we will and should continue to rely to implement complex programs and projects. Many of the current arguments among public and private sector actors operating in complex contingency environments—like Iraq—involve absence of trust, logistical conflicts, and lack of information sharing that we must resolve through common education and training programs.

In sum, we must continue our intense pursuit of peaceful resolution of the critical, "front-page" issues that dominate the Middle East, especially the Israeli-Palestinian dispute, Iraq, and Iran. However, we must also recognize that viewing the region only as a function of these specific issues, and concentrating our scarce resources only on them, will ignore the many sources and drivers of conflict there. A "crisis-focused" approach overlooks the broader impediments to economic, social, and political development that will sustain stability, growth, and global engagement for the region. The best way for the United States to contribute to this latter vision is to recruit, hire, and train the experts needed to perform inherently governmental functions in foreign affairs; to re-create the ground-level, person-to-person programs that enhance America's credibility; and to integrate all instruments of national power in effective pursuit of the U.S. national interest in the Greater Middle East, which include stability, economic interdependence, and good governance.

Notes

1. Defined as Egypt, Israel, Jordan, Syria, Lebanon, Turkey, Saudi Arabia, Kuwait, Bahrain, Qatar, UAE, Oman, Yemen, Iraq, and Iran.
2. Murrow actually said, "To be persuasive we must be believable; to be believable we must be credible; credible we must be truthful."
3. Typified by the September 2007 incident involving the former Blackwater security company in Baghdad's Nusoor Square.
4. The author personally was aware of such incidents, and the difficulties they posed for U.S. policy in Iraq.
5. See President Obama's January 27, 2009, interview with Hisham Melhem on Al-Arabiyah television network, and his address to the Turkish Parliament, April 6, 2009. Since the Ruger Workshop, he has delivered a major address to the Muslim world in Egypt on June 4, 2009.
6. According to the author's personal recollections of discussions with Arab leaders and U.S. ambassadors in the region, there has been deep frustration with the tendency of senior Washington-based U.S. officials to hold meetings with Gulf heads of state, sometimes in several different countries in one day. This was especially irritating when these U.S. officials arrived in country, purporting to seek the view of the country in question, but instead unfolded a U.S. strategy, in which the country was expected to play a specific part, usually giving money, without listening in advance to the head of state's opinion. Our official then flew off to the next country, usually declining the request to stay overnight or some other demonstration of hospitality. The message to the Gulf leaderships was clear: this relationship is purely transactional, not strategic, enduring, or involving strong personal ties. We then wondered why we failed to achieve our objective.
7. Released by the White House March 19, 2009.
8. See in particular Secretary Clinton's interview with NBC's Andrea Mitchell on March 31, 2009, following Ambassador Holbrooke's casual meeting that day with Iranian senior diplomat Mehdi Akhunzadeh in The Hague.
9. The paragraphs on Iran were updated after the Workshop, in light of the dramatic events following the Iranian elections of June 12, 2009.
10. From the State Department's Web site (www.state.gov), "The Department's Diplomatic Readiness Initiative represents a new strategic human capital plan designed to ensure the availability of adequate human resources through a more streamlined and aggressive hiring process from recruitment to intake."
11. The most articulate case for enhancing U.S. public diplomacy in the Middle East remains Advisory Group on Public Diplomacy for the Arab and Muslim World, *Changing Minds, Winning Peace: A New Strategic Direction for U.S. Public Diplomacy in the Arab & Muslim World*, report prepared for the House Committee on Appropriations, 108th Cong., 1st sess., 2003.
12. American Academy of Diplomacy and the Henry L. Stimson Center, *A Foreign Affairs Budget for the Future: Fixing the Crisis in Diplomatic Readiness,* October 2008, pp. 18–19.

13. See U.S. Army Special Operations Command, http://www.soc.mil.
14. *A Foreign Affairs Budget for the Future.*

Engaging the Muslim World beyond al Qaeda

Dr. Marc Lynch
Associate Professor of
Political Science and International Affairs
The Elliot School of International Affairs
George Washington University

> The United States is not, and will never be, at war with Islam. . . . In fact, our partnership with the Muslim world is critical not just in rolling back the violent ideologies that people of all faiths reject, but also to strengthen opportunity for all its people.
>
> I also want to be clear that America's relationship with the Muslim community, the Muslim world, cannot, and will not, just be based upon opposition to terrorism. We seek broader engagement based on mutual interest and mutual respect.
>
> —Barack Obama, address to Turkish Parliament, April 6, 2009[1]

President Obama's call in Turkey for an engagement with the Muslim world beyond al Qaeda offers the prospect for a dramatic and long-overdue shift in the American approach to the "war of ideas" that since 9/11 has occupied a central place in American national security policy. Over the last seven years, a wide consensus has emerged about the importance of such a "war of ideas." Stunned by the catastrophic decline in international approval of the United States in a series of global public opinion surveys, an Advisory Commission on Public Diplomacy's 2003 report urged a dramatic increase in U.S. engagement with the world. The 9/11 Commission report defined the "war of ideas" as a generational challenge facing the United States. The Bush administration's 2006 National Counterterrorism Strategy declared that "[i]n the long run, winning the War on Terror means winning the battle of ideas." This rhetorical consensus has been met with a commensurate commitment of resources, at least on the military side. But the nature, objective, and appropriate means of this "war of ideas" remain ill defined and poorly conceptualized.

After an initial period focused on marketing American values and trying to improve American favorability ratings in public opinion surveys, the Bush administration came to define the "war of ideas" primarily in terms of counterterrorism, counter-radicalization, and "combating violent extremism." While this does indeed represent an important component of any serious approach to the problem of al Qaeda, it represents a dangerously narrow focus for America's engagement with the Islamic world. The focus on violent extremism as the primary mission of engagement privileges and reinforces al Qaeda's conception of the nature of the confrontation, ironically at a time when al Qaeda is weaker than it has ever been as a political force in the Arab world. This requires a move beyond the "counter-radicalization" and "violent extremism" formulation. The United States should not be in a tacit dialogue with al Qaeda on its own terms, entertaining its fantasies of a global caliphate or offering any sustenance to its conceptions of an essential clash of civilizations between the West and Islam. Al Qaeda should be marginalized, recognized for the radical fringe movement that it is, and not allowed to dominate our vital dialogue with the mainstream of the Arab and Muslim worlds.

Al Qaeda is relatively weak today, at least in the Arab world. But the spirit of "resistance" to American hegemony—a mass-based, political resistance rather than a fringe, religious radicalism—is strong and rising. The focus of U.S. engagement with the Muslim world must be to reframe and transcend the binary oppositions that fuel the appeal of the advocates of resistance. That means focusing far less upon al Qaeda or upon grand ideological rhetoric, and more on practical issues related to core diplomatic agendas: building broad support for American foreign policy goals, establishing long-term foundations of trust and mutual respect, supporting engagement with potential adversaries, and addressing the political issues that provide the sustenance for the rhetoric of resistance.

Al Qaeda thrives only when it can hijack other, more popular projects—the popular vehicles of Arab and Muslim anger and concern such as the Palestinian issue, Iraq, or fury with Arab authoritarianism or American hegemony. Iran today is attempting the same gambit, appropriating more popular issues and seeking to be the standard-bearer of resistance. The old-school version of the war of ideas, which lumped together all forms of Islamic activism under one global jihadist label, served to reinforce and strengthen that narrative. So does the current effort in some quarters to present Iran as the epicenter of radicalism in the region, with Hamas and Hezbollah simply proxies for Iranian malevolence. If the key for American strategy against al Qaeda must be to expose it for the marginal, radical fringe that it is, the key against a popular, politically oriented, mass-based sense of resistance must be to address those political issues that give its appeal resonance. But in both cases, the key is to disaggregate rather than to aggregate, to split the problem rather than lump it together into a single threat, and to deny the adversary the advantage of being viewed as the primary alternative to the United States.

This does not mean entirely a return to traditional "public diplomacy." The global media landscape and information environment have radically changed, and America's engagement with the world must change to reflect this. Instead, engagement must take the form of an ongoing dialogue across multiple levels—what James Glassman has called a "great conversation"—with insights gleaned from this conversation integrated directly into the policy formation process at the conceptualization, not implementation, phase. American engagement with the Islamic world cannot fall into the "marketing trap," attempting to sell policies formed in splendid isolation from the Islamic world's interests and concerns.

Conceptualizing the "War of Ideas"

It is now a cliché to say that the most important war of ideas isn't about us, but rather a war inside the Muslim world. But the United States has generally not been able or willing to internalize the implications of this fully. Attitudes toward al Qaeda and toward the United States are only very weakly related to one another, and it is entirely possible for al Qaeda to lose without America winning. If the war of ideas is taking place within the Muslim world—and it is—then what is the appropriate role for America?

This question should take into account the reality that the more intensely that the United States is involved, the worse the outcome is likely to be. This is because of the deep-rooted fears in the Islamic world of American hegemony and intrusion, and the much-mentioned but nevertheless real "kiss of death" (which can be lessened when America pursues a more respectful and effective foreign policy than that on display in the last eight years, but is unlikely ever to go away given the realities of power and vulnerability in today's international system). It is also because America's involvement inevitably turns the struggle into one between the United States and Islam—precisely the "clash of civilizations" narrative preferred by al Qaeda and that American public diplomacy should be striving to undermine.

There are at least three different conceptions of the "war of ideas," all of which move beyond the traditional conception of public diplomacy as explaining American policy and values to foreign publics.

The first, which emerged powerfully in the later portion of the Bush administration, is rooted in counterterrorism, and involves a narrow campaign to marginalize al Qaeda and delegitimize violent extremism. As the most influential proponent of this perspective, former under secretary of state for public diplomacy James Glassman, put it,

> While educational exchanges and other such efforts seek over the long term to encourage foreigners to adopt more generally favorable views of the United States, the war of ideas today should have a different, specific focus. The aim must be to ensure that negative sentiments and day-to-day grievances toward the U.S. and its allies do not manifest themselves in violence. We want to create an environment hostile to violent extremism, especially by severing links between al Qaeda and like-minded groups and their target audiences.[2]

Glassman is very clear about the difference between the "war of ideas" and traditional public diplomacy: "the aim of the war of ideas is *not* to persuade foreign populations to adopt more favorable views of the United States and its policies. Instead, the war of ideas tries to ensure that negative sentiments and day-to-day grievances toward the United States and its allies do not manifest themselves in the form of violent extremism."[3] This is the appropriate domain of "strategic communications," but should be only a minor, secondary sliver of the larger portfolio—because, most fundamentally, al Qaeda could be completely defeated without the United States winning in the areas that matter. Below, I argue that this campaign is going rather well, only partly because of anything the United States has done.

The second is the vastly ambitious campaign to spread liberal values through the Islamic world, bringing about fundamental changes in Arab and Muslim political cultures and promoting Western civilization. While I am broadly sympathetic to the aspirations of many advocates of this campaign, I also believe that it is well beyond the capabilities of the U.S. government, can only be conceived as a long-term campaign rather than a short-term political campaign, and must be carefully circumscribed to avoid triggering defensive reactions. The United States, whether the Pentagon or the State Department, is almost uniquely poorly positioned to "reform Islam," to "promote moderate readings of the Qur'an," to "combat *salafi* interpretations of Islamic tradition," or any of the other ideas often on offer in the "war of ideas" industry—and trying to do so is likely to discredit the approved carriers of the message and to ignite fierce opposition. At any rate, the foundations upon which to build such a strategy are far too weak today to be effective—the less than 10 percent who in surveys say that sharia should not play a role in law, the "secular Muslims" featured in many conferences—and are a fool's errand for any concrete strategic policy. Rather than try to choose sides aggressively in intra-Muslim debates, this level should involve a long-term campaign aimed at building foundations of civil society, tolerance, and public freedoms—not as part of the war of ideas or war on terror, but for its own sake. It is not clear that government should take the lead on this, and certainly not the Pentagon. The strategic imperative should be to ensure that this is not perceived as a war on Islam, and that it be viewed as a partnership with Islam to strengthen and support the aspirations of Muslims and Arabs.

Third, and by far the most important, is broad-based engagement across the mainstream of Arab and Muslim political societies with the goal of explaining American policies, building support where possible while building networks and relationships of mutual respect. This should involve sustained and productive dialogue with those with whom we disagree, whether states or publics, secularists or Islamists. It also requires serious changes in policy to give substance to the dialogues—full engagement on the Israeli-Palestinian conflict, careful management of the withdrawal from Iraq, and openings to Syria, Iran, and others. As discussed below, the Obama administration has made a strong beginning, but will require the full-scale support of the bureaucracies for what has thus far been primarily a White House initiative. The State Department and not the Pentagon should have the lead, while the National Security Council should play a strong role in coordinating and guiding the activities of both to ensure that the resource advantage inevitably enjoyed by the Defense Department does not warp the mission unintentionally.

The Challenge: From Radicalism to Resistance

The dominant approach to the war of ideas after 9/11 took the form of what I would call "lumping": conceptualizing the Islamist threat as a single, undifferentiated challenge in which internal divisions are primarily over tactics. The lumpers tend to be attracted to cultural explanations for "why they hate us," and to see all Islamist groups, whatever their surface differences, as engaged in a single jihad. "The crux of the debate between al Qaeda and the Muslim Brotherhood [MB] is not over the ends," such analysts argue, "but rather the means by which to realize the greater goal of Islamic governance throughout the Muslim world."[4] Even differences as obvious as Sunni-Shia divisions pale for the more extreme lumpers, for whom Islamist extremism is its own sufficient category. For lumpers, the appropriate response is a full-bore confrontation with Islamism in all of its manifestations. The advantage of the lumping approach is that it focuses attention on the very real sharp antagonism between American foreign policy and most Islamist movements, and the essential shared attitudes of many of them. The disadvantage is that it squanders opportunities to divide and conquer, obscures very real and crucially significant differences among them, magnifies the challenges, and unintentionally strengthens the hand of our most radical adversaries. Liberal or "secular" trends in the Arab and Islamic world are exceedingly weak; what sense does it make to build a grand strategy on a base that might be measured in public opinion surveys as a residual category?

The second approach, which has gained increasing acceptance over the last few years, might be called "splitting"—seeing the internal differences among Islamist groups as extremely significant both analytically and politically. In this view, the differences among Islamist groups go beyond tactics, and extend to fundamentally different approaches to politics, society, and the use of violence. The advantage of the splitting approach is that it undermines al Qaeda's polarizing worldview, takes advantage of the intense internal conflicts between al Qaeda and its Islamist adversaries, and opens up new opportunities for policy. Hamas and Hezbollah are problems, but they are very different problems from al Qaeda and are not reducible to Iranian interests—and it does no good to group them together under a common label.

Indeed, taking seriously the cliché that the real battle is inside of the Islamic world virtually requires a "splitters" approach. Indeed, even lumpers claim to be splitters—they just draw the lines on one extreme, and view groups like the MB as part of the problem ("nonviolent extremists") rather than as part of the solution. There is no question about the Brotherhood's deep conservatism and commitment to Islamizing the public realm, as well as its enthusiastic support for Hamas and for the insurgency in Iraq. But for splitters, a wide range of conflicts capture attention—the MB's embrace of democratic participation, its rejection of violence against civilians in the West, and its ideological rejection of al Qaeda's extreme brand of salafi jihadism. The disadvantage is that it may lead the United States to support groups that could prove problematic in the future, or strengthen cultural or political trends that conflict with American liberal values.

Generally speaking, the shift from "lumping" to "splitting" within the U.S. government probably roughly coincided with the experience in Iraq, where an undifferentiated "al Qaeda" enemy gave way to a much more nuanced understanding of the internal competition and ideological differences among insurgency factions and tribal groupings that facilitated the "Awakenings" strategy, turning "former" insurgents against the hardest-line al Qaeda factions. The Iraq experience showed graphically the tactical value of careful exploration of the lines of division within Islamist movements. Years of undifferentiated warfare against an insurgency seen as monolithic and infused with radical extremist ideology only strengthened that insurgency, while the decision to work with the "Awakenings" and to cooperate with "former" insurgents proved far more effective (at least in the short run). The same logic could be applied to the "war of ideas": differentiating carefully among different groups, splitting where possible, and forming tactical alliances to marginalize and defeat the most dangerous adversaries. If such an approach worked in Iraq, why not try it not only in Afghanistan but in the Islamic world as a whole?

Al Qaeda: The Case for Its Decline

While al Qaeda appropriately remains a primary focus of counterterrorism, and remains strong in some ways, as a political force in the Arab world it has lost considerable ground over the last few years. This is in large part due to its declining ability to claim effectively the mantle of generic resistance to the United States or to promote its unitary, clash-of-civilizations narrative.

Al Qaeda is still an active and dangerous organization, capable of doing harm. But in the Arab world, at least, it is virtually unrecognizable from its post-9/11 profile. It has become more South Asian in its orientation, which for Arab audiences might as well be on the moon. It has become more ideologically pure, at the expense of its mass appeal, while its "near enemy" attacks have mobilized outrage in virtually every instance. Its media units produce more and more material, to less and less effect, while the fragmentation and competition of the Arab mass media deny it access to the kind of unified public that it seized after 9/11. Iraq has gone from an unmitigated blessing to a serious problem, with the backlash against its attacks on Shia civilians and the confusion sown by its falling-out with other Sunni insurgency factions and tribal groups. And finally, it has found itself on the wrong side of virtually every major issue for Arab publics in the last few years—with its anti-Shiism putting it in stark opposition to those who admire Iran and Hezbollah, and its feuds with Hamas and the Muslim Brotherhood putting it at odds with virtually all of mainstream Arab public opinion.

While U.S. strategic communications efforts may have helped along these trends, for the most part they were the product of the Arab world's own internal dynamics—as Arab regimes such as the Saudis and popular Arab movements like the Muslim Brotherhood and Hamas turned against al Qaeda out of their own self-interest. Al Qaeda thrived after 9/11 by hijacking the popular issues about which mass Arab publics cared—Palestine, Iraq, Arab despotism—but over the last few years has lost its ability to claim this mantle of generalized resistance. The key for American strategy against al Qaeda must be to expose it for the marginal, radical fringe that it is—while demonstrating, as we did in Iraq, that the United States is able and willing to work respectfully with less radical forces even if they are Islamists.

The Muslim Brotherhood Question

The Muslim Brotherhood, the largest and most influential mass-based Islamist movement in the Arab world, poses a unique challenge to efforts to combat al Qaeda and like-minded groups. It is one of the key sources of Islamist thought and political activism, with organizations in almost every country in the world and a sophisticated political and social infrastructure.[5] It plays a crucial role in promoting Islamic consciousness and organizing political activism in a wide range of countries, particularly in the Arabic-speaking world. At the same time, the MB has consistently denounced al Qaeda's ideology and terrorist activities, and offers a significantly different vision of an Islamic state from that favored by salafi-jihadist groups. As an Islamist movement with global reach and a message that resonates widely with Arab publics, the MB represents the strongest challenger to al Qaeda and like-minded groups within Islamist politics. Its leaders speak the language of democracy, reject extremism and *takfir*, and advocate peaceful political participation, yet the MB remains deeply committed to spreading a conservative vision of Islamic society and its cadres are deeply hostile to Israel and to American foreign policy.

The long-latent conflict between the MB and al Qaeda has emerged over the last few years as a central cleavage in Islamist politics, driven by intense disagreements over Iraq, Palestine, the Shia question, and the legitimacy of participation in democratic elections. Al Qaeda leaders from Ayman al-Zawahiri to Abu Omar al-Baghdadi and Abu Hamza al-Muhajjir have sharpened their long-standing animus against the MB into a more global critique. In a series of tapes and writings, al-Zawahiri savaged Hamas and the Egyptian MB for their participation in elections and public life. Al-Baghdadi and al-Muhajjir identified the MB as the driving force behind the setbacks of the jihad in Iraq, pointing not only to the Iraqi Islamic Party but also to a wide range of other Islamist adversaries lumped together into the MB label.

The MB therefore poses the question very directly: should the United States seek to take advantage of this intra-Islamist conflict, or should it seek to combat all forms of Islamism? If the goal is marginalizing al Qaeda, then the MB is a tacit ally as it wages its ideological and organizational battles. If the goal is to promote secularism and liberalism, then the MB is a tenacious adversary as it seeks to promote cultural conservatism. And if the goal is to transcend the rhetoric of "resistance," then the MB should be ground zero of the campaign to persuade, engage, and compete.

"Resistance": The Real Challenge

The shift to "resistance" and the Muslim Brotherhood's pivotal role both point to the new challenges the Obama administration confronts. Al Qaeda's decline doesn't mean that support for American foreign policy is rising. It has always been the case that al Qaeda can lose without the United States "winning" with the mainstream publics that most matter. Indeed, despite some optimism over Obama's election and appreciation of his outreach efforts, thanks in large part to Israel's recent war with Gaza the spirit of "resistance" is strong and rising. Responding effectively to that requires different tools and conceptual frameworks from those that were appropriate for the struggle against al Qaeda.

Al Qaeda offered a radical religious ideology that sought to hijack popular political issues to broaden its appeal, and primarily drew upon a small, marginal fringe of Arab and Muslim societies. It had no political demands that could be addressed. But today's discourse of resistance is mass based rather than concentrated in a small radicalized fringe, and is fundamentally political rather than religious. Anger over the invasion of Iraq, outrage over Guantánamo, and outrage over Israel were widespread, majority attitudes with mass publics. That means a political response, not a response focused on delegitimizing violent extremism, and a public diplomacy oriented toward mass publics rather than strategic communications oriented toward a concentrated, marginal niche.

The focus of our engagement with the Muslim world must be to reframe and transcend the binary oppositions that fuel the appeal of the advocates of resistance. That engagement must be oriented not toward "counter-radicalization" but toward public arguments about the political issues about which mass publics and elites care—whether through traditional means such as broadcasting and appearances on satellite television or massively ramped-up exchange programs, or through Internet-based new media technologies. Either way, such a "great conversation" will have to tackle head-on the major political issues—above all, the Israeli-Palestinian conflict, where the Obama administration should be and is taking a very public lead role in pushing for a just two-state solution—and cannot simply be about combating "radicalism" or even about "Islam."

America's public engagement in this environment should—and will—focus less upon al Qaeda and more on building broad support for American foreign policy goals, establishing long-term foundations of trust and mutual respect, supporting engagement with potential adversaries, and moving beyond the counterproductive binary oppositions and threat inflation that have blocked progress for so many years.

Framing the region's politics as a binary choice between Israel and the United States versus Iran would repeat the mistakes of the early post-9/11 years—inflating the strength of the adversary and choosing unnecessary battles by making it about "us." The lessons of the last few years should be that the better approach is to take away the appeal of "resistance" by reframing the confrontation, disaggregating the challenge, and dealing pragmatically with the political issues rather than engaging in rhetorical wars of ideas. Riding the tiger of anti-Iranian sentiment would be counterproductive—increasing rather than decreasing Iran's appeal in the region, strengthening its most repressive and autocratic forces, and sharply conflicting with President Obama's vision for the region.

How can Obama defuse the "resistance" discourse and genuinely transform the political contours of America's engagement with the Middle East? Obama's election and successful early outreach—including his much-anticipated speech to the Islamic world scheduled for June 5 in Cairo—helps, but only so much. To change attitudes will require a new form of engagement that adopts a genuinely different approach, as outlined in this paper. And above all, it requires serious changes in policy to give substance to the dialogues. It is here where the Obama administration has made a strong beginning: high-level engagement on the Israeli-Palestinian peace process, announcement of a withdrawal from Iraq, direct outreach to the Islamic world, the closure of Guantánamo and renunciation of torture. The new approach to engagement and public diplomacy should be directly and forcefully integrated into the policy process to ensure that they work together to the vital strategic goal of reducing the political appeal of the "resistance" discourse with Arab and Muslim publics and offering a constructive, positive alternative vision of partnership.

Notes

1. "Remarks by President Obama to the Turkish Parliament" (Ankara, Turkey, April 6, 2009), available at http://www.whitehouse.gov/the_press_office/Remarks-By-President-Obama-To-The-Turkish-Parliament/.
2. James K. Glassman, "How to Win the War of Ideas," *Wall Street Journal,* June 24, 2008, p. A19.
3. James Glassman, "Winning the War of Ideas" (speech, Washington Institute for Near East Forum Special Policy Forum, July 8, 2008).
4. Lydia Khalil, "Al-Qaeda & the Muslim Brotherhood: United by Strategy, Divided by Tactics," *Terrorism Monitor* 4, no. 6 (March 23, 2006), available at http://www.jamestown.org/single/?no_cache=1&tx_ttnews[tt_news]=714.
5. Mona El-Ghobashy, "The Metamorphosis of the Egyptian Muslim Brothers," *International Journal of Middle East Studies* 37 (2005), pp. 373–395; Carrie Rosefsky Wickham, *Mobilizing Islam* (Columbia University Press, 2001); Bruce Rutherford, *After Mubarak* (Princeton: 2008); Marc Lynch, *The Brotherhood's Dilemma,* Middle East Brief 25 (Crown Center for Middle East Studies, Brandeis University, January 2008).

Why the United States Should Engage Islamists

Dr. Heidi E. Lane
Associate Professor of Strategy and Policy
Greater Middle East Area of Study Coordinator
Naval War College

It is time for the United States to launch an initiative aimed at direct and focused engagement with Islamist organizations. Although such a proposition may seem counterintuitive and at odds with some aspects of our national security policies, opening channels of dialogue with Islamist organizations would have far-reaching benefits for U.S. objectives in the Middle East region and would, therefore, contribute to our larger strategic vision. First, engagement would increase our knowledge about nonstate actors and how they function and survive under hostile political conditions. It would also increase our understanding of the nexus among nationalism, ideology, and religious belonging. As the well-respected Jordanian journalist and commentator Rami Khouri has pointed out, "The United States does not have a problem with Islam other than it does not understand how religion, nationalism, politics, and resistance fit into each other."[1] Finally, engagement with Islamists could begin to chip away at the pervasive sentiment in the region that U.S. objectives are nothing more than cultural hegemony, dominance over natural resources, and the weakening of Islam itself. The United States should also engage Islamist organizations because the collective social and political capital that Islamists now possess in Arab and Muslim-majority nations has grown steadily over the past twenty years and these organizations will have a vote in the future of the region. Both radical and moderate Islamists are now considered by many in their target publics as a viable alternative (even if not an ideal one) to ineffectual and repressive governments and aging autocrats. Some Islamist organizations have gained immense popularity because of charismatic and populist leaders, while others are known for providing social services to citizens where local governments have repeatedly fallen short. Many of these organizations have also engaged in different forms of violence and coercion and have, in a few cases, directly threatened incumbent regimes or preyed on the populace in failing or failed states. The United States should engage both the radical and the moderate among these organizations with an eye toward defanging the most violent and empowering the most moderate among them. But we should also maintain reasonable expectations. Islamist organizations, regardless of type, will not magically transform themselves into the moderate democrats of the future nor will they allow themselves to be transformed by outside forces. These organizations and their leaders may or may not respond to a new U.S.-led initiative, but the United States must be proactive and committed to setting in motion new and innovative standards of political dialogue and public diplomacy if it wishes to have a constructive role in helping shape the future of the Middle East.

Five Good Reasons Why the Time Is Now

There are five good reasons that this is now a propitious time to pursue a new type of engagement with Islamists. These reasons emanate as much from the present political and social environment in the United States as they do from sociopolitical conditions in the Middle East. The first reason is that the entrance of the new American administration onto the world stage has produced both real and imagined "windows of opportunity" for reshaping the image of the United States, especially in areas of "soft power" such as public diplomacy. Both at home and abroad, there exists a hopeful expectation that the new American administration will make a clear break with the policies carried out under the former administration. This hope, however, will not last indefinitely. There are many who believe that the Obama administration represents a political sea change—the chance to build on this positive global public mood must not be squandered. This is not to say that the United States should expose itself to unnecessary risk, but if it were to lead the way to a new type of engagement with Islamist organizations, this would be a brave and valuable first move on the part of a powerful nation. It would signal to both our enemies and our friends that the old era is over and that the United States does not fear leaving it behind in favor of an innovative and forward-looking strategy. The first step the United States can make in this direction is to accept the fact that Islamists matter and that Islamism is the dominant political ideology of our day and will be part of the political fabric of the region for the foreseeable future. Simple and direct acknowledgment of nonstate actors as an integral part of the political culture of the region would be an important and meaningful first step. And on such a first step, a more concrete path to engagement could be built.

The second reason that we should engage Islamist organizations stems from the fact that the American public is better prepared to take this step than ever before. Americans may never have been more informed about the intricacies of a foreign culture and its internal political and religious divisions than they are today about "political Islam." The American public opinion has arguably digested more information about the Middle East, Islam, and Islamic extremism than about any other single foreign policy issue since President Franklin D. Roosevelt embarked on a "strategic communication" campaign to inform and prepare the U.S. public for entry into WWII. Though the relationship between the U.S. government and its citizens is fundamentally different than it was during WWII, both the current and former U.S. administrations have been instrumental in attempting to remind and educate Americans about the gray areas in the debate over religious extremism and its relationship to the fundamentals of Islam. The American public has struggled mightily (along with our government) to grasp what separates those who advocate political Islam from those who advocate terrorism, where the term "moderate" begins and ends, and how societal ills such as poverty, lack of political rights, and failed states are believed to contribute directly to the growth of violence carried out under the banner of religious extremism. Our collective public discourse is now peppered with terms like "salafi," "jihadi," and "fatwa"—and other complex concepts that prior to 9/11 would have only interested a small group of regional specialists. The result of this "baptism by fire" is that Americans have now come to an uneasy understanding of "why they hate us" while at the same time slowly accepting, if not embracing, the mantra that "the United States is not against Islam." As the policies of the "war on terror" are dismantled and recast in new terms, it should be remembered that our government will likely remain committed to the same basic and shared objectives: protecting the homeland, precluding the growth of extremism, and assisting other nations in ways that promote regional and global security. Though Americans have neither "forgiven" nor "forgotten" the trauma of being attacked on 9/11, the lessons of this trauma have encouraged a new political maturity that will prove to be a valuable asset for embarking on the next logical step of engagement with Islamists. As Fareed Zakaria recently argued, we must learn to live with some aspects of radical Islam.[2] In many respects, Americans are well on their way to deciding for themselves which aspects they can live with and which they cannot. Now the U.S. government should take the next step, which includes direct engagement with Islamist organizations. These organizations represent a disproportionately influential (albeit numerically small) percentage of the populace in the Middle East, but their collective political attitudes, social preferences, and religiously grounded ideologies mirror the larger majority populace. One important path to reaching the majority is through the Islamist minority.

The third reason that the United States should engage Islamists has to do with the political attitudes and perceptions of those who make up the proverbial "Arab street." President Obama and the new administration have already made a valuable initial investment in recasting America's image in the region. This was evident in the president's June 2009 speech in Cairo, but such efforts must be ongoing if they are ultimately to make a difference. The United States could continue this positive trend by publicly addressing the failure of its democratization policies over the past decade. It is abundantly clear that the "Freedom Agenda" did not produce the outcomes it was designed to procure. The dominance of counterterrorism as a component of the Freedom Agenda inadvertently encouraged some governments in the region to reconsider (and sometimes renege) on prior commitments toward political and economic liberalization. Instead of democratizing, many regional governments, including those that had previously shown the greatest willingness to support liberalization, began to develop allergic responses to what they perceived as unsustainable pressures brought to bear on them by divergent U.S. policies. This pressure was intensified and exacerbated by the growth and popularity of Islamist organizations that these states feared would become the next Algerian FIS (Islamic Salvation Front) or Palestinian Hamas. The regrettable outcome for the United States is that the policies associated with the war on terror paradoxically undermined those of the Freedom Agenda. Dissonance and conflicting priorities emanating from the United States therefore severely impaired U.S. credibility in two important audiences. First, it damaged the relationship between the United States and regional governments who rely heavily on a complex domestic security apparatus for state survival. Second, it further alienated those at the level of civil society where Islamist organizations have gained a firm and popular foothold.[3] In this respect, both regional governments and their disenfranchised constituencies developed a strong mistrust of the spirit and actions of the United States on issues of democratization and reform. Indeed, in the most recent Annual Arab Public Opinion Survey undertaken by the University of Maryland with Zogby International, 46 percent of those polled reported that their view of the United States is "very unfavorable."[4] Unpacking why views were so unfavorable, the survey found that respondents report that "protecting Israel" and "controlling oil" were what most believed motivated U.S. policies in the region.[5] Although the above results are an improvement over those of 2008, when the percentage of respondents who viewed the United States as "unfavorable" was a full 64 percent,[6] U.S. policy in the region continues to be seen, at best, as an insincere gesture on the part of a superpower. At worst, U.S. policy is viewed as part of an insidious plan to remake the region in our own self-indulgent image. Alleviating such a high degree of mistrust may be a long process, but that process should begin as soon as possible. The United States should begin drawing a clear line between those policies that are intended to address terrorism and insurgent groups and those that are intended to facilitate democratic values and liberalization more broadly within local governments and civil society organizations. Our engagement efforts must be at once comforting to governments who feel more vulnerable than ever and challenging to Islamist organizations who have grown more emboldened in recent years. The way to challenge Islamist organizations is to impress upon their supporters that the United States is open for new ways of doing business.

The fourth reason that the United States should engage Islamists is because the rapid growth and evolution of Arab satellite television now offer a public and transparent medium that is viewed as increasingly credible by its broad and diverse consumer base. Any program of public diplomacy on the part of the United States must take into account this meaningful change in the way that information is disseminated in the region. Many have called the boom of satellite television and other new media a true information revolution. President Obama chose to take a bold step by granting his first public interview to the Arabic satellite network al-Arabiya.[7] By doing so, he spoke directly to inhabitants in the region through a medium that the people of the region trust as a news source. Not surprisingly (and although many critics argued that al-Jazeera would have been the best venue for this interview) President Obama's interview was widely viewed in the region as a public relations success. Such steps are meaningful when made at the right time and place. According to recent polling results, 64 percent of respondents watch their international news on either al-Jazeera or al-Arabiya networks.[8] This new media realm is both responsive to and driven by an ever-growing and increasingly sophisticated consumer base whose appetite for more rather than less news is, as yet, insatiable. If the United States were to open up avenues of dialogue with Islamists through this medium, our efforts would reach a critical audience of hundreds of thousands of viewers overnight.[9] And, via the medium of satellite television, our actions would be amplified, hotly debated, and fully dissected by talk shows, pundits, and the viewing public. Even if Islamist organizations rejected our initiatives for engagement, their actions and reactions to a new U.S. policy directed at the Middle East would be forced into the public eye, where they would be debated, dissected, and evaluated by both supporters and detractors.

The fifth and final reason that the United States should engage Islamists is because both moderate and radical Islamist organizations have begun to show signs of failure. Moderates have seldom succeeded in breaking into formal political circles through "playing by the rules" of their respective governments. Known as "participating Islamists,"[10] these moderate Islamist organizations have found that their progress has instead been regularly blocked through clever manipulation of the existing political system. This has meant that the costs of Islamist participation in a manipulative system have often exceeded the benefits—Islamist organizations cannot boast more than cosmetic changes to government process or miniscule (and often reversible) gains in political access as a means of increasing their own constituencies. Instead, ruling governments have been especially adept at managing and mitigating the threats that they perceive emanate from an Islamist opposition. For the most liberally oriented Islamist organizations, this means that they lose disenchanted members of their prospective support bases to other, often more radical organizations. Especially bitter for democratically oriented Islamists (and other secular and social reform movements in the region) is that survey data indicate that large percentages of the Arab public are supportive of the idea and ideals of democracy.[11] Continued political alienation therefore continues to feed the impoverished state of political pluralism in the Middle East, both within and outside formal political channels.

On the other side of the spectrum are radical Islamist organizations. Contrary to popular myth and propaganda disseminated widely by regional governments themselves, most of these groups do not possess the popular support to lead a revolution against or undertake the violent overthrow of their respective regimes. Most would be unprepared to govern even if they could somehow grab hold of the reins of power. This problem was evidenced by Hamas's electoral victory in 2006 and the organization's subsequent inability to govern effectively. Many sympathetic observers would argue that Hamas's failure has been caused by a host of other issues associated with Israeli occupation and international rejection of their legitimacy, but Palestinians themselves now increasingly question whether Hamas would have succeeded even in a permissive environment. Since January 2009, Hamas has suffered a decline in popularity. Recent polls show that the percentage of Palestinians who would vote for Hamas in an election has declined from 27.7 percent to 18.8 percent.[12] Since hard-line Israeli policies that have previously guaranteed Hamas an extra amount of sympathy are as strong as ever, Hamas's loss of popular support is an even greater signal that Palestinians have begun to question Hamas's ability to act as a governing body.

That leaves those Islamists and their supporters who are somewhere in the middle between participating and excluded, moderate and radical, manageable and unmanageable. It is precisely this opaque center that makes up a growing and dynamic, but unknown, percentage of Islamist activism. Moreover, this is the aspect of Islamist politics that has proved to be most transnational and, therefore, unfettered by the institutional norms of its more veteran predecessors such as those of the Muslim Brotherhood. These cross-generational and technologically sophisticated Islamists are mobile and unmotivated by typical nationalist sentiments. They guarantee themselves popular support in places where government services, public order and security, and other basic necessities are denied or become contested. It is in precisely these environments, such as Iraq and Pakistan, where Islamist successes are based on one part "popularity" and one part coercion and fear. In these "tipping states" a proactive program of public diplomacy by the United States may prove a most valuable investment. This will not be the case because such groups will be inclined toward "dialogue" with the United States, but rather because their strategies depend on being able to market themselves without interference. The United States would be wise to interfere by challenging the legitimacy of these new Islamist brand names before they become household brands.

Realistic (and Unrealistic) Expectations

The United States stands to benefit from a variety of engagement strategies with Islamist organizations. Far more difficult is the task of assessing how such an invitation for engagement would be received by Islamist organizations themselves. It is critical that the United States remain realistic in its expectations. The first reaction we should fully expect is that many Islamist organizations, for fear of colluding with a sworn "imperialist enemy" and supporter of the "Zionist state," will shun any gesture made by the United States to open a dialogue or create the basis for engagement. But initial rejection should not impede our efforts. It will take time before these investments will yield tangible returns. Islamist organizations are numerous, ideologically diverse, and politically dynamic and are therefore inextricably linked to the larger strategic picture. Even initial rejection of a U.S. overture of direct engagement will alter the status quo and open up new opportunities for U.S. diplomacy. Even negative reactions should be regarded as opportunities on which to build further. If there is any doubt on this point, one need only point to the recent U.S. "dialogue" with Iran, which, though cautious and slow on both sides, replaces a long period of no contact at all between U.S. diplomats and our Iranian counterparts.

The United States must also remain realistic, but firmly resolute, with regard to reactions among our regional allies and coalition partners. Their responses to engagement with Islamist organizations will range from mild irritation to acute panic. Many will voice their protests against what they believe is the legitimization of Islamist and other nonstate actors. We must remind them as gently as possible of the oft-repeated retort that "one man's terrorist is another man's freedom fighter" without giving credibility to those groups that use violence as their main tool. This, like all diplomacy, will be a delicate dance. We should also expect that some regional partners will offer to act as interlocutors between the United States and Islamist organizations on the grounds that there is already an established and natural interaction between these organizations and their regional governments. The United States should certainly carefully weigh such overtures, but remain ever mindful that the most important function of public diplomacy is that it be public and transparent.

One important caveat must be heavily emphasized here: This program of engagement with Islamist organizations should never be interpreted as a change in the fundamental principles that the United States has upheld for itself. It should be made abundantly clear to our own public and others that the United States continues to view terrorism and the use of indiscriminate violence against civilians (no matter what the political cause) as abhorrent. The United States must reiterate this as the common denominator between our government and the U.S. public, our allies and partners in the region, and citizens in the Middle East even when other issues are in dispute. The United States must work hard to overcome the prevalent belief that it is content to sing the praises of the philosophical benefits of freedom and democracy, but is reluctant to wade into the muddy and gray waters where these philosophical principles are tested and contested at the most basic levels of human society.

The Stakes

In his 2002 book *The Stakes,*[13] University of Maryland professor and Middle East expert Shibley Telhami argues that

> public diplomacy is not the same as propaganda, and there are limits to what can be achieved through this means. The United States needs to explain its policies and to disseminate information about American culture, values, and aims. But public diplomacy must be present at the inception of any policy and must include dialogue and feedback: If the aim of policy is to send messages to others or to generate particular responses, it cannot succeed without understanding those others' aims, aspirations, priorities, and sensitivities.[14]

The United States can continue to keep its distance from Islamist organizations and hope that the challenges they represent will resolve themselves. This is, however, an unlikely outcome. More likely, moderate and reform-minded Islamists will slowly disappear from the political arena and be replaced by their radical counterparts whose constrained strategic choices have convinced them that violence and coercion are the most expedient ways to obtain their objectives. Before this happens, the United States should challenge Islamist organizations with a policy of engagement and extend it even to those organizations whose anti-U.S. rhetoric has been distasteful or incendiary. It should be remembered that engaging Islamist organizations would not be unprecedented. The United States has done so in the past through indirect and back channels, and other governments, such as the British, have privately and publicly debated the pros and cons of a similar program of engagement. [15] The reasons that the United States has not engaged such organizations more frequently are many and valid, but even carefully crafted policies can, without proper review, become fixed moral positions rather than effective strategies intended to produce a specific outcome. It is time to break old taboos and reexamine our logic. Indeed, if a battle-hardened Israeli public can tolerate its government's imperfect "negotiation" process with Hezbollah, so too can the American public adapt to the necessity of and potential benefits from engaging Islamist organizations in pursuit of greater U.S. interests.

U.S. public diplomacy must now take a bold and confident step forward and rest upon a clear set of objectives that will further our long-term objectives in the region. We will undoubtedly benefit from furthering a solid foundation of in-house knowledge about Islamist organizations, but there is still a great deal we do not know about the aspirations, competing strategies, and ideological flexibility of such groups. Some have argued that the United States and other Western nations missed the opportunity to mitigate the growth of violent religious extremism after the conclusion of the first U.S.-Iraq war in 1990–1991, when there was a groundswell of public dissent in the region against returning to the "status quo" of repressive and unrepresentative governance.[16] This may or may not be accurate, but it is true that engagement with Islamist organizations after 9/11 has been largely of a kinetic sort and has lacked an appreciation for diversity and dynamism within groups, as well as for the deep societal roots that many organizations have cultivated within and across states in the region. The United States and other Western powers have too often started with the premise that Islamist political objectives and strategies are ideologically impoverished, as well as antithetical to democracy and political development. The United States should avoid tainting our ability to break away from old paradigms in international relations by avoiding this judgment at the outset. We must begin with the possible premise that although we eschew violence and extremist behavior, misguided strategies on the part of these organizations do not always imply misguided policies and it is time for the United States to increase its ability to discern between types. The United States should pioneer a program of direct and focused engagement with Islamists as an outgrowth of a new commitment to "soft power"—investing in the future of the region by engaging Islamists entails less risk than does allowing this opportunity to slip away for yet another generation.

Notes

1. Rami Khouri, "America, the Middle East, and the Gulf: An Arab View of Challenges Facing the Next U.S. Administration," CIRS Brief 1 (Georgetown University School of Foreign Service in Qatar, 2008), p. 7.
2. Fareed Zakaria, "Learning to Live with Radical Islam," *Newsweek,* March 9, 2009.
3. Heidi Lane, "Let's Make a Deal: Walking the Line between Counterterrorism and Political Liberalization in the Arab and Muslim World" (draft manuscript, under revision for publication). See also Dalia Dassa Kaye et al., "More Freedom, Less Terror? Liberalization and Political Violence in the Arab World" (RAND Corporation, 2008); Tamara Cofman Wittes, *Freedom's Unsteady March: America's Role in Building Arab Democracy* (Brookings Institution Press, 2008); Marina Ottaway and Julia Choucair-Vizoso, eds., *Beyond the Façade: Political Reform in the Arab World* (Carnegie Endowment for International Peace, 2008).
4. This should be compared to the opposite end of the spectrum, which yielded 3 percent of respondents who had a "very favorable" view of the United States. See *2009 Annual Arab Public Opinion Survey* (University of Maryland with Zogby International, 2009). The full PDF version of this poll is available at http://www.brookings.edu/events/2009/0519_arab_opinion.aspx. Polls for previous years can be found at http://sadat.umd.edu/surveys/index.htm.
5. Ibid., slide 19.

6. Ibid. These numbers represent an overall improvement from 2008 to 2009 of 18 percent and include Egyptian respondents who have tended to weight the results in a negative direction for this particular issue.
7. The interview took place on January 26, 2009, with al-Arabiya Washington bureau chief and veteran journalist Hisham Melhem. For a full transcript, see http://www.america.gov/st/texttransenglish/2009/January/20090127161320xjsnommis0.705578.html.
8. These responses were summarized in the *2009 Annual Arab Public Opinion Survey*, section entitled "Media," slide 59.
9. Another interesting discussion on this subject can be found in Marwan Kraidy, "Arab Media and U.S. Policy: A Public Diplomacy Reset," Policy Analysis Brief (Muscatine, IA: The Stanley Foundation, January 2008).
10. Marina Ottaway and Amr Hamzawy, "Islamists in Politics: The Dynamics of Participation," Carnegie Paper 98 (Washington, DC: Carnegie Endowment for Middle East Peace, November 2008).
11. See Amaney Jamal, "Reassessing Support for Islam and Democracy in the Arab World? Evidence from Egypt and Jordan," *World Affairs* (Fall 2006).
12. "Poll No. 68—June 2009: Press Release" (Jerusalem Media and Communication Centre), http://www.jmcc.org.
13. Shibley Telhami, *The Stakes: America and the Middle East* (Boulder: Westview Press, 2002).
14. Ibid., p. 179.
15. British debate on this issue is briefly summarized in an article by Martin Bright, "Talking to Terrorists," *The New Statesman,* February 20, 2006.
16. One of the most comprehensive analyses of this period remains James Piscatori's edited volume entitled *Islamic Fundamentalisms and the Gulf Crisis* (American Academy of Arts and Sciences: The Fundamentalism Project, 1991).

Panel V: The Greater Middle East

Summary of Discussion

Dr. Hayat Alvi-Aziz
Associate Professor of
National Security Affairs
Naval War College

The Greater Middle East panel discussion centered on current political and military/security crises in the Middle East region, and analyses of U.S. policies and strategies pertaining to them. The panel participants included Ambassador David Litt, Professor Marc Lynch, and Professor Heidi Lane. The Greater Middle East panel generated vigorous discussions and raised numerous questions during the question and answer session, as the topics and issues discussed are timely, changing on practically a daily basis, and pose critical challenges and potential threats to U.S. interests and security, especially since the 9/11 attacks, the U.S. invasion of Iraq, and the U.S. campaign in Afghanistan. The participants' discussions concerned aspects of the following subjects and issues:

- Strengthening U.S. public diplomacy in the region;
- Strategic relationships in the region and the need for collaborative efforts;
- Resolving the Arab-Israeli conflict;
- Stabilizing Iraq and Afghanistan;
- Dealing with Iran and its nuclear ambitions;
- The State Department's role: more diplomats and Foreign Service Officers needed;
- The Bush and Obama administrations' distinctions in policies and strategies;
- The current status of the global war on terrorism (GWOT);
- Islamic extremism, Islamic organizations, and political dialogue: should the United States talk to "radical" elements?
- The role of Arab media in shaping public opinion;
- Obama's outreach efforts toward the Muslim and Arab worlds;
- The popularity of the discourse of "resistance;"
- Prioritizing issues and problems: what should the United States do first?

The premise of Ambassador Litt's presentation was the need to improve U.S. credibility in the Middle East region, and the need to change the approach to public diplomacy. Damaged U.S. credibility helps the enemies more, he argued, and we need more collaboration with other states in developing strategic relations in the region. Personal relations are especially important to Gulf Arab leaders.

Ambassador Litt also suggested that the United States needs better policies and assertive efforts for the Arab-Israeli peace process, and with a new administration in place, there are some positive signs of measures being taken in the right direction in terms of outreach and diplomacy. He went on to say that true reconciliation is needed in Iraq, and that such reconciliation must be deeply rooted so as to prevent ethnic and sectarian disputes from continuing. Although the United States would have a limited role in this process, the United States certainly can urge resolution of the allocation of oil resources, as well as advocate nonpartisanship in Iraqi politics. In addition, USAID and State Department projects should continue in Iraq. It would also be wise to seek help from other Arab states to help stabilize Iraq.

We need to have experts in the Middle East at the ground level, with private-sector civilian programs to help improve the quality of life and good governance in the region. "Front page issues are not everything." However, Ambassador Litt mentioned a caveat: Yemen is a crisis on the horizon.

Professor Marc Lynch began his discussion by mentioning General Mike Herzhog's speech about the Israeli vision of creating an Arab-Jewish alliance against the Persians (i.e., Iran, Syria, and Hezbollah). Professor Lynch argued that this is "a very wrong approach," because the future strategy for dealing with the region cannot afford to polarize and isolate players. He pointed out that "Islamism was conflated upon the 9/11 attacks," and now we need to address each challenge individually. Falling prey to the "marketing trap," as he explains in his paper, would be very detrimental to U.S. interests.

General Herzhog's view supports the unchanging status quo, which is counterproductive. The configuration and variables are changing on the ground, and in order to devise appropriate strategies for dealing with the region, we cannot afford to be out of touch with the (regional) public. Professor Lynch made several suggestions for adjustments to approaches to a new strategic vantage point relative to the Middle East:

1. We need to diverge from the GWOT focus on al Qaeda, because al Qaeda is now quite marginalized in the Arab world. Today, the discourse of "resistance" is growing in popularity. With regard to this, Obama's outreach and public diplomacy endeavors in the region are positive steps, and they should continue to target the mainstream mass public.
2. The issue of sequencing is extremely important: Which issue should the United States tackle first, Iran? The Palestinian-Israeli conflict? It is very likely that the United States will talk to Iran after the election in June.

The premise of Professor Lynch's discussion was the need to depolarize the region, or else risk facing the prolongation of the status quo. The key question to ask, in order to devise a strategic engagement with the Muslim world, is the following: "If the war of ideas is taking place within the Muslim world—and it is—then what is the appropriate role for America?" It is also equally important to note that "al Qaeda's decline [in the region] does not mean that support for American foreign policy is rising."

Professor Heidi Lane began her discussion by posing a question: should the United States talk to Islamist organizations? Professor Lane contended that "opening channels of dialogue with Islamist organizations would have far-reaching benefits for U.S. objectives in the Middle East region and would, therefore, contribute to our larger strategic vision."

Professor Lane explained that the United States suffers an image problem in the Middle East region, and that the region's populace often views the U.S. government's refusal to talk to Islamist organizations as an affront against Islam. Professor Lane expressed that "engagement with Islamists could begin to chip away at the pervasive sentiment in the region that U.S. objectives are nothing more than cultural hegemony, dominance over natural resources, and the weakening of Islam itself."

The increasing popularity of Islamist organizations throughout the region cannot be dismissed or ignored, and diplomatic opportunities for outreach should not be missed. In addition, the growing prominent role of nonstate actors needs to be acknowledged.

Professor Lane listed five reasons that the U.S. government should talk to, but not necessarily negotiate with, some Islamist organizations (except al Qaeda):

1. "Can't look like more of the same"; that is, there need to be distinctions between the Bush and Obama administrations;
2. The public has consumed a lot of information about Islam, Islamism, and the like; thus, there is a need for understanding the parameters of the U.S. role in the region;
3. We need to clarify our "freedom agenda"—i.e., do our counterterrorism programs provide an excuse for states to push back on domestic reforms? This needs to be addressed;
4. We need to leverage the Arab media more, because this is fertile ground and reaches a vast public in the region;
5. There is a need for clarifying the classifications of Islamic "moderates," "radicals," and "extremists," and we need to understand and interact accordingly with civil society in the region. There are critical perceptions to address, mainly how to deal with Hamas and Hezbollah.

In short, the United States needs to engage with Islamist organizations, which would be "a new commitment to 'soft power'—investing in the future of the region by engaging Islamists entails less risk than does allowing this opportunity to slip away for yet another generation."

During the question and answer session, the participants asked numerous questions about the effectiveness of strategic communications; public diplomacy, outreach, and dialogue; and Iran's nuclear ambitions and what to do about them. The discussants offered a range of responses.

Questions included: How can we improve strategic communication, cultural understanding, and public diplomacy? What is the significance of choosing Egypt as the location of President Obama's speech? Respondents suggested that the more efficient and effective use of modern information technology, such as satellites, cyberspace, and the Arab news media to reach vast audiences in the region, should be embraced.

Another discussant remarked, regarding the location of Egypt for the president's speech, that we should "wait and see what he says; give him a chance." The concern of many in the region is that the selection of Egypt is a default endorsement of an authoritarian regime, and counters the promotion of democracy. The discussant expressed that President Obama "wanted it to be in an Arab country, because he had already spoken in Turkey, and he's going to go to Indonesia later."

In response to a question about outreach to radical or moderate Islamist elements in the region, a discussant responded, "We [the United States] should be persuasive about the United States, who we are, what we believe with respect to the Middle East, and not get into discussions of who is moderate, and do we support moderates versus radicals." Another question asked was about how to restart the Arab-Israeli peace process, and how to deal with Iran and its nuclear ambitions. A discussant responded that Iran is really the focus of attention for Israel right now. "This may be an issue between the U.S. Jewish community and Israel, and within the U.S. Jewish community." The discussant went on to say that "it's impossible to get anywhere on Palestinian negotiations if you don't have some kind of effective Palestinian unity government. If you're going to have an effective unity government, it's going to have to include Hamas somehow, and I think that's what everybody is arguing about pretty vigorously at the moment."

The Middle East region contains ongoing conflicts and political and military/security dilemmas in which the United States is directly and indirectly involved. These realities have major security implications for the region and also for the global community. The consciousness of the importance of the region made for a profoundly interesting and informative panel discussion.

Panel VI

Europe and Russia

Dr. Sharyl Cross
Professor and Director of Studies, Program in Advanced Security Studies, George C. Marshall European Center for Security Studies

Dr. R. Craig Nation
Professor of Strategy and Director, Eurasian Studies, U.S. Army War College

Moderator:
Ambassador (Ret.) Mary Ann Peters
Provost, Naval War College

Advancing a Strategy for Constructive Security Engagement: "Resetting" the U.S./NATO Approach toward Russia

Dr. Sharyl Cross
Professor and Director of Studies
Program in Advanced Security Studies
George C. Marshall European Center for Security Studies

Introduction

Russia's incursion into Georgia in August 2008 shocked Western officials and triggered a fundamental reassessment of Moscow's intentions toward neighboring nations and the wider international community. Tensions among Russia, its neighbors, and the West had been escalating for some time. Russia's intervention in Georgia was entirely predictable. Moscow had consistently expressed opposition to NATO enlargement, especially possible fast-track membership for Ukraine and Georgia. Russia's objection to the proposed placing of missile defense assets in Poland and the Czech Republic, and U.S. and other-Western-partner recognition of Kosovo's new independence also exacerbated tensions. The negative climate in U.S.-Russian relations was clear at the Munich Conference in 2007 when Russia's President Vladimir Putin took the podium to charge that today's world was witnessing an "uncontained hyper use of force," and that "democracy" was not compatible with a "one master" world order.[1] U.S. defense secretary Robert M. Gates responded observing that Putin's remarks resurrected sentiments of the East-West confrontation, but also cautioned that "[o]ne Cold War was quite enough."[2]

By January 2009, when Barack Obama assumed the presidency, the state of U.S.-Russian relations was more strained than at any period during Russia's post-Soviet experience. At the Munich Security Conference in February 2009, U.S. vice president Joseph Biden signaled early on that the new administration sought to "press the reset button" with Moscow, suggesting there are "many areas" where the United States "can and should be working together with Russia."[3] One of the major objectives in the "Declaration on Alliance Security" issued at NATO's sixtieth anniversary aimed to build a strong and cooperative relationship with the Russian Federation.[4]

The United States and NATO will undoubtedly require Russia's cooperation for meeting the priority transnational security challenges of the twenty-first century. The Obama administration must not only foster an improved climate with Moscow, but also manage the often divergent perspectives among NATO member nations. As NATO's Secretary General Jaap de Hoop Scheffer notes:

"It is no secret that when it comes to Russia, there are a wide range of views within NATO, from the very cautious to the forward-leaning. . . . Until we narrow the range it will be difficult to engage Russia effectively. . . ."[5] This paper explores the foundations for building a U.S./NATO-Russia common security agenda, as well as defines major challenges for developing an enduring security partnership with Russia.

Initial Euphoria, Sobering Reassessment

The dramatic transformations throughout the former Soviet bloc ushered in an initial period of euphoric expectations in Western capitals with respect to early post-Soviet Russia. However, pledges of "partnership" and even "friendship" on the part of both President Bill Clinton and President Boris Yeltsin were not matched by substantive progress in the U.S.-Russian relationship through the decade of the 1990s. For most Russians, the Yeltsin period of so-called democratic transition was experienced as "chaos" or "lawlessness" resulting in complete lack of stability or normalcy in ordinary life. Many Russian citizens lost their entire modest life savings during the 1998 ruble devaluation, a crisis situation commonly perceived by Russians to have been brought about as a result of the influence of Western economic advisers.

Although some have argued that the United States will not be able to influence Russia's domestic transition or relations between Russia and the West, U.S./NATO actions certainly did have a dramatic impact in early post-Soviet Russia's domestic and international development. From Moscow's perspective, two developments were most significant in explaining the sobering reassessment that took place over the decade of the 1990s regarding the potential for security partnership with the United States/NATO. First, despite U.S./NATO attempts to downplay Russia's opposition to NATO enlargement, every Russian president has expressed sustained objections to expanding the alliance. Russia's political-party platforms left, center, and right have been unified in opposing NATO enlargement. The former Soviet president, Mikhail Gorbachev, claims to have had a "gentlemen's agreement" with officials of the first Bush administration that removal of Soviet forces from Eastern Europe and German unification would not be followed by NATO expansion.[6] As early as 1994, Russia's President Boris Yeltsin warned of a "Cold Peace" falling over Europe, stating that "plans for expanding NATO" would "create new divisions" and "sow seeds of distrust."[7]

The U.S./NATO air war in Kosovo also delivered a major blow that deflated high expectations for cooperation with the United States and Western nations. Dr. Victor Kremenyuk, long-time specialist on Russian-American relations in Moscow, captured the impact of the U.S./NATO air campaign among Russia's foreign policy community: ". . . [T]he world after 1999 will never be what it was immediately after the end of the Cold War: the reconciliation of former foes, hopes for partnership, democratization of international affairs, growth of cooperation. . . ."[8] Following the Kosovo war, Russian official foreign and security documents reflected reassessment of the potential for cooperation with Western nations. For example, Russia's first *Foreign Policy Concept,* of 1993, was decidedly pro-Western, stating that Russia and the West possess "common understanding of the main values of world civilization and common interests," while the same document issued in 2000 warned of the "establishment of a unipolar structure of the world with economic and power domination of the United States."[9] Russia's *National Security Concept* issued in 2000 notes that "NATO's transition to the practice of using military force outside its zone of responsibility and without UN Security Council sanction could destabilize the entire global strategic situation."[10]

Russians tended to conclude that NATO enlargement and the U.S./NATO Kosovo air war demonstrated that Western countries would take advantage of Russia's transitional domestic turmoil and weakness. After Kosovo, Russians, more than ever, desired a leader who could display resolve in resisting Western pressures, and restore Russia's prestige in the world community.

Drifting toward Resurgent Cold War?

Vladimir Putin's pragmatic orientation toward realizing Russia's international objectives led to willingness to resume relations with NATO following suspension during the Kosovo war. Vladimir Putin and George Bush established a good personal rapport at their first summit, held in Ljubljana in June 2001.

The attacks on 9-11 at the World Trade Center and Pentagon generated genuine sympathy on the part of Russia's president and public. Putin was the first among world leaders to call President Bush to express support following the September 11 terrorist attack, and Moscow offered intelligence and basing access in Central Asia that was invaluable for executing the first stage in Bush's war on terrorism. Putin hoped to gain greater support for the Chechen campaign by casting the 9-11 incident and other terrorist attacks as emanating from the same international source, to include al Qaeda and its affiliates.

The visit of NATO Secretary General George Robertson to Moscow in February 2000 marked the beginning of the restoration of the Russian-NATO relationship after Kosovo. In March 2000, Putin surprised audiences in the West when he responded to a question posed concerning whether Russia might join NATO, replying "Why not?" Putin stated that he was "open to more profound integration with NATO," and that while not ruling out Russia's membership in NATO, this would be conditional on taking Russia's views into account as an "equal partner."[11]

The early overtures toward placing the U.S./NATO-Russia relationship on a more positive track began to deteriorate with the U.S. military intervention in Iraq. Russia and many other nations could accept the justification offered by Washington for the first phase of Bush's war on terrorism, which aimed at removing the Taliban and disrupting and dismantling al Qaeda's base in Afghanistan. Russia's reaction to the Iraq intervention could be described as calmer and more calculated than in the case of the response to the implosion of former Yugoslavia, but still Russia's commercial and strategic interests in Iraq outweighed other factors in leading Putin to join France, Germany, and other nations of the world community in opposing the U.S. response to Saddam Hussein's failure to comply with UN resolutions.

Putin's *Vertikal Vlasti*, reversing democratic transformation on multiple fronts and rendering presidential leadership increasingly unaccountable, led to sharp criticism and pressure from the West. Accusations from the United States/Europe identifying reversals in Russia's democratic project yielded little more than resentment from Moscow. Alleged support by the West in the colored revolutions of Ukraine and Georgia was perceived in Moscow as further meddling in the internal political dynamics of neighbors, rather than any genuine commitment to encourage democratic transformation.

Russia's concerns with respect to encroachments in the post-Soviet space intensified in response to proposed plans for placing missile defense assets in Poland and the Czech Republic. Putin offered a counterproposal, suggesting that the United States and Russia cooperate in using the Gabala radar station in Azerbaijan, but Bush administration officials rejected the offer, concluding that the proposal would not suffice in terms of technical adequacy.

While Putin recognized the importance of the West to fulfilling Russia's security and economic aspirations, his posture toward the United States/NATO became increasingly combative in the latter stage of his presidency. Reflecting on NATO at the Munich Conference in 2007, Putin charged that ". . . it is obvious that NATO expansion does not have any relation with the modernization of the Alliance itself or ensuring security in Europe. On the contrary, it represents a serious provocation that reduces the level of mutual trust. . . ."[12] After failing to comply with the terms of the Conventional Forces in Europe (CFE) Treaty requiring Moscow to withdraw forces from Moldova and Georgia, Putin announced in July 2007 that Moscow would no longer participate in the CFE Treaty, which had been regulating the deployment of troops and weapons systems in Europe since 1990. Moscow has adamantly opposed recognition of Kosovo's independence, and suggested that claims for independence on the part of South Ossetia and Abkhazia hold greater historical legitimacy than does the case of Kosovo.

Putin's designated successor, Dmitri Medevedev, confronted serious conflicts with the West from the outset of his presidency. In addition to Moscow's diplomatic isolation and condemnation for the disproportionate use of force in the Georgian war, Medvedev advanced a new European Security System proposal that would ultimately replace NATO. Russia again threatened to suspend energy supplies to neighboring nations and Europe during winter 2008–2009. Moscow appears to have provided considerable economic incentive to the Kyrgyz leadership to encourage the closure of the U.S. air base at Manas that would jeopardize vital U.S. supply/support lines for Afghanistan. Russia conducted naval exercises in the Caribbean and deployed TU-160 strategic bombers to Venezuela in fall 2008, and Venezuela's anti-American president Hugo Chávez claims to enjoy a "strategic partnership" with Russia. All of these factors provided evidence for several in the U.S., European, and Russian academic communities to argue that it was obvious that a new Cold War had developed or would inevitably define the U.S./European-Russian relationship.

"Resetting" the U.S./NATO Relationship with Russia

At least for the foreseeable future, the United States and NATO will encounter significant obstacles attempting to build a relationship with the Russian Federation on the basis of common values. Some believe that Vladimir Putin's successor, Dmitri Medvedev, would be more inclined to seek democratic liberalization. In his November 2008 address to the Federal Assembly, Medvedev called for reversing some of the restrictions on civil society that had been instituted by his predecessor.[13] Notably, Medvedev granted the first interview of his presidency with *Novaya Gazeta*, the Russian newspaper source known for controversial reporting, such as coverage exposing human rights abuses in Chechnya by famous staff journalist Anna Politkovskaya, who was murdered in 2006. In the interview, Medvedev underscored the point that Russians would not sacrifice "rights and freedoms in exchange for economic well-being and stability," and argued that "[w]e've had democracy, we have it, and we will have it."[14] Medvedev recently launched a forum on a LiveJournal blog featuring vibrant political debate, encouraged public square discussions for expressing political views, and opened discussion with human rights advocates.[15] However, Medvedev still enjoys the backing of the same group that surrounded Putin, supporters who reject most features of Western democracy.

Contemporary Russia and the West are no longer divided by ideological conflict, but there are undeniable differences in culture and values. Western criticism of Putin's restrictions on political parties, media, and nongovernmental organizations was dismissed by Moscow's leadership and much of the broader society as nothing more than inappropriate intrusive interference in Russia's domestic affairs that could again lead to destructive "chaos" as in the 1990s. Placing emphasis on the obvious shortcomings of Russia's domestic political development is still likely to be counterproductive and damaging to cooperation at this stage.

The intention of the new U.S. administration to "reset" the relationship with Russia should not be dismissed as a "naive hope," but seen as recognition that the United States and Russia share vital interests. With continued engagement and integration, over a period of decades rather than years, transference of values or greater openness on the part of Russian society toward adopting European/Western values could take place. In the meantime, building a relationship based on engagement in areas of shared interests is the most viable option for achieving long-term democratic transition in Russia. We should promote cooperation in areas of common interests, and offer a unified Western voice in seeking to resolve differences on issues where serious conflicts of values/interests emerge. The United States/European nations are more likely to contribute to a positive end state for Russia with an approach based on setting a desirable example and partnership, rather than amplifying criticism of the inadequacies of Russia's complex and difficult transition.

U.S./NATO Interests

Russia's vast territory and geographic reach adjoining Europe, the Middle East, and Asia make the nation a major player for regional and global security. Russia retains a nuclear arsenal approximating that of the United States. Both countries possess capabilities far superior to other nuclear weapons states. As a net energy exporter, Russia boasts combined gas/oil resources exceeding those of any nation, including the United States. Russia's status as a permanent member of the UN Security Council with veto authority, and membership in other major international organizations (Organization for Security and Co-operation in Europe [OSCE], Group of Eight, Shanghai Cooperation Organization [SCO]), provides Moscow with the diplomatic gravitas for exerting considerable influence in shaping the world security agenda and determining vital collective security responses. Russia's longstanding diplomatic ties and networks in the Middle East, Asia, and beyond can be important for the United States and European partners in addressing the most difficult security challenges, including settlement for the Israeli-Palestinian conflict, and many other problem areas.

Russia's cooperation is needed for almost every priority security concern for the United States and NATO nations. The Strategic Arms Reduction Talks [START I] agreement will expire in 2009, and both the Obama and Medvedev administrations recognize that continued progress in arms control and proliferation of weapons of mass destruction (WMD) will require U.S.-Russian engagement and cooperation. Russia's support will be crucial in preventing the advancement of nuclear programs for both Iran and North Korea, and for strengthening the Nuclear Non-proliferation Treaty and arresting the further proliferation of WMD.

The United States and Russia share varying commitment to countering the ambitions of al Qaeda and its affiliates. Barack Obama has affirmed that the threats from al Qaeda and the Taliban remain the highest security priorities for the United States. Russia's counterterrorist strategy has tended to focus domestically on threats emanating from Chechnya and the predominantly Muslim regions of the northern Caucasus. Nevertheless, both nations recognize the threat posed by the potential access of terrorist groups to WMD and the importance of countering the ideological appeal of al Qaeda and its affiliates.

The United States, NATO, and Russia share the common objective to prevent the Taliban from returning to power in Afghanistan. While the alliance has been seriously strained as a result of differing levels of support for a long-term commitment in Afghanistan, all NATO countries and Russia will benefit by secure development for Afghanistan. The United States/NATO and Russia also share interests in ensuring the stability of nuclear-armed Pakistan and strengthening that nation against the destabilizing influences of militant extremism, and promoting a secure future for Iraq.

NATO has recently included energy among the priorities for the alliance's security agenda. The dependence of Europe/Eurasia on Russian gas enables Moscow to use energy as a potential "geopolitical weapon" in advancing interests with respect to these countries. The United States/NATO and Russia would all stand to benefit by fostering stable energy regimes providing reliable access and markets for Europe/Eurasia.

Long-term stability in Eurasia and the broader Euro-Atlantic community will require that Russia recognize the territorial integrity and sovereignty of its neighbors. Without a functional relationship with Russia based on mutual respect, continued democratic transition and security in these newly formed nations will continually be jeopardized.

Ensuring reliable communication channels and clear understanding of objectives on the part of all actors is essential for avoiding misperception that holds the potential for escalating into a crisis situation between Russia and the West. In the case of Georgia, the West failed to anticipate the extent of Russia's resolve in responding to secessionist impulses from South Ossetia and Abkhazia. The Georgian president appears to have miscalculated when he failed to hear Washington's messages cautioning restraint.

The United States will increasingly require other major powers, including Russia and China, to assume a greater share of the burden in managing transnational security challenges, particularly in those areas contiguous to their nations. U.S. resources and capabilities are not infinite, and thinking more about how to ensure the constructive development of U.S. and European relationships with both Russia and China would certainly serve U.S./NATO long-term interests.

Priorities for the Russian Federation

While Russia shares several similar critical security interests with the United States/NATO, there are other unique driving motivations that will factor in Moscow's behavior that can be even more decisive in determining responses. First, and above all, the new Russia seeks respect and prestige in the world community. From a national psychological perspective, Russia's leadership and society will never be able to come to terms with diminished status following the collapse of the Soviet empire. The Russian national identity is largely based on a perpetual "great power" image and a "special destiny" for Russia in world society. Moscow officials have frequently complained that they could not accept a "junior partner" status in NATO. The leadership in Moscow would not rule out the possibility of joining NATO, but they could never reconcile their sense of national great-power status with standing behind new member countries of East-Central Europe, Ukraine, or Georgia in seeking admission to the alliance. Russians will often claim that the United States affords China greater respect than Russia, even though China and the United States are sharply divided ideologically. References to the need to "reassert" a "strong" and "self confident" Russia contained in the *Review of Foreign Policy of the Russian Federation* released in 2007 reflect this priority/motivation.[16] In Moscow's foreign policy discourse, repeated references to the influence of the "Cold War winner's syndrome" reflect the sense Russia's leadership must demonstrate significant presence and independence in world politics.

A second related major factor concerns Russia's relationship with neighbors or post-Soviet nations. Elements within Russia's foreign and security apparatus are threatened by the diminished capacity to exert influence among border nations that previously constituted part of the Soviet empire, while at the same time these neighboring countries are forging stronger associations with the United States and other European nations. Dmitri Medvedev has continued with the tradition established by his predecessor, suggesting that Russia enjoys a "zone of privileged interests" in the post-Soviet space. The Russian leadership displays a sense of entitlement in relationships with their neighbors that derives from a combination of factors including history, shared cultural ties, and security and economic dependencies. Igor Zevelev has made the point that while the United Kingdom "possessed" a distant empire, Russia "was" an empire where no borders divided Russia from its neighbors.[17] For Zevelev, this explains the fact that Russia's behavior toward its neighbors in the post-Soviet space can often be emotionally driven, rather than consistently calculated to reflect Russia's national interests. The Obama administration should anticipate that Russia will continue to attempt to assert influence on its periphery, and not always in ways that are most advantageous even from Russia's perspective.

In addition, stimulating economic growth and improving the society's living standards are among the highest priorities. Russians must overcome the industrial and infrastructure obsolescence resulting from the decades of the inefficient Soviet centralized economic management. Burgeoning capacity in the energy sector fueled Russia's renewed influence on the world stage, and the society seeks to achieve a level of economic opportunity and living standards approximating that of the leading European nations. The bulk of foreign direct investment in Russia comes from the European Union, with Germany being Russia's most important trading partner. U.S.-Russian economic ties are quite less significant than U.S.-Chinese economic dependence, but Russia was still severely impacted by the global economic crisis, which originated in the U.S. subprime mortgage debacle. Stabilization of the world economy and further integration with Western nations is critical for Russia to achieve the stated objective of becoming one of the world's top-five economies and bettering conditions for a population that suffered such hardship during the communist era.[18] Should the financial recovery be delayed, socioeconomic turmoil and resulting political instability could further complicate choices for the Medvedev leadership. The need for diversification of Russia's economy, declining birthrates, and poor health care conditions also present a unique set of demands on the Russian leadership.

Russia's economic climate deteriorated significantly in the aftermath of the Georgian intervention. Foreign investors immediately began to withdraw capital from the Russian market, and the main Russian stock market index plunged sharply. Projections indicate that Russia's gross domestic product is likely to contract 4.5 percent in 2009, with escalating unemployment and other socioeconomic hardships for society.[19] As Russia's society suffers the adverse repercussions of the world financial downturn, it has become more obvious than ever to Russia's leadership, financial community, and society that Russia's fate is linked to stabilization of the U.S. and world economy, and that overcoming the crisis will require unprecedented levels of cooperation among nations.

Strategy for Building a Common Security Agenda

The United States and Euro-Atlantic community should work cooperatively to forge a constructive, consistent, and coherent strategy toward Russia on the basis of long-term strategic vision. Robust engagement and eventual integration of Russia in the Western security and economic community will benefit all involved.

Greater attention must be devoted to fostering transatlantic unity in building strategy toward Russia. Over the past decade, Putin and Medvedev have sought to divide the United States and European nations, exploiting differing priorities. To the extent that the United States/European nations can speak with a single positive and constructive voice with respect to Russia, the potential for effective engagement with the Russian Federation will be far greater. The United States, together with other long-standing NATO members, must reaffirm existing security commitments to new alliance member nations of East-Central Europe, while reinforcing or shifting focus on the importance of successful engagement with Russia in advancing common security objectives.

Although Barack Obama said little about Russia during the campaign, the U.S. presidential transition has generated high expectations in Moscow for improving U.S.-Russian relations. In fewer than 100 days, one can conclude that at least the tone in the U.S.-Russian relationship has improved, though considerable pessimism remains on both sides.

A resurgent Cold War is not inevitable. The United States/NATO and Russia recognize the imperative need for security cooperation, and no one would like to witness a return to an adversarial East-West struggle. Western countries fully recognize the fact that Russia could become a spoiler or work against vital U.S./NATO interests in the future, and all appreciate the importance of reversing the negative climate with Russia. Today there is no ideological orientation among Russia's leadership committed to confrontation with U.S./Western civilization.

Reversing the deterioration in Russia's relationship with the West must be among the highest priorities for the new U.S. administration. This will require sustained commitment and patience; it will not be easy. The two prior administrations initially enjoyed excellent personal presidential relationships, but both attempts to build a common Western security agenda or strategic partnership with Russia failed and the Clinton-Yeltsin and Bush-Putin eras ended with great disappointment on both sides of the Atlantic. The sense of urgency today among the U.S., European, and Russian policy communities for finding new ways of trying to cooperate stems from the fact that all recognize that past attempts have fallen short of success, and there is so much at stake. If Obama and Medvedev fail to set the relationship on a more positive track, this third attempt might have been the final opportunity.

Presidential investment is absolutely essential, but perhaps there was too much reliance on personal U.S.-Russian presidential ties in the past. Greater focus should be placed on building sustained institutional ties for the United States and Russia at other levels of government and society. The overtures by Senator Mikhail Margelov, chair of Russia's Foreign Relations Committee, for opening dialogue with U.S. congressional counterparts on a range of security issues and committing the Russian side of the joint legislative working group to improve bilateral relations represent the type of broader-based cooperation that should be encouraged.[20]

Bold new initiatives from the West commensurate with Russia's Euro-Atlantic and global importance will be needed to help to set relations on a new track. The United States and European partners must demonstrate a willingness to work with Russia on equal terms. Any perception that one side is dominating the agenda without a mutual willingness to make concessions toward shared interests will doom the relationship to repeat past failures in efforts to build strategic partnership.

In terms of specific recommendations, resuming the NATO-Russia Council and seeking renewed military-to-military ties should be a first priority. After a brief suspension of the NATO-Russia Council in response to the conflict in Georgia, the NRC resumed consultations in March–April 2009. NATO and Russia collaborated under the auspices of the NRC on a number of security issues, and Russia was invited to participate in NATO exercises, such as Active Endeavor, involving unprecedented access. Still, the NRC has thus far failed to serve as an effective mechanism for fully realizing the potential for security cooperation between Russia and NATO. With the resumption of dialogue in the NRC, Russia has agreed to permit NATO transfer of nonmilitary supplies to Afghanistan from Russian territory. The United States/NATO and Russia share common interests in a stable and secure outcome in Afghanistan, but both sides must also remain deeply committed to working through other sharply contentious issues that have impeded full-scale cooperation in the past.

In his first foreign policy address of June 2008, Dmitri Medvedev introduced his proposal for a new European security architecture to encompass Vancouver to Vladivostok. Medvedev's ambitious proposal is a clear indicator that Moscow is completely dissatisfied with NATO, the OSCE, and the entire post–Cold War security regime in Europe, which, from its perspective, do not hold a proper place for Russia or serve its interests. Russia's ambassador to NATO, Dr. Dmitri Rogozin, suggested in response to questions about Medvedev's security proposal in November 2008 that the current security system is "unstable" and "unhealthy," and that we need to discuss "new concepts" to eliminate "bloc politics" approaches.[21] Rogozin said that it was a mistake for Russia not to be asked to join the alliance early on, and that the NATO members would have been "wise" to overcome the "inertia of mind" to invite Russia to join NATO.[22]

American and European leaders mostly dismissed Medvedev's European security system proposal, suggesting that the document lacked specifics. However, the Moscow leadership intended to offer a proposal that was deliberately vague in order to invite further discussion with Western nations in elaborating parameters for such a new security configuration.[23] The West should give serious consideration to the proposal, and open dialogue or support convening meetings for evaluating the effectiveness of existing security structures and exploring options for better engaging and integrating Russia within the Western security community. The suggestion to replace NATO is unrealistic and there are unquestionably significant practical obstacles to eventual Russian membership in NATO (common values, defense transparency and military reform, committing to defense of Russia's vast territory). However, the current system is obviously not working from Moscow's perspective, and Western nations should be willing to discuss these objections and explore viable alternatives.

In addition, if the purpose of integration of Ukraine and Georgia into NATO was to enhance Euro-Atlantic/Eurasian security, the entire objective is undermined if, in the process, the U.S./European relationship with Russia deteriorates and diminishes regional stability and cooperation on priority critical global security issues. The actual prospects for NATO membership for Ukraine and especially for Georgia, given domestic instability particularly in the aftermath of the Georgian intervention, have rendered early membership for these countries unlikely and appropriate probably for only academic discussion. In the coming years, eventual membership for Ukraine and Georgia should remain open, but the U.S./NATO relationship with Russia's neighbors should advance in concert with deepening security ties and cooperation with the Russian Federation.

The potential for another crisis situation involving Russia and the United States/NATO over Georgia and Ukraine can hardly be dismissed. The Russo-Georgian war in August 2008 demonstrated that interstate war in the Euro-Atlantic community could still occur, and anxieties are still reverberating among new NATO members of East-Central Europe with respect to Russia's intentions. In April 2009, Moscow demanded that NATO call off planned peacekeeping exercises in Georgia, indicating that holding the exercises would constitute a "provocation" that could prevent the rebuilding of security ties with the West even before the process begins.[24] The exercises are scheduled from May 6 to June 1 and involve some 1,300 troops from nineteen countries. The buildup of Russia's forces on the Georgian border and military presence in South Ossetia and Abkhazia create circumstances for potential miscalculation and confrontation. Given the existing tensions, it seems NATO could have considered rescheduling the exercises for a more appropriate time.

The United States, European Command, and NATO should explore all options for renewing military-to-military ties and practical security cooperation at the bilateral and multilateral levels with the Russian Federation. This will require considerable sustained support and engagement from the West, and there will surely be significant barriers on the Russian side. During the Kosovo conflict, as diplomatic ties broke down, the U.S./NATO-Russian military-to-military channels provided the only fully functional lines of communication. The United States/NATO and Russia established a foundation for successful military collaboration in peacekeeping in former Yugoslavia.[25] The accomplishments of former adversaries working together, in spite of political differences, were impressive and should continue. Such joint task-oriented cooperation in areas of mutual interest contributes to breaking down barriers to trust and improves capacity to cooperate/react in meeting common security challenges. Russia's participation with Western nations in joint threat assessments and security-issue-focused working groups can strengthen capacity for anticipating and reacting to crisis security situations. An educational institution such as the Marshall Center can offer an appropriate venue for Russian military officers and defense sector civilians to come together with the U.S., European, Eurasian, and wider security communities to discuss the origins and solutions for critical shared security challenges.

Instead of fueling the prevailing perception among several post-Soviet nations that they must choose between either Russia or the West, the United States/European nations should attempt to cultivate a more constructive Eurasian climate emphasizing the importance of contributions that both Russia and the West can bring to these societies. Nations of the Caucasus, Central Asia, Ukraine, and Moldova will never enjoy secure development without Russia's cooperation. All these nations share the same neighborhood, with long-standing ties and dependencies that must be acknowledged and respected. The challenge will be for the United States/Europe to contribute to shifting the focus among Russia's neighbors beyond historical fears and grievances toward the aspirations they all share for a better future. Episodic conflicts as geostrategic interests and ethnic forces collide must be expected, but this does not have to lead to a permanent return to confrontation between Russia and the West. The Obama administration should try to diffuse concerns that we seek to undercut or jeopardize Russia's interests among neighbors. The role of the United States in reacting to progress and reversals in democratization among Russia's neighbors, the influence of the United States and European nations in peacekeeping or brokering conflicts in neighboring nations, and ongoing U.S. military presence in Eurasia will be significant. We simply can't allow tensions in these regions stemming from Russian sensitivities to thwart cooperation with Russia in areas of higher security priority.

The United States should make clear that while we support the reform aspirations of the post-Soviet nations, U.S. capacity in Eurasia faces significant constraints. Above all, the Georgian conflict demonstrated the limits to U.S./European commitment in these countries, and at the same time made clear that conflict resolution in the region requires the full cooperation of the United States, Russia, and Europe. The United States/NATO, Russia, and Russia's neighboring nations must seek mechanisms for dialogue so that intentions are clear and miscalculation that could lead to potential confrontation can be avoided.

A logical area for beginning to "reset" the U.S.-Russian relationship is in arms control and weapons proliferation. The initial steps taken by the presidents of the United States and Russia in April 2009 to resume cooperation in arms control are encouraging. The two presidents have announced the intention to pursue an arms agreement package that would cut nuclear warheads below the levels previously agreed in 2002, committing both sides to cutting arsenals to 1,700–2,200 warheads by 2012.[26] The United States and Russia should also find common ground in cooperating to enhance the nuclear nonproliferation regime.

In addition, the United States/NATO and Russia have remained at an impasse over suspension of the CFE Treaty providing for regulations of movement of conventional armed forces, inspections, and other confidence-building mechanisms. Russia and Western countries could benefit by resumption of the CFE, but both sides will have to be willing to reconsider the terms of the agreement in light of changing force structures in Europe and commit to complete timely compliance and transparency.

Another more promising area for moving past the deadlock at the end of the Bush administration is the issue of missile defense. The willingness of the Obama administration to reconsider or temporarily slow plans for deploying missile defense assets in Poland and the Czech Republic opens the possibility for finding a mutually acceptable compromise with Russia. In November 2008, Medvedev announced that he would order the stationing of Iskander missiles in the Kaliningrad enclave targeting Poland and the Czech Republic to counter U.S. missile defense plans, but then, perhaps signaling willingness to seek a compromise solution, suggested that the missiles would not be deployed. The United States/NATO should try to work with Moscow in finding a solution that is acceptable for all parties. Proceeding with missile shield deployment plans lacking Russia's buy-in is likely to create more adverse challenges in the security sphere than these costly systems will solve.

The Obama administration has made initial overtures that may be intended to link Russia's cooperation in halting development of Iran's nuclear program or access to nuclear intercontinental missiles with U.S. willingness to seek alternatives acceptable to Russia in missile defense. Iran has been a traditional ally for Moscow, and Russia's commercial interests benefited by supporting development of the Bushehr nuclear power station and supplying Tehran with S-300 ground-to-air missiles. Whether Moscow would increase pressure on Iran in the UN in support of economic sanctions or suspend the provision of air defenses to Tehran remains to be seen. However, the fact that the Obama administration will explore options with Russia in these areas could help to reverse the progressive worsening of relations over the past several years.

In addition, the United States, NATO/European nations, and Russia should take this transitional opening between administrations to advance significant new proposals for addressing a range of transnational security challenges where the United States/NATO–Russia share common interests and can build greater cooperation for the future. The United States/NATO will need to engage the Russian leadership more effectively in the energy security area. In April 2009, Dmitri Medvedev announced that Moscow would soon release a new regional energy proposal.[27] While the Russians have tended to resist discussion of energy in the NATO-Russia Council, Moscow's energy proposal provides a good opportunity for Western nations to engage Russia further on this issue. Charges that Russia monopolizes energy supply lines for political purposes must be addressed, and developing a sustainable energy security regime for Europe-Eurasia will be important for the coming decades. Continuing to seek diversification of suppliers and alternative sources could provide European/Eurasian nations greater leverage in working with Moscow in this area.

Similar discussions should be initiated with Russia in both bilateral and multilateral forums to tackle the transnational threats of terrorism, crime, drug trade, human trafficking, piracy, climate change, poverty, disease/pandemics, and other issues. Further expansion of the security agenda beyond the traditional challenges of arms control and weapons proliferation will benefit the cultivation of sustainable cooperation with Russia, and improve capacity for responding effectively to the most serious global threats confronting the world community in the coming decades.

U.S./European-Russian transnational security initiatives must reach beyond the transatlantic community to build partnerships throughout the world that include China, India, and major actors in the Middle East and elsewhere. There is no reason that NATO should not open dialogue mechanisms with the nations of the SCO on topics of common transnational security interest such as countering terrorism, weapons proliferation, etc. The most pressing security challenges will only become increasingly global in the coming decades, and a strong Euro-Atlantic community to include Russia could provide a foundation for reaching out to build partnerships throughout the world to address these challenges.

The United States and Europe/European Union commercial communities can play a central role in setting relations with Russia on a more positive course for the future. The United States and European partners should support Russia's progress toward membership in the World Trade Organization. Outdated trade restrictions in the United States, relics of the Cold War such as the Jackson-Vanik Amendment, should be removed/graduated, and the new leadership in Washington and Moscow should involve the private sectors of both nations and Europe in seeking means for advancing economic ties. The infusion of expertise, technology, investment, and human support provided by the Western private sector will continue to be important to Russia, and should be fully tapped in support of security objectives.

Finally, the United States and NATO must devote greater focus and resources to improving mechanisms for communication and security dialogue with Russia at every level. In the aftermath of the Georgian conflict, Russian defense expert Dr. Dmitri Trenin of the Carnegie Foundation in Moscow claims that Moscow views new NATO members as "little more than U.S. satellites, ready to act as platforms to launch American armed forces."[28] This is a genuine reflection of perceptions of many experts and officials of Russia's foreign policy community, even to include those with a more receptive attitude toward the West. U.S. strategic communication or soft/smart-power initiatives directed toward countering distorted narratives and enhancing understanding of U.S. values, objectives, and intentions are sorely needed. The newly instituted Russian-American dialogue group cochaired by Henry Kissinger and Yevgeny Primakov involving bipartisan representation to encourage discussion with Russians on strategic issues is promising, and promoting additional similar exchanges at every level can contribute to better understanding and defining common approaches for potential collaboration.

Persistent Challenges for U.S./NATO-Russian Engagement & Integration

There are several persistent challenges to long-term U.S./NATO-Russian engagement. The unwillingness of Moscow to provide transparency necessary for fostering greater cooperation with Western military forces will continue to impede cooperation. Russian officials were prepared to sacrifice the lives of sailors trapped in the *Kursk* submarine catastrophe of August 2000, rather than to accept assistance from the United States, which would have exposed the inadequacies in capabilities. Both Yeltsin and Putin placed military reform among the highest priorities, but neither was able to achieve the progress desired in revamping Russia's military-industrial complex. Dmitri Medvedev and his reform-oriented minister of defense Anatoliy Serdyukov will still have to overcome the inertia of the Soviet mind-set remaining in Russia's senior officer staff to implement their plans for far-reaching transformation of the armed forces.[29]

NATO's engagement with Russia will depend on the capacity of the alliance itself to overcome differences. Division among the United States and the old and new alliance members in Europe with respect to Russia and other issues will continue to present difficulties in developing a common, consistent, and constructive approach toward Russia. Former Soviet bloc nations of East-Central Europe and Eurasia would undoubtedly benefit from improved relations among the United States, NATO, and Russia, but it remains quite difficult for these emergent nations to overcome their historical memories of fear and distrust of Russia and Russian intentions.

In addition, the future of the U.S./NATO-Russian bilateral security relationship is likely to hinge to a great extent on Russia's relationship with its neighbors and resolution of frozen conflicts in Eurasia, especially to include South Ossetia/Abkhazia, Transdniestria, Crimea, and Nagorno-Karabakh. U.S. vice president Joe Biden's point that the United States will not recognize "any nation having a direct sphere of influence" stands in direct contrast to Moscow's entitlement to a "zone of privileged interests" with respect to its neighbors.[30] Russia's *Foreign Policy Concept* issued in July 2008 established the pretext for further Russian intrusions in post-Soviet neighboring nations by including a provision for protecting Russians living outside its borders.[31]

The development of Russia's national identity and progress in evolving increasingly to a free, open society based on a democratic value system will undoubtedly be critical for deepening Russia's relationship with the United States and European countries at every level. The fact that so much uncertainty remains in resolving Russia's identity makes it difficult to build a shared vision with Western nations. At the most fundamental level, the NATO alliance coalesced on the basis of adherence to common values. Surely the United States and European and NATO nations must exercise patience with respect to Russia's transition and abandon unrealistic expectation that the process of reform would require less than a period of decades. At the same time if Russia's development takes on increasingly authoritarian aspects in the future, then the potential for full integration of Russia into the community of Western nations will never materialize.

Finally, while it is difficult to move beyond the mind-set/stereotypes of the Cold War, these perceptual legacies must be overcome in order to build a more constructive U.S./NATO-Russian relationship suited to meeting the challenges of the twenty-first century. Almost twenty years after the collapse of the Soviet Union and the Warsaw Pact, mutual suspicions remain. Both NATO nations and Russia must put to rest the vision of a bifurcated European continent and realize the full potential for NATO-Russian cooperation on countless priority security issues. Otherwise, NATO's existence and activities are likely to continue to thwart relations with Russia. Remaining locked in the past recreating an East-West zero-sum rivalry will not serve the interests of the United States, Russia, Europe, or the broader community of nations.

Notes

The views expressed do not necessarily reflect the official policy or position of the Marshall Center, the U.S. Department of Defense, the German Ministry of Defense, or the U.S. or German governments.

1. Vladimir V. Putin, "Speech at the 43rd Munich Conference on Security Policy" (February 10, 2007), available at http://www.securityconference.de/konferenzen/rede.php?menu_2007=&menu_konferenzen=&sprache=en&id=179&.
2. Linda D. Kozaryn, "World Much More Complex Than during Cold War, Gates Says," DefenseLink, February 11, 2007, http://www.defenselink.mil/news/newsarticle?id=3021.
3. Ross Colvin and Noah Barkin, "Biden Vows Break with Bush Era Foreign Policy," Reuters, February 7, 2009.
4. NATO, "Declaration on Alliance Security, Issued by the Heads of State and Government Participating in the Meeting of the North Atlantic Council in Strasbourg/Kehl," news release, April 4, 2009.
5. Steven Erlanger, "NATO Struggles with Conflicting Views on Russia," *International Herald Tribune,* April 10, 2009.
6. Mikhail Gorbachev, discussion with the author, Gorbachev Foundation, Moscow, Russia, April 1999, originally cited in Sharyl Cross, "Russia and NATO toward the Twenty-first Century: Conflicts and Peacekeeping in Bosnia-Herzegovina and Kosovo," *Journal of Slavic Military Studies* 15, no. 2 (June 2002).
7. Boris Yeltsin, remarks (CSCE Summit, Budapest, December 5, 1994), cited in "Peace and Security: Defining the Roles of the U.N., NATO and CSCE," *Foreign Policy Bulletin: The Documentary Record of United States Foreign Policy* 5, nos. 4–5 (1995), p. 11.
8. Victor A. Kremenyuk, "Rossiya-SShA: Pervye Yroki Balkanskovo Krizisa," *SShA,* no. 1 (January 2000), p. 6.
9. *Foreign Policy Concept of the Russian Federation* (June 1993); and *Foreign Policy Concept of the Russian Federation* (June 2000).
10. *Rossiyskaya Gazeta,* January 18, 2000.
11. "Putin Says 'Why Not' to Russia Joining NATO," *Washington Post,* March 6, 2000; "Putin Says Russia Could Join NATO," *The Independent,* March 6, 2000; "Russia Could Join NATO If Terms Equal," Reuters, March 5, 2000; and "Putin Brings Passion, Earnestness to Summit," Reuters, March 5, 2000.
12. Putin, "Speech at the 43rd Munich Conference."
13. Dmitry Medvedev, "Address to the Federal Assembly of the Russian Federation" (Moscow, November 5, 2008), available at http://www.kremlin.ru/eng/speeches/2008/11/05.
14. "Russian President Says No Trade-Off between Prosperity, Democracy," Radio Free Europe/Radio Liberty (RFE/RL), April 15, 2009.
15. See "Medvedev Reaches Out to Russia's Unruly Bloggers," Agence France-Presse, April 22, 2009; "The Internet as Collective Organizer, Global Network Gives Russian Democracy a New Opportunity," *Nezavisimaya Gazeta,* April 21, 2009.
16. *Review of Foreign Policy of the Russian Federation* (Ministry of Foreign Affairs of the Russian Federation, March 27, 2007).
17. Igor A. Zevelev, lecture (Marshall Center, via VTC from Washington, DC, March 11, 2008).
18. See "Concept of the Long Term Socioeconomic Development of the Russian Federation to 2020" (Ministry of Economic Development and Trade, Moscow, September 2007); and Andrew C. Kuchins, Amy Beavin, and Anna Bryndza, *Russia's 2020 Strategic Economic Goals and the Role of International Integration* (Center for Strategic and International Studies [CSIS], July 2008).
19. See Zeljko Bogetic, "Global Crisis and Its Impacts on Russia: Refocusing Policy on Households," *International Economic Bulletin,* April 2009.
20. Mikhail Margelov, "Starting Fresh with Obama," *The Moscow Times,* January 20, 2009; and "Creating a New Reset Button," *The Moscow Times,* April 17, 2009.
21. Ambassador Dmitri Rogozin, remarks in Brussels, Belgium, at a conference session hosted by the East-West Center, New York, November 25, 2008.
22. Ibid.

23. For further perspective on Medvedev's proposal see Sergei Lavrov, "Russia and the World in the 21st Century," *Russia in Global Affairs,* no. 3 (July–September 2008).
24. "Medvedev Says NATO Exercises in Georgia 'Open Provocation,'" RIA Novosti, April 30, 2009; "Russia Demands NATO Call Off Exercises in Georgia," Reuters, April 16, 2009; and Pavel Felgenhauer, "Russian Forces Deploying for Possible Action in Georgia," *Eurasia Daily Monitor* 6, no. 73 (April 16, 2009).
25. See Cross, "Russia and NATO toward the Twenty-first Century."
26. "Obama, Medvedev Agree to Pursue Arms Deal," RFE/RFL, April 1, 2009.
27. "Medvedev to Float European Energy Security Proposal," RFE/RL, April 20, 2009.
28. Dmitri Trenin, "To Be Free from America," *Newsweek,* September 1, 2008.
29. See "Challenges to Russian Military Reform," STRATFOR, February 10, 2009; and "Medvedev Reforming a Military in Crisis," *Eurasia Daily Monitor* 6, no. 61 (March 31, 2009).
30. Colvin and Barkin, "Biden Vows Break with Bush Era Foreign Policy."
31. *The Foreign Policy Concept of the Russian Federation* (July 15, 2008).

Resetting U.S.-Russian Relations

Dr. R. Craig Nation
Professor of Strategy and Director, Eurasian Studies
U.S. Army War College

Introduction

The Five-Day War between Georgia and its breakaway province of South Ossetia and the Russian Federation waged during August 2008 bore many of the marks of a war by proxy. Since his accession to power in the "Rose Revolution" of 2003 President Mikheil Saakashvili had gone out of his way to align his country with the United States and attract security cooperation assistance. Saakashvili was regarded in Moscow as a U.S. protégé, and his ebullient personality and assertive local agenda were both resented and feared. By thwarting Georgia's effort to reestablish its territorial integrity by restoring control over South Ossetia, massively damaging Georgia's armed forces, and humiliating its leader, Russia seemed to be sending a message to the United States as well. The Five-Day War was not only a setback for the Republic of Georgia. It was also "a post–Cold War nadir for U.S.-Russian relations."[1]

Near-panicked reactions to the Russian incursion by Western observers may have been cathartic.[2] In the aftermath a concerted effort has been launched by both the Russian Federation and the United States to back away from confrontational policies and, in the phrase originally coined by U.S. vice president Joseph Biden, "press the reset button" in order to place relations on a more stable and businesslike foundation.[3] The apparent thaw in U.S.-Russian relations that has followed seems full of hope, even if to date it has been short on substance. But major challenges lie ahead. The U.S.-Russian relationship remains a significant axis in world politics, with considerable interests and complex issues at stake. Short-lived flowerings of goodwill have withered in the past. If the current attempt at normalization is to succeed where others have failed it will need to rest upon a more sophisticated understanding of larger trends working to shape U.S.-Russian relations both for good and for ill.

From Partnership to Confrontation

U.S. relations with the Russian Federation have moved through two broad phases since the collapse of the USSR in 1991. The first phase corresponded to the period of optimism inspired by the end of the Cold War, captured intellectually by Francis Fukuyama's influential "end of history" argument, which equated the Soviet defeat to the universal and enduring triumph of Western institutions and values.[4]

On the basis of such presumptions and in the wake of the Soviet implosion, a policy of *strategic partnership* was articulated by the administrations of William Clinton and Boris Yeltsin. This policy rested on a series of assumptions on both sides. The United States assumed that Cold War rivalry between the West and the USSR was essentially ideological; that the sources of strategic friction had disappeared along with Soviet communism; that the new Russian Federation was well-placed to make an efficient transition to democratic norms, the market economy, and the rule of law; and that this new Russia would share a wide range of common interests with the United States and be capable of functioning as a legitimate partner.[5]

On the Russian side, as articulated by Yeltsin's foreign minister Andrei Kozyrev, it was assumed that Russia was indeed an integral part of the historical West, that the new Russia was committed to the pursuit of a far-reaching process of transition, that substantial moral and material support from the advanced countries of the West led by the United States would be required to effect that transition, and that it was in Russia's best interest to subordinate its short-term national goals to the overriding interest of association with its inescapable post–Cold War sponsor.[6]

These assumptions were quickly exposed as flawed. The new Russia was too weak, troubled, and poorly led to function as a "partner" in any positive sense. The United States lacked significant motivation to offer concessions to a former rival perceived to be on the ropes. The impact of Yeltsin's failed economic policies, combined with his deference to U.S. international priorities, conjured up a powerful domestic reaction to the entire transition agenda. Russia's fragility, and unreliability as an international partner, was brutally revealed by the armed confrontation between president and parliament in October 1993 and subsequent elections that made the Liberal Democratic Party, led by the demagogic populist Vladimir Zhirinovskii, Russia's leading political force in voting by party list.

Though the rhetoric of partnership lingered, from 1994 onward the Clinton administration moved toward a policy of *selective engagement* toward Russia. The illusions of alliance were abandoned, but it was presumed that despite its unreliability Moscow could still be a useful associate in specific areas of mutual interest. From the U.S. side this meant engaging Russia in regard to questions such as nonproliferation and regional stability on the Russian periphery, while being careful not to dilute the U.S. strategic agenda in deference to Moscow's preferences.[7] That agenda included initiatives to which Russia strongly objected, including NATO enlargement, the Partnership for Peace program projecting NATO-sponsored security cooperation relationships deep into Eurasia, and an interventionist policy in the Yugoslav wars of secession. From the Russian side, the replacement of Kozyrev as foreign minister by Evgenii Primakov in January 1996 was accompanied by the articulation of a more distanced and competitive international posture based on a rather traditional definition of national interests.[8] The ultimate decider was not Primakov, however, but rather Yeltsin and his corrupt "Family," for whom self-interest, rather than national interest, was the ultimate rationale. Yeltsin worked on the presumption that Russia could not afford to defy the "sole remaining superpower," and consistently deferred to U.S. priorities when forced to choose.

In office as president from 2000 to 2008, Vladimir Putin brought Primakov's deputy Igor Ivanov with him as new foreign minister, and many of Primakov's priorities as well. Buoyed by windfall profits from the oil and natural gas sector, under Putin Russia experienced a dramatic economic revival accompanied by expanding international self-confidence. The disorder of the Yeltsin years gave way to an agenda for rebuilding state authority under strong central direction. In fact, Russia was in the process of becoming the kind of state which the United States might seriously have envisioned as a strategic partner in 1992. Meeting with Putin in Ljubljana, Slovenia, in June 2001, U.S. president George W. Bush struck up what would prove to be a positive personal relationship with his Russian counterpart, famously remarking that he "looked the man in the eyes" and "was able to get a sense of his soul."[9]

A more substantial foundation for strategic partnership seemed to come into place following the terrorist attacks of September 11, 2001. Putin moved quickly to express solidarity with the United States in its newly coined global war on terrorism (which he likened to Russia's ongoing war in Chechnya), gave a green light to the establishment of U.S. military bases in post-Soviet Central Asia, cooperated in meaningful ways (using leverage with Afghanistan's Northern Alliance and intelligence sharing) with the U.S. military campaign in Afghanistan, partnered with the Proliferation Security Initiative, and opened a renewed energy security dialogue. Putin's initiatives in the wake of 9/11, imposed in the face of significant domestic opposition, have been described as a "strategic choice" in favor of enhanced cooperation with the West and away from a traditional paradigm of geopolitical competition.[10] If so, the choice was not reciprocated. In crafting its war on terror, notes Lilia Shevtsova (no friend of Putin's Russia), the United States "took Russia for granted, overlooked the issues that divided them, and put U.S. strategic interests first, without paying attention to Moscow, in just the same way that it paid scant attention to its Atlantic allies."[11]

The promise of renewed strategic partnership proved to be a mirage. Russian-American cooperation in Afghanistan and elsewhere was short-lived, while sources of disagreement and confrontation (NATO enlargement, the American military presence in Central Asia, frozen conflicts on the territory of the former USSR, the Kosovo question, access to the oil and natural gas resources of the Caspian basin, the premises of Russian governance, U.S. military action in Iraq, the missile defense program in central Europe, etc.) proliferated. A kind of rhetorical low point was reached in the course of 2006–2007. In May 2006 U.S. vice president Richard Cheney, during state visits to Lithuania and Kazakhstan, publicly denounced the Kremlin for backsliding on democratization and using oil and natural gas exports as "tools of intimidation and blackmail," supported the construction of oil and natural gas pipelines that would bypass Russia to the south, and specifically described Russia as "not a partner or trusted friend."[12] The *New York Times* grudgingly acknowledged that the "Bush administration was right to re-examine its stance on Russia" while in the Russian press the vice president's remarks were being compared to Winston Churchill's 1947 "Iron Curtain" speech and described as a new declaration of cold war.[13] In his annual speech to the Russian Federal Assembly, Putin shot back with a call for enhanced military readiness and a curious reference to the United States as "Comrade Wolf, who eats in silence, and knows who to eat."[14] At the annual Munich *Wehrkunde* in February 2007 Putin upped the ante in a ferocious speech that characterized the United States as an aspiring global hegemon seeking to impose its will by military force by constantly intervening outside its borders where its actions "only made things worse," whose excessive ambitions risked to destroy it from within.[15] U.S. secretary of defense Robert Gates responded temperately, noting the habit of "old spies" to "speak bluntly," and insisting that "one Cold War was quite enough."[16] But Russian evaluations were perhaps more apropos, underlining that the exchange of remarks at the conference "does not fit into the framework of a dialogue between partners" and calling attention to the apparent readiness of both sides "to take steps leading the two countries towards confrontation."[17] These remarks seemed to be validated in August 2008 by the events of the Five-Day War. The origins of the conflict remain disputed, and its sources were clearly complex, but the possibility of Russian-American strategic friction contributing to the outbreak of armed conflict on the volatile Russian periphery was made unmistakably clear.

The Foundations of Strategic Rivalry

The failure to reestablish a meaningful Russian-American partnership in the wake of the 9/11 attacks was to some extent the result of choices made by leaders on both sides, for whom national priorities took precedence over a bilateral relationship whose relevance seemed to lie in the past. But the sources of strategic rivalry between Russia and the West also have a structural foundation. Russia and the United States are different kinds of polities with distinct strategic cultures and sometimes sharply contrasting priorities. A history of rivalry has led to the accumulation of considerable mistrust. If the challenge of recasting bilateral ties is to be addressed more effectively in the future it is important to grasp the systemic forces working to reshape the roles that both states can aspire to play in international affairs.

Resurgent Russia

The history of the independent Russian Federation divides rather neatly into parts. The first, under the direction of Boris Yeltsin, dates from 1991 to 1999. It was characterized by a historically unprecedented national decline. The Soviet Union was declared out of existence as a result of a cabal-like encounter between three men (Yeltsin, Stanislav Shushkevich, and Leonid Kravchuk representing Russia, Belarus, and Ukraine respectively) in the forests of Belovezhskaya Pushcha on December 8, 1991.[18] The decision sealed Western "victory" in the Cold War, but worked against Russia's traditional national interests. The demise of the USSR reduced the state's territory by 20 percent and its population by nearly half. Soaring inflation eliminated savings overnight, wage arrears became chronic, unemployment leaped upward, and social ills such as alcoholism, drug abuse, criminality, and increased mortality rates became national plagues. From 1991 to 1999 Russian GDP per capita declined by about 50 percent and 90 percent of the population experienced sharp reductions in income.[19] Against the foundation of popular disaffection occasioned by these conditions, political instability became chronic as well—in all his years in office Yeltsin never achieved a stable and supportive parliamentary majority. The low point arrived in the summer of 1998 with a catastrophic financial crisis leading to the collapse of the ruble and a major default. In December 1999 an increasingly ill and dysfunctional Yeltsin drew the appropriate political conclusion, resigning on behalf of his self-selected successor, the still virtually unknown Vladimir Putin.

Putin's terms in office produced very different results. The new president conveyed an image of competency that contrasted sharply with Yeltsin's painfully visible decline, while economic performance was dramatically reversed. From 1999 to 2008 the Russian economy grew at an annual rate of over 7 percent, and cumulatively by over 65 percent. The driving force was a dramatic increase in oil prices (at $10 per barrel in 1998 and peaking at over $140 per barrel in the summer of 2008), but the energy sector served as a stimulus to other parts of the national economy as well.[20] With revenues in excess of expenditures by over 8 percent Russia was able to pay down outstanding international debts and build up imposing currency reserves. Russia remained a highly inegalitarian society, but the social benefits of national revival were notable. By 2007 the number of Russians living below the officially designated poverty line had fallen from 42 million to below 20 million and unemployment had declined from 10 percent to 6 percent. Economic recovery was accompanied by a dramatic, though perhaps exaggerated, revival of national self-confidence. Yeltsin's Russia was perceived as a weak state in precipitous decline. Putin presided over its transformation into a strong and stable polity well aware of its great-power tradition and determined to reassert itself as a factor in world affairs. This "new" Russia was in fact a partial reincarnation of the traditional Russian and Soviet great powers—inevitably an assertive and difficult international actor.

Authoritarian Drift

Economic revival under Putin was accompanied by a concentration of political authority in the hands of the state. The Putin leadership never sought to downplay its intentions in this regard. The Yeltsin period was portrayed as a phase of destructive chaos and anarchy. Russia, it was claimed, needed strong central direction and a political regime that suited its character and traditions, idealized by notions such as the "dictatorship of law," "sovereign democracy," and the "verticality of power" (*vlastnaia vertikal'*).[21] In practice, this emphasis contributed to the drift away from democratic norms that has become a major item on the Western bill of indictment against the potential of the new Russia as an international partner. The critique has a degree of validity. Putin oversaw the elimination of direct election of Russia's regional governors, dropped regional constituencies in favor of election by party list with a high 7 percent bar for representation designed to discourage opposition, used blatant judicial manipulation to intimidate or destroy potential rivals, and applied state assets purposefully to advantage his own dominant "party of power" *United Russia*.[22] His regime championed state buyouts of major media outlets and in general imposed greater uniformity in media coverage, and imposed state supervision on the economically critical oil and natural gas industries as well as other so-called national champions. According to some, Putin's power actually rests upon a state within the state consisting of well placed former associates drawn from the state security apparatus (the so-called *siloviki*).[23] Most ominously, Putin seems to have looked on while a virtual campaign of intimidation and murder was waged against regime critics.[24] These trends have affected foreign policy decision making as well. Assertive nationalism and an emphasis upon a distinct and ineffable Russian national idea gained greater salience in foreign policy discourse. Policy formation was concentrated within the presidential administration and remained quite opaque, if not unfathomable, to external observers.[25] George Bush's second inaugural address highlighted the promotion of democracy as a basic goal of U.S. policy, in sharp contrast to what seemed a developing Russian reality.[26]

Russia's reputation as a state with a "distinct Russian approach to politics . . . based on xenophobia, authoritarianism, historical revisionism, and exceptionalism" contributes significantly to its identification as an undeserving, unreliable, and menacing international actor.[27] This too, of course, is nothing new—Russia's distinctiveness has always set it apart. The challenge of bridging the cultural and institutional gap and engaging Russia on a basis of respect and equality has been and will continue to be considerable.

Decline of the West?

Since 2003 the United States has been engaged in two simultaneous and open-ended regional conflicts. These wars have included abuses that have cast international opprobrium on the United States and its armed forces. Strategic overextension, the perception of miscalculation and failure, and associated loss of international stature and respect, have placed limitations upon the U.S. ability to impose its will. Accumulated debt and increased economic dependency have likewise called U.S. economic leadership into question. To the extent that there ever was a post–Cold War "unipolar moment" it may be said to have passed. The relative decline of U.S. power is arguably a reality that encourages a more assertive policy on the part of aspiring regional influentials, including the Russian Federation.[28]

During the 1990s the European Union (EU) sought to develop a more dynamic policy toward the Russian Federation and the so-called Common Neighborhood along their interface from the Baltic to Black Seas and into the Caucasus.[29] In conception this policy was to have been built upon expanding cooperation with Moscow institutionalized by a Partnership and Cooperation Agreement in force from 1997 and a "Common Strategy" seeking to define common spaces for collaboration developed in 1999.[30] Ironically, the process of EU enlargement worked to undermine this intent in two different ways. Bringing former members of the Soviet bloc into the EU had a significant impact upon institutional priorities, creating an anti-Russian lobby leery of expanded cooperation with a Russia hypothesized as a historical threat. Polish opposition would eventually scuttle attempts to renew the EU-Russia Partnership and Cooperation Agreement on schedule.[31] Rapid growth also had the effect of stimulating "enlargement fatigue" among many European citizens and increased skepticism about the larger agenda for ever-closer union. Such perceptions contributed to the defeat of projects for EU constitutional reform in referendums in France and the Netherlands in 2005 and again in Ireland in 2008.

As a result the EU's aspiration to redefine itself as a coherent strategic actor has lost momentum, and its ability to impose meaningful constraint on a recalcitrant Russia has been reduced. Russia continues to function as a kind of free agent in international affairs, tied to but not bound by the priorities and concerns of the West.

A Moscow-Beijing Axis?

During the Putin years Russia succeeded in establishing a strategic partnership with the People's Republic of China with considerable more substance and dynamism than its flagging relationship with the United States. Russo-Chinese relations rest upon a firm foundation of mutual interests—increased commercial relations, energy security agreements, significant Russian arms transfers, shared commitments to regional security and a reduced U.S. presence in Inner Asia, and political support in international forums regarding issues of high national priority. Both parties have consistently opposed a presumed U.S. unilateralism and hegemony on behalf of a "multipolar" alternative.[32] The relationship is institutionalized in the Shanghai Cooperation Organization, which since its origin in 2001 has gradually become a more significant Eurasian forum with the potential to develop and grow.[33] The principals doubtless have instrumental reasons for association, but their relationship almost certainly represents more than an "axis of convenience."[34] The calculus of power is still a significant foundation for interstate relations. Balancing against the intrusive role of the United States in post-Soviet Eurasia has been a logical gambit for both Moscow and Beijing, one that has served their interests well and is likely to remain in place as a stabilizing factor in the politics of the region for the foreseeable future. Russia is empowered by positive association with its powerful Chinese neighbor, and better positioned as a result to assert its prerogatives regionally and globally.[35]

The Geopolitics of Eurasia

The Soviet collapse created a geopolitical vacuum along the Russian periphery that the Russian Federation and the Western security community have both sought to fill, with sharp geopolitical competition, sometimes characterized as a "new great game," the result. Under Kozyrev's direction Russia briefly aspired to "join" the West on a basis of equality.[36] Under pressure domestically, this priority was quickly conditioned by the need to cultivate an autonomous political sphere of influence within the boundaries of the former USSR. Already in 1993 a new Russian military doctrine defined the Commonwealth of Independent States as a region of "vital interests," and a new foreign policy concept spoke of Russia's "special responsibilities" in the former Soviet space.[37] The transition from Kozyrev to Primakov marked a significant watershed. Henceforward the project of association with the Western security community would be rejected as both impracticable and undesirable.[38] This turn away from the West was demonstrated in a literal way by Primakov's dramatic decision, made during March 1999 while en route to Washington for a meeting with U.S. vice president Albert Gore, to turn his plane around in mid-air and return to Moscow in protest against the U.S. decision to go to war in Kosovo.[39]

Resistance to what is perceived as Western encroachment upon Russia's self-styled area of special responsibility in the new Eurasia (in the form of NATO enlargement, the EU Common Neighbourhood Policy, the U.S. missile defense program, democracy promotion initiatives, sponsorship of anti-Russian regional forums such as GUAM—the Organization for Democracy and Economic Development, etc.) has become a major source of strategic friction.[40] From the Baltic region, through the central European corridor, along the Caucasus and beyond the Caspian Sea into Inner Asia, the Russian periphery has become the site of some of the world's most intractable regional conflicts. Post-Soviet Eurasia has become a contested strategic space where the struggle for leverage and influence is defined by both sides as a zero-sum contest with little room for flexibility or compromise. In this environment, the familiar call to "contain" Russian expansion has arisen as if by reflex.[41] In the wake of the armed conflict over South Ossetia hostile rhetoric has on occasion become inflammatory, with at least one influential U.S. commentator openly calling for U.S. support for secessionist movements inside the Russian Federation itself.[42]

Many factors contributed to the failure to preserve the spirit of Russo-American cooperation briefly manifested in the immediate aftermath of the 9/11 terrorist attacks. The Russian and American leaderships may both be faulted for an unwillingness to engage in good spirit in a process of cooperation. Building a foundation for Russo-American rapprochement, however, has never been an easy or self-evident endeavor. On both sides, the legacy of the Cold War is still alive—Russophobia and anti-Americanism remain powerful forces among respective policy elites. Shifting power balances encourage rivalry. The U.S. determination to sustain its status as the sole remaining superpower has generated the creation of countervailing coalitions inspired by the ideal of multipolarity, with Russia as an enthusiastic partner. A contrasting discourse of values has proven to be extraordinarily divisive. Russo-American enmity may not be written in the stars, but there are significant structural factors at work that make strategic partnership a challenging goal.[43]

Pushing the Reset Button

If it is not difficult to explain the difficulties and inconsistencies that have defined the U.S.-Russian relationship in the post-Soviet period, it has become increasingly difficult to justify them. Global economic recession has created new strains that threaten to promote destabilization and transform latent conflicts into open confrontations. Russia's fragile recovery has been pushed off the rails. The U.S. role as global economic leader has been called into question. Both countries confront pressing problems that have little to do with the legacy conflicts that divide them. Under these circumstances, U.S.-Russian strategic rivalry has great potential to do harm, but little capacity to serve positive goals. The competitive political environment that has been allowed to develop, harshly revealed by the Five-Day War, is arguably a luxury that neither side can afford in the challenging geostrategic environment produced by a process of globalization that has yet to be mastered, combined with increased economic stress.[44]

Global recession has set the stage for a renewed effort to stabilize U.S.-Russian relations. Rather than view the Russo-Georgian conflict as a watershed leading toward renewed hostility, both sides seem to have understood it as a warning shot highlighting the risks of uncontrolled strategic rivalry. New leadership on both sides has reinforced these conclusions. The administration of U.S. president Barack Obama has clearly articulated the goal of re-engagement.[45] Russian president Dmitri Medvedev has sought to cultivate a public image as a socially conscious reformer and champion of international stability on a foundation of international law and organization.[46] The role of Prime Minister Putin in the new administration is a subject of some dispute, but need not be equated with a hard-line alternative. The U.S.-Russian Strategic Framework of April 2008, with its assertion that "the era in which the United States and Russia considered one another an enemy or strategic threat has ended," and rejection of "the zero-sum thinking of the Cold War" can serve as a positive foundation for new thinking.[47] The problem becomes how to translate abstract principles into practical initiatives. The much-publicized meeting between Secretary of State Hillary Rodham Clinton and Foreign Minister Sergei Lavrov in March 2009, where Lavrov accepted the gift of a "reset" button as a symbol of a "new start" in bilateral relations, seemed to be an auspicious inauguration.[48] The ball has now passed into the courts of Presidents Obama and Medvedev, who reiterated assertions of goodwill during their first meeting as heads of state in April 2009.[49] The atmospherics of Russian-American relations had clearly changed for the better, but the most difficult sources of division and rivalry remained to be confronted.

What needs to be done to address the systemic sources of Russo-American rivalry and prevent the current phase of engagement from going the way of other short-lived attempts at improved understanding? Recognizing the considerable weight of inherited hostility and the magnitude of the task ahead is a good starting point. Beyond that, some more practical suggestions linking a reconceptualization of the relationship to positive policy initiatives may be in order.

The Discourse of Values

Chronic criticism of Russia's domestic regime based upon a purported clash of values between Russia and the West has become an independent variable exerting a divisive and unhelpful influence on the U.S.-Russian relationship. There is of course nothing wrong with calling attention to the many imperfections in the current Russian variation on democracy. However, the United States has little or no effective leverage over issues related to Russian domestic governance. Moreover, the tenor of U.S. criticisms is often exaggerated or misplaced. It is popular to label Russia as an "authoritarian" state as though authoritarianism and democracy were existential states of being. Democracy is a process, and though Russia's institutions are clearly imperfect they may fairly be described as the most open and democratic in spirit in the country's millennial history. To speak of a "war of values," in the telling phrase of Edward Lucas, as "the biggest question" in relations between Russia and the West borders on irresponsibility.[50] To make democratization on Western terms a litmus test for political reliability, as many urge, is not sound policy.[51] The United States does not apply such a litmus test to its relations with other key international actors. Russia's leaders reiterate the conviction that "historically, Russia has always been a part of European civilization."[52] To define Russia's relationship with the West in terms of a purported clash of civilizations is suspect analytically and destructive politically. The United States in particular is no longer in a position (as it may have been until quite recently) to *dictate* the terms of its relationship with Moscow. Respect, reciprocity, and empathy for the partner's situation and concerns will be necessary components of the relationship looking ahead—not rewards for good behavior but preconditions for businesslike interaction.

The Discourse of Interests

Russia and America share a large number of coinciding interests that provide a strong foundation for practical cooperation.

- *Strategic Stability and Nonproliferation:* Together Russia and America control over 90 percent of the world's nuclear weapons and are the only countries that can realistically attack one another and effect decisive damage. They have a powerful vested interest in regulating strategic competition, maintaining transparency, and sustaining a minimal deterrent posture. The strategic pillar of the relationship has been badly neglected in the past, to no one's advantage. Renegotiating the START Accord is a high priority for 2009, hopefully as a step toward a more comprehensive rethinking of respective nuclear postures and doctrines. The U.S. agenda for deploying missile defense systems in Poland and the Czech Republic, negotiated outside the NATO context and in defiance of Russia's repeatedly expressed concerns, without convincing guarantees of the reliability of the systems in question, in response to a hypothetical Iranian threat that has yet to materialize, poses a major barrier to expanded cooperation in these domains. Russia has been briefed repeatedly on the program's intentions and limitations, and itself confronts potential intermediate range missile threats from both China and Iran, but persists in viewing it as a potential threat to the integrity of its strategic deterrent force. In view of the numerous assets that the United States possesses in strategic competition with the Russian Federation, these sensitivities should be respected. Ideally this agenda would be rethought and recast in such a way as to engage Russia as a partner in the context of a larger, reanimated arms control dialogue—a win-win solution that should not be outside the realm of the possible.[53]

  Both parties likewise have a related interest in blocking the proliferation of weapons of mass destruction and reinforcing a robust nonproliferation regime. The currently egregious cases of Iran and North Korea are salient examples of the extent to which cooperation between the world's leading nuclear powers has become an essential foundation for meaningful counter-proliferation policies.

- *Counterterrorism*: Washington and Moscow share a common perception of the threat of catastrophic terrorism emerging from radical jihadist movements. Indeed, Russia has been much more congenial to U.S. definitions of the global terrorist threat than have many key U.S.

European allies. Twenty percent of Russia's population is Muslim and the state's southern boundary corresponds to a major Huntingtonian "fault line" with the Islamic world. Russia's geopolitical situation and security assets (including intelligence capability, military, and police forces) position it to play the role of a useful ally in a "long war" against the terrorist threat.

- *Energy Security*: As the world's largest consumer and producer of energy resources, respectively, the United States and Russia should have a vested interest in regulating global energy markets to their mutual advantage. Stability of supply and stability of demand are not mutually exclusive categories. In this regard the "battle of pipelines" that currently categorizes strategic competition in the Caspian basin, driven by geopolitical considerations rather than commercial logic, represents yet another luxury that neither Russia nor the West can really afford. Russia has been successful in warding off Western advances and sustaining a competitive position in the Caspian region, but at the price of aggravated regional tensions and chronically troubled relations with its Western rivals.
- *Confronting the Global Crisis*: The new Russian Federation is a fully converted participant in the global capitalist market economy. Its economic fortunes have an important influence on systemic stability and it possesses important economic assets. The United States should move with alacrity to eliminate the Jackson-Vanik Amendment, a Cold War relic that benefits no one, and use all its influence to bring Russia into the World Trade Organization (WTO) as soon as possible. The conditions for WTO membership will stimulate positive economic reform domestically, and the more Russia is integrated into world markets the more closely our mutual interests in economic stabilization will be aligned.
- *Eurasian Security:* The Eurasian landmass has become the venue for many of the world's most threatening regional conflicts. Finding a way to escape from the zero-sum logic of U.S.-Russian competition in the new Eurasia is an essential basis for meaningful rapprochement. Important vested interests are at stake, however, and the challenge is more easily evoked than resolved. From a Western perspective the momentum of NATO enlargement into the post-Soviet space needs to be slowed, frozen, or if possible put off until the Greek calends. NATO enlargement is not an inherently destabilizing dynamic. Viewed objectively the alliance does not threaten the Russian Federation and really is in the business of exporting the kind of stability that should work to Russia's advantage. But there is a large consensus within the Russian elite on all sides of the policy spectrum to the effect that NATO as a military alliance and historic rival represents an objective threat whose will to absorb strategically and culturally sensitive regions such as Ukraine and Georgia is an intolerable affront. The common assertion that sovereign states "have a right" to choose their associations is formally correct but

politically irrelevant. Under current circumstances NATO membership for Ukraine or Georgia is not commensurate with an agenda for resetting U.S.-Russian relations—and might well become the prelude to a really existing new Cold War.

Further NATO enlargement also risks creating an illusory sense of security among new members who see the alliance first of all in its traditional configuration as a forum for collective defense aimed against a real and present Russian threat. It is not at all clear, however, that important alliance members have any real intention of stepping into the breach should push come to shove. Alliance reactions to what appear to have been Russian-sponsored cyber attacks against Estonia in the wake of the "bronze soldier" fiasco were concerned but not forceful.[54] Western responses to Russia's military punishment of Georgia were remarkably tepid. In Georgia the Russian Federation demonstrated both the will and the capacity to use military force to impose itself against weak states on its periphery.[55] Although Georgia had been cultivated for years as a U.S. protégé a forceful response was not forthcoming. The Russian military has suffered terribly in the context of post-Soviet transition, but it remains the world's third or fourth largest land force equipped with thousands of tactical nuclear weapons and a military doctrine that commits to their use if the integrity of the Motherland is threatened. In the context of the Cold War the United States was prepared to risk nuclear war in defense of its alliance commitments in Europe. Whether such a risk makes strategic sense under current conditions might be considered by some to be an open question.

The best way to work around the contrasting perceptions that make these issues so difficult to resolve cooperatively would be to reanimate NATO-Russian dialogue and reconstruct a more robust and dynamic NATO-Russia Council. NATO may have lost its original raison d'être, but it has proven to be a remarkably adaptive and resilient organization. With its unique combination of political and military instruments and substantial legitimacy, the alliance should be capable of playing a positive role as security provider in the new Eurasia. But so long as it is constituted against rather than with the Russian Federation it will be in the business of reproducing the security architecture of the Cold War. Russia does not oppose NATO as such, and President Medvedev has spoken in support of "full fledged, competent, and mutually beneficial relations" with the alliance.[56] Russia's current proposal for a new, pan-Eurasian security architecture seeks to look beyond the outmoded logic of an Iron Curtain, though it may not be realistic as currently defined.[57] A process whereby NATO reaches out to engage Moscow in expanding security cooperation, works to "enlarge the range of questions on which Russia can join the allies on a basis of equality for discussions,"[58] institutionalizes that cooperation in a more firmly grounded NATO-Russia Council, and perhaps establishes a more formal relationship with Moscow's preferred Eurasian security forum, the Collective Security

Treaty Organization, may provide a context for associations that bring to life some of the benefits of the Russian proposal and clear the way for closer association with countries like Ukraine and Georgia outside the zero-sum context of siding "with us or with them."[59]

If meaningful progress is to be made in recasting U.S.-Russian relations it will have to be inspired by a new and expanded conception of security itself. This is not an abstraction derived from the formal study of international relations but a practical challenge critical to sustaining global stability in the face of twenty-first century challenges. The most pressing security concerns of our time no longer relate to the threat of armed aggression by Napoleonic neighbors. Economic instability, social inequities, environmental disintegration, depletion of energy sources, pandemic disease, mass casualty terrorism, the proliferation of weapons of mass destruction, and many other really real and present dangers are challenges that can only effectively be addressed through cooperative initiatives on a global scale.[60] During the Bush and Putin years both Russian and American thinking in this regard seems to have regressed, with renewed emphasis on hard power and geopolitical competition. And yet in terms of conventional threats both countries probably enjoy the most benign security environments that they have ever experienced in their long national histories. Neither Russia nor America poses a direct threat to the other's vital interests. The objective foundations for pragmatic cooperation on behalf of shared interests and mutual security are already in place. The effort to "reset" derives in part from this realization, but will have to wade through an imposing legacy of inherited hostility and systemic rivalry. The alternative conceptual framework provided by a mutual security paradigm can make an essential contribution to pushing an agenda for practical cooperation forward.

Conclusion

A strategic reassessment of the Russian-U.S. relationship that places renewed emphasis upon areas of mutual interest and builds upon an enlarged concept of security offering practical solutions to twenty-first century threats has been long overdue.[61] Such an assessment is now under way. It is by no means certain that it will lead to positive results, but the effort is undoubtedly worth making. We have arrived at a historical juncture where the driving force of global strategy need no longer be great-power rivalry—the kind of rivalry that drove the world into two global wars in the course of the twentieth century. With the partial exception of the threat of mass casualty terrorism, in conventional terms the great powers are probably more secure today than ever before. Reversion to a situation dominated by aggravated great-power competition or a "new cold war" would be a major step backward. Taking advantage of a relatively benign security environment and bringing Russia, China, India, and other emerging powers out from the cold and into a functional great-power concert represents a strategic opportunity. Fixing what is broken in the U.S.-Russian relationship would be one good place to start.

Notes

1. Jeffrey Mankoff, *Russian Foreign Policy: The Return of Great Power Politics* (Lanham, MD: Rowman & Littlefield, 2009), p. 104.
2. For contrasting perspectives see Ryan O'Donnell and Sally McNamara, "The Return of History: Confronting the Russian Bear after the Georgian War," *WebMemo*, no. 2056, September 9, 2008, and Dimitri K. Simes, "Russian Roulette," *The National Interest Online*, October 30, 2008.
3. Craig Whitlock, "'Reset' Sought on Relations with Russia, Biden Says," *Washington Post*, February 8, 2009.
4. Francis Fukuyama, *The End of History and the Last Man* (New York: Penguin, 1992).
5. Michael Cox, "The Necessary Partnership? The Clinton Presidency and Post-Soviet Russia," *International Affairs* 70, no. 4 (1994), pp. 635–658.
6. Andrei V. Kozyrev, "Russia and Human Rights," *Slavic Review* 51, no. 3 (Autumn 1994), pp. 287–293.
7. Detailed in Michael McFaul and James Goldgeier, *Power and Purpose: American Policy toward Russia after the Cold War* (Washington, DC: Brookings Institution Press, 2003).
8. See the text of the Russian National Security Blueprint issued in 1997 in *Rossiiskaia gazeta*, December 26, 1997, pp. 4–5.
9. Caroline Wyatt, "Bush and Putin: Best of Friends," *BBC News*, June 16, 2001.
10. Dmitri Trenin, "Osennii marafon' Vladimira Putina i rozhdenie Rossiiskoi vneshnepoliticheskoi strategii," Carnegie Moscow Center Briefing Paper, November 15, 2001, p. 6.
11. Lilia Shevtsova, *Russia—Lost in Transition: The Yeltsin and Putin Legacies* (Washington, DC: Carnegie Endowment for International Peace, 2007), p. 230.
12. Christine Lagorio, "Cheney: Russia is Limiting Rights," *CBS News*, May 4, 2001; Ilan Greenberg and Andrew E. Kramer, "Cheney urges Kazakhs to Bypass Russia," *International Herald Tribune*, May 6–7, 2006; and Lloyd de Vries, "Cheney Seeks Oil in Kazakhstan," *BBS News*, May 5, 2006.
13. "From Cheney, with Nerve," *New York Times*, May 8, 2006; Sergei Iur'ev, "Cheini poliubil rossian bol'she, chem amerikantsev," *Izvestiia*, May 11, 2006.
14. "Poslanie Prezidenta Rossii Federal'nomy vystupleniia," *Komsomol'skaia pravda*, May 11, 2006.
15. "Vystuplenie prezidenta Rossii Vladimira Putina na Miunkhenskoi konferentsii po voprosam politiki bezopoasnosti 10 fevralia 2007 goda," *Nezavisimaia gazeta*, February 13, 2007.
16. "Secretary Gates Remarks at the Munich Conference on Security Policy in Munich, Germany," *U.S. Department of Defense Transcripts Archive*, February 11, 2007, available at http://defenselink.mil/transcripts/archive.aspx.
17. Iuliia Petrovskaia, "Kholodnaia voina bez islishnevo politesa," *Nezavisimaia gazeta*, February 12, 2007.
18. The decision embodied in the resultant Belovezh Accords or Minsk Agreement was apparently determined in the course of discussion between the three leaders without wider consultation. See Mikhail Gorbachev, *On My Country and the World* (New York: Columbia University Press, 2000), pp. 149–168.
19. Shevtsova, *Russia—Lost in Transition,* pp. 27–28.
20. Andreas Goldthau, "Resurgent Russia? Rethinking Energy Inc.," *Policy Review*, no. 147 (February/March 2008), pp. 53–63. Russia's barrels-per-day oil production also nearly doubled over the period.
21. Vladislav Surkov, "Suvernitet vazhnee demokratii," *Nezavisimaia gazeta*, July 13, 2005; Vladimir Ovchinskii, "Diktatura zakona: promezhtochnye itogi," and Ivan Sukhov, "'Vertikal' vlasti' i samosozananie natsii," *Rossiia v global'noi politike,* no. 2 (March/April 2008).

22. The argument is summarized in Lilia Shevtsova, *Putin's Russia*, rev. ed. (Washington, DC: Carnegie Endowment for International Peace, 2005).
23. Ian Bremmer and Samuel Charap, "The Siloviki in Putin's Russia: Who They Are and What They Want," *Washington Quarterly* 30, no. 1 (Winter 2006–2007), pp. 83–92. The phenomenon of the *siloviki* ("strong men") is detailed in Olga Kryshtanovskaia, *Anatomiia Rossiiskoi elity* (Moscow: Zakharov, 2004), pp. 262–288.
24. "Revisiting Putin's Soul," *New York Times*, December 4, 2006; and Michael Spector, "Kremlin, Inc.," *New Yorker*, January 29, 2007.
25. Dmitri Trenin and Bobo Lo, *The Landscape of Russian Foreign Policy Decision-Making* (Washington, DC: Carnegie Endowment for International Peace), 2005.
26. George W. Bush, Second Inaugural Address, January 20, 2005, available at http://www.batleby.com/124/pres67.html.
27. Edward Lucas, *The New Cold War: Putin's Russia and the Threat to the West* (New York: Palgrave Macmillan, 2008), p. 120.
28. R. Craig Nation, "National Power," in *U.S. Army War College Guide to National Security Issues: Vol. II—Theory of War and Strategy*, Boone Bartholomees, ed., 3rd ed. (Carlisle, PA: Strategic Studies Institute, 2008), pp. 163–175.
29. Karen E. Smith, "The Outsiders: The European Neighbourhood Policy," *International Affairs* 81, no. 4 (2005), pp. 757–773.
30. European Union, "Common Strategy of the European Union of 4 June 1999 on Russia (1999/414/CFSP)," *Official Journal of the European Union*, No L 157, June 24, 1999.
31. Efforts to renew the Agreement are ongoing. See Vladimir Pankov, "Options for the EU-Russian Strategic Partnership," *Russia in Global Affairs*, no. 2 (April–May 2008).
32. Jenny Clegg, *China's Global Strategy: Toward a Multipolar World,* (New York: Palgrave Macmillan, 2009); and "Russia Won't Accept Unipolar World—Medvedev," *Russia Today*, September 1, 2008.
33. E. N. Orazalin, *ShOS-osnovy formirovaniia, problemy i perspektivy: Puti sovershenstvovaniia mekhanizmov sotrudnichestva* (Almaty: Institut mirovoi ekonomiki i politiki pri Fonde Prezidenta RK, 2007).
34. See the argument in Bobo Lo, *Axis of Convenience: Moscow, Beijing, and the New Geopolitics* (London: Chatham House, 2008).
35. Aleksandr Lukin, "Usilit' aziatskii vektor," *Rossiia v global'noi politike* 7, no. 2 (March–April 2009), pp. 82–95.
36. A. V. Kozyrev, *Preobrazhenie* (Moscow: Mezhdunarodnye otnosheniia, 1995).
37. See Andrei Tsygankov, *Russia's Foreign Policy: Change and Continuity in National Identity* (Lanham, MD: Rowan & Littlefield, 2006), pp. 78–81.
38. Mankoff, *Russian Foreign Policy*, p. 62.
39. Remarks by Oxana Gaman-Golutvina, Chairperson, Scientific Council, MGIMO-University, Moscow, at the conference The EU, Russia and the Global Crisis, April 18, 2009, Forlì, Italy.
40. Russia's 2007 Foreign Policy Concept identifies the territory of the former USSR as an area of Russian "special responsibility." *Obzor vneshnei politiki Rossiiskoi Federatsii* (Moscow: Foreign Ministry of the Russian Federation, 2007), p. 28.
41. Yuliya Tymoshenko, "Containing Russia," *Foreign Affairs* 86, no. 3 (May/June 2007), pp. 69–82.
42. Janusz Bugajski, *Expanding Eurasia: Russia's European Ambitions* (Washington, DC: Center for Strategic and International Studies, 2008), p. 173.
43. See Gary Hart and Dimitri K. Simes, "The Road to Moscow," *National Interest Online*, May/June 2009.
44. The potential for socioeconomic discontent to spark political instability was a major theme in Dmitri Medvedev's surprise interview with the critical newspaper *Novaia gazeta* (The New Gazette) in April 2009. "Deklaratsiia Medvedeva. God 2009," *Novaia gazeta*, no. 39, April 15, 2009.

45. Meeting with Russian Foreign Minister Sergei Lavrov in Xinhua, China, Obama remarked with characteristic optimism: "We have an excellent opportunity to reset the relations of the U.S. and Russia on a whole host of issues." "Obama: U.S.-Russia Can Overcome Differences," *China View*, May 8, 2009.
46. Speaking in the wake of Russia's incursion into Georgia, Medvedev outlined a vision for a "new" world order based on five principles defined as international law, multipolarity, rejection of isolation, protection of citizens, and legitimate spheres of influence. Paul Reynolds, "New Russian World Order: The Five Principles," *BBC News*, September 1, 2008.
47. "U.S.-Russia Strategic Framework Declaration," *Council on Foreign Relations: Essential Documents*, April 6, 2008.
48. Nikita Krasnikov, "Lavrov i Klinton 'perezagruzilis,'" *Komsomol'skaia pravda*, March 11, 2009. The "reset" button was marked in English on one side, and with the incorrect Russian word "*peregruzka*" ("overload") (in the Latin alphabet) rather than "*perezagruzka*" ("reset") on the other. Lavrov was quick to point out the error. The good humor that accompanied the incident seemed a positive augury.
49. Peter Fedynsky, "Russia Hails First Obama-Medvedev Meeting," *VOA News*, April 2, 2009.
50. Lucas, *The New Cold War*, p. 213.
51. This is an important theme in the major study coordinated by John Edwards and Jack Kemp, *Russia's Wrong Direction: What the United States Can and Should Do: Independent Task Force Report No. 57* (Washington, DC: Council on Foreign Relations, 2006).
52. Dmitri Medvedev, "La Russie, la crise mondiale et la sécurité européene," *Politique étrangère*, 2008, no. 4, p. 739.
53. Stephen J. Blank, *Russia and Arms Control: Are There Opportunities for the Obama Administration?* (Carlisle, PA: Strategic Studies Institute, March 2009).
54. Martin Ehala, "The Bronze Soldier: Identity Threat and Maintenance in Estonia," *Journal of Baltic Studies* 40, no. 1 (March 2009), pp. 139–158. The Estonian decision to remove a Soviet World War II memorial from central Tallinn in April 2007 led to days of rioting and a series of cyber attacks against Estonia suspected to be of Russian origin.
55. George T. Donovan, *Russian Operational Art in the Russo-Georgian War* (Carlisle, PA: Strategy Research Paper, 2009).
56. "Russia-NATO Ties Must Work, Not Just Exist on Paper—Medvedev," RIA Novosti, May 11, 2009.
57. "The Role of Russia in Europe's Security Architecture," *Russian Analytical Digest*, no. 55 (February 18, 2009).
58. Thomas Graham, "Les relations États-Unis/Russie: une approche pragmatique," *Politique étrangère*, 2008, no. 4, p. 755.
59. Adam Weinstein, "Russian Phoenix: The Collective Security Treaty Organization," *Whitehead Journal of Diplomacy and International Relations* 8, no. 1 (Winter/Spring 2007), pp. 167–180.
60. R. Craig Nation and Giampiero Giacomello, *Security in the West* (Ithaca, NY: Cornell University Press, forthcoming in 2009).
61. The Right Direction for U.S. Policy toward Russia: A Report from the Commission on U.S. Policy toward Russia (Washington, DC, March 2009).

Panel VI: Europe and Russia

Summary of Discussion

Ambassador (Ret.) Mary Ann Peters
Provost, Naval War College

There was considerable agreement between Sharyl Cross and R. Craig Nation on the state of U.S.-Russian relations and what needs to be done to "reset" relations. In "Advancing a Strategy for Constructive Security Engagement: 'Resetting' the U.S./NATO Approach toward Russia," Sharyl Cross explains why Russia's cooperation is important for "meeting the priority transnational security challenges of the twenty-first century," and provides specific recommendations to enhance cooperation. Both the United States and Russia share areas of common interest, including progress in arms control and counterproliferation of weapons of mass destruction (WMD), prevention of nuclear proliferation in Iran and North Korea, countering the Taliban in Afghanistan, stability in Pakistan, and stable energy supplies and markets in Europe/Eurasia. The importance of dialogue is stressed to ensure an understanding of common interests and objectives. "The United States and Euro-Atlantic community should work cooperatively to forge a constructive, consistent, and coherent strategy toward Russia on the basis of long-term strategic vision." And it is imperative that the United States and its European partners "demonstrate a willingness to work with Russia on equal terms." The "stereotypes of the Cold War" must be left behind to build a more cooperative and constructive relationship with Russia.

In "Resetting U.S.-Russian Relations," R. Craig Nation suggests that "[i]f the current attempt at normalization is to succeed where others have failed it will need to rest upon a more sophisticated understanding of larger trends working to shape U.S.-Russian relations both for good and for ill." The areas of disagreement between the United States and Russia are considerable and include "NATO enlargement, the American military presence in Central Asia, frozen conflicts on the territory of the former USSR, the Kosovo question, access to the oil and natural gas resources of the Caspian basin, the premises of Russian governance, U.S. military action in Iraq, the missile defense program in central Europe, etc." He further explains that there is a structural foundation for the "strategic rivalry between Russia and the West." This includes "different kinds of polities with distinctive strategic cultures and sometimes sharply contrasting priorities." He discusses the "systemic forces" of a resurgent Russia, an authoritarian drift, a potential decline of the West, a potential Moscow-Beijing axis, and the geopolitics of Eurasia. As a first step in resetting relations, he suggests that "[r]ecognizing the considerable weight of inherited hostility and the magnitude of the task ahead is a good starting point." And "[c]hronic criticism of Russia's domestic regime based upon a purported clash of values between Russia and the West has become . . . a divisive and unhelpful influence on the U.S.-Russian relationship."

Russia and the United States share common interests that provide the basis for "practical cooperation." Both nations have strong common interests concerning moderating strategic nuclear competition, enhancing arms control and counterproliferation of WMD, dealing with terrorism, ensuring energy security, managing the global economic crisis, and escaping from "the zero-sum logic of U.S.-Russian competition in the new Eurasia." Finally, "recasting U.S.-Russian relations . . . will have to be inspired by a new and expanded conception of security itself." "Taking advantage of a relatively benign security environment and bringing Russia, China, India, and other emerging powers out from the cold and into a functional great-power concert represents a strategic opportunity. Fixing what is broken in the U.S.-Russian relationship would be one good place to start."

The panel discussion opened with an extended consideration of Russia and energy. The first questioner asked, "Is there anything that we can or should do about Russia playing the energy card towards Europe? I would hold that one of the reasons we didn't say very much about the incursion [in Georgia] last August was because of Germany's position, which was perhaps largely driven by energy. Is this just a fact of life that we have to deal with, or can we do something about it?" Another participant questioned whether energy is an area of common interest with Russia. "It seems that the major U.S. interest in energy is to have the energy markets be open to the world and be smoothly functioning. And as the question suggests, that is not necessarily how the Russians see their interest."

A respondent acknowledged that this is a critical obstacle and added that we need to engage with the Russians on energy in the context of the range of other security interests where we have common objectives and interests. Medvedev has proposed a new energy proposal. NATO has energy on its agenda. While the Russians have resisted discussion with NATO on this topic, the respondent believed that we could push them to talk about the issues of ensuring reliable access and the use of energy as a geopolitical weapon but we should encourage diversification of supply in Europe as well.

Another participant asked what is wrong with pipelines moving across Russia, and why that is threatening. The Soviet Union/Russia actually has been a very reliable energy supplier for decades. The Russians tend to interpret this "war of the pipelines" in the Caspian Sea as another form of Western encroachment on what for them is an area of vital interest. To the Russians this seems to have no other point than to weaken Russia's position globally. The European Union has not been able to define itself effectively as a strategic actor, to develop coherent policies, or to speak with one voice on this or other issues. There really isn't an alternative to dependency on Russian sources of energy. The reality is that for the next several generations, the European states are going to be considerably dependent on Russian sources of oil and natural gas—and Russia will be dependent on the European markets. So it is a fact of life that we must live with. The question whether it needs to be treated as a geopolitical battle was raised. If we want to treat it that way, we are losing. Russia has been very competitive in sustaining and expanding its infrastructure, developing new markets in Asia, and negotiating long-term bilateral deals in Central Asia that make our alternative projects questionable in commercial terms. Then there is the question of Iran. If we could normalize relations with the Iranians and bring them into a cooperative regime to exploit Caspian energy that would change the rules altogether.

A participant asked whether energy security should be on NATO's agenda as this was not a military issue. If not on NATO's agenda, then where does it belong? A respondent said that it would be useful to engage the Russians on energy in the context of the NATO-Russia Council. Energy and security issues are interrelated. Another respondent pointed out that putting energy on the NATO-Russia Council agenda would actually put it on the State Department's agenda, which is not a bad place for it.

Another participant underscored the importance of this energy discussion, citing Michael Klare's earlier presentation on the "three e's," and his recent book *Rising Powers, Shrinking Planet: The New Geopolitics of Energy,* which hypothesizes that a major driving force behind nations' actions is their need for access to energy. Klare divides the world into net consuming nations and net supplying nations. There is a tremendous transfer of wealth occurring from the consumers to the producers. That is what has given Russia strength. The recent drop in oil prices is only temporary. By 2050 the United States and China will be the two largest consumers of energy and the Europeans will not be able to survive without energy sources from Russia and Central Asia. So over the long term we will have to deal with the fact that energy is and will be a significant driving force. What is going on in Africa is another example of a big scramble for energy. This also explains then-president Putin's rationale for renationalizing Russia's energy industry. Coming from a free market capitalist system we tend to view this negatively. However, Putin felt energy was such a powerful source of wealth that the nation-state had to be in control. After ten years of economic decline and deprivation it made eminent sense for a new leader to say, "I have to take charge and get us on a path to where the state has 'command' of its resources." Over 80 percent of all known petroleum reserves are controlled by national oil companies, and a number of governments are authoritarian.

A respondent commented that Putin never made any pretense and has never apologized for the recent realization of authority in Russia. It has these ideological cloaks: the dictatorship of law, sovereign democracy, and the verticality of power. For Russia, the argument is very consistent. Russia is a special case—"We have our traditions and history. We're a vast country. We need strong central authority to be functional. We collapsed in a period of chaos and anarchy in the 1990s. Consolidating the state is the task at hand and this is our way." In the energy sector in particular, this has the potential to work to Russia's long-term disadvantage. Russia needs more openness. It needs more investment if it's going to exploit its resources more effectively. Its relatively closed system does not allow it to attract such investment.

A respondent returned to the larger important question of NATO and Russia. NATO is here to stay. It has powerful assets, political and military, and it can play a very positive role as a security provider in the new Eurasia, so long as it is not configured nor perceived to be configured against Russia—that would just repeat the security architecture of the Cold War. So how do you square the circle? Any NATO initiatives such as naval patrols in the Black Sea or energy security initiatives in the Caspian are valid but they will just regenerate geopolitical competition, if the larger relationship is competitive. The really important piece of this agenda of "resetting" is making the NATO-Russia Council more robust. We made a mistake by going after the low-hanging fruit like search and rescue at sea, the areas where we felt we could cooperate. So this NATO-Russia organization never took on the substance that made it really valuable. And then there is the "curse of oil," the risk of dependence on a single commodity and the corruption that comes from a funding source delinked from the population.

A participant asked how we should prioritize the range of issues. Unless we believe that each is of equal urgency, how do we go about identifying which are potentially the most important? How do you make choices—do it bureaucratically? Is there an argument for a sequence?

A questioner turned to the issue of Iran, and asked what it means to be more flexible with Russia in regard to Iran. What kind of prescription does that mean, and what could Russia contribute?

Another participant stated that if there is one thing that Moscow could do that would get America's attention and change U.S. perceptions of Russia's intentions, it would be cooperation on the Iranian issue, but noted there seems to be a suggestion that Russia, for a variety of reasons, is constrained to help.

A respondent stated that all the issues are really important. When the Russians see us willing to really move in these areas that have been important to them, then we can have the give-and-take across the issues. The respondent was not so optimistic that Russia is going to be able to help us so much with the Iranian issue, but suggested we can talk about it in the context of missile defense. We have to shift the frame of reference for discussion so both sides understand that the development of Iranian nuclear potential can affect Russian interests equally.

A participant said we have this horrible habit, especially with respect to Arab countries, of going in and claiming we're going to engage in a dialogue, and then telling them what's going to happen and what they ought to do. We do the same thing with Russia concerning Iran—"You, Russia, have to impose sanctions. Here is what you have to do." Have we ever sat down and listened to the Russians, asked, "What do you think should happen?" A respondent said they would appreciate that tremendously. This raises the issue of joint threat assessment; we have to talk about it together. Russia has to see it is in its interest to proceed. Another participant then asked what happens if Russia decides it is not in its interest. A third participant interjected, "Then we are no further back then we already are."

A respondent noted that Russia has contradictory interests in its relationship with Iran. The relationship is important commercially and it reinforces Russia geopolitically. But the respondent did not think Russia has any interest, whatsoever, in seeing Iran as a nuclear-weapons power. Russia can do a lot. It can buy into the sanctions regime, it can reduce arms transfers to the Iranians, and it has already offered programs for recycling and assisting the Iranians in their energy program.

Another participant stated that the logic of what the last panel said on Iran —although it did not exactly say this—was that we may just have to learn to live with a nuclear Iran. The world is full of countries that would prefer that Iran not be a nuclear-weapons state. The disagreement is over what to do about it. The previous suggestions about what Russia could do indicate that Russia has more influence over this process than any of the people who have actually been involved so far. Regarding sanctions, we actually know something about how sanctions work. They can work over a whole lot of time and if there is a whole lot of support, meaning countries that will actually put their backs into implementing them. And then we must be prepared to wait ten, fifteen years.

A respondent came back to the issue of priorities, saying the priorities have already been laid out. The administration wants to change the tone. We want to emphasize arms control; that is the agenda for 2009. In effect, we de facto have frozen the agenda for NATO enlargement with respect to Ukraine and Georgia. That is a pretty good number of steps toward this "resetting."

A participant asked about domestic links between what's happening within Russia and what Russia is doing externally, noting that the phrase "the foreign and defense policy establishment says" has been used. Can we step back to identify the underlying dynamics within Russian society or economy and pull out a few features that will drive Russian behavior? What is helping to inform this "establishment" and influencing what it decides to do? Toward a related question, the participant noted criticism of the notion that the United States should be engaged in any way inside the Russian black box. The suggestion is that it just simply won't work effectively to focus on the freedom agenda or values. When one looks at countries, much time is spent worrying what's happening inside those countries. So should we be worried about what is happening in Russia?

In response, a participant said it is not that we do not worry about what is happening domestically. Both U.S. and European engagement, especially through the private sector, working with Russia at multiple levels, is very important. But what has been counterproductive is the criticism identifying Russia's shortcomings.

Another respondent suggested that our foreign policy preparation process fails to do all the things that the previous question suggests. It is fairly ideological. Russia is an authoritarian state. Therefore, this is the way it is going to behave. It is remarkable how little classical standard foreign policy analysis is devoted to the Russian Federation, to the things we look at in any other state. For example, who are the actors, what does the policy process look like and how does it impinge upon decisions, what is the ideological framework, what is the role of the public in foreign policy formulation, what are Russia's national interests and how does it see its interests, what courses of actions does it use to pursue its interests, and so on? There is a very underdeveloped literature that addresses these questions. It is not easy since the Russian policy process is very opaque. That is why people tend to fall back on this not particularly helpful phrase, "the foreign security policy establishment." This may be as close as we can get to what is going on inside the walls of the Kremlin.

A discussant stated that most of the important factors identified as causes of conflict with Russia are also true of our relationship with China. China is a rising power, with a history of enmity and a desire for prestige, surrounded by flash points and by U.S. allies and alliances that we're beefing up. If we are prioritizing on what we're going to do then we really have to address whatever we decide are the fundamental causes of the conflict.

It was suggested that Americans are culturally more comfortable with China than Russia. This is what Russophobia rests upon. The discussant replied there is a big difference in economic importance. The respondent agreed that is a totally different relationship that we simply do not have with Russia—we do not have powerful invested economic interests.

Another respondent explained that Russians complain that the United States is more disrespectful toward them than it is toward China even though there isn't an ideological conflict with the Russian Federation—maybe it is the history of anti-Russian sentiment. The Chinese are certainly ambitious, but they are sort of modest in the way they present themselves in the world.

A participant interjected that the Chinese are not a defrocked superpower. The Russians are, and that psychological edge is palpable to Russian elites.

Another participant stated that the Chinese are much more confident, have developed much better relationships, while the Russians seem to lack confidence and approach the world suspiciously. The Chinese are much more open to engage in a variety of different ways, not that we always agree.

A respondent stated that one of the major differences was that expectations (following the end of the Cold War) were so high initially, and the Russians were disappointed with the way everything developed with the West. There is a lack of trust as they work to come back. There is a psychological damage that has to be overcome to proceed in a constructive way.

A participant stated that last fall Russia sent a number of navy ships to the Caribbean, made port calls in Venezuela and Cuba, and held maneuvers. About the same time, Prime Minister Putin may have said that he would like to consider having some overseas bases. Do we see any circumstances that could lead to a persistent military presence in the Western Hemisphere by the Russians?

A respondent remarked this is an example of the lack of confidence we just talked about. It is bravado—pounding the chest to call attention to itself. It is very different from the Chinese use of its navy in the Pacific, which is much more strategically pointed.

A final question related to perceptions in our society that this region is not as important, and what that does to our ability to handle a crisis that comes up rapidly, like our lack of regional and language specialists after 9/11. How do we grow knowledge in government and elsewhere to keep up just in case?

A respondent said there is danger in deciding what part of the world is important and what part is not. When people dismiss the Russian Federation, that is a mistake. We need to see it in a new way—see the United States, European nations, NATO, and the Russian Federation serving as a strong transatlantic community reaching out to deal with global issues. We need to approach Russia with recognition that it is a major player, that we understand its importance. If we concentrate only on one contentious issue we will get bogged down. We need an agenda that allows patient engagement.

Another respondent said that in the Cold War we spent a lot of money training civilian and military specialists on the Soviet Union. We do not have to repeat that level of effort, but these cultural understandings are important. It also works the other way; they need to understand us as well. This is where this notion of a new cold war is not productive.

A final respondent noted that we can grow language specialists. An understanding of key languages allows for learning of similar languages.

Panel VII

Africa

Ambassador (Ret.) Princeton N. Lyman
Adjunct Senior Fellow for Africa Policy Studies, Council on Foreign Relations

Dr. Peter J. Schraeder
Professor and Graduate Program Director, Department of Political Science, Loyola University, Chicago

Moderator:
Dr. Stephen A. Emerson
Associate Professor of National Security Affairs, Naval War College

Africa in U.S. Foreign Policy

Ambassador (Ret.) Princeton N. Lyman
Adjunct Senior Fellow for Africa Policy Studies
Council on Foreign Relations

Placing Africa in a Strategic Perspective

For a long time, Africa was seen as outside the strategic interests of the United States, and nowhere more so than within the U.S. military. As a long-time Department of Defense (DOD) official, charged with Africa policy, put it in the 1980s, "DOD sees Africa as a place to fly over, not stop there." This perspective was reinforced by the U.S. military intervention in Somalia in 1992–1993. The purpose of that intervention was humanitarian, to prevent massive starvation caused by drought on the one hand and the total breakdown of central government on the other. What began as a humanitarian intervention, however, morphed into a fight with a dominant warlord. That in turn led to the tragedy of "Blackhawk Down," the loss of eighteen servicemen, and the withdrawal of American troops shortly thereafter. It would be nearly twenty years before American troops would again be committed to a long-standing commitment on the continent.

In foreign policy circles, Africa was seen as largely a humanitarian interest, especially following the Cold War. During the Cold War, Africa served as a proxy area of competition between the United States and the Soviet Union, in the Horn—Ethiopia, Somalia, Sudan—and in southern Africa—Angola, and to a lesser extent Mozambique, Namibia, and South Africa. But with the collapse of the Soviet Union, attention to Africa receded to the province of humanitarian nongovernmental organizations, African American constituencies, and some interested congressmen. The debate over U.S. policy toward South Africa throughout the 1980s, which did draw national attention, pitted moral forces against only vaguely defined strategic interests, and when the Soviet Union collapsed, U.S. policy readily took the moral high ground and placed itself firmly on the side of liberation.

The argument in foreign policy circles for a proactive Africa policy thus became, "We must do something for those poor people in Africa." Then the discussion would shift to "serious" foreign policy issues elsewhere. U.S. assistance to Africa during the 1990s drifted downward, falling for a while below the once "bottom line" of $900 million. Intelligence assets were sharply curtailed in Africa. USAID closed numerous missions, especially in the Sahel region and similar small states. The State Department's resources similarly declined, with large numbers of vacancies even in major countries like Nigeria, and the closing of three embassies and the only consulate in north Nigeria, where more than 60 million Muslims live. Still smarting from the debacle in Somalia, the United States failed to respond, or to enable the UN to respond, to the genocide that erupted in Rwanda in 1994. Afterward there was much hand-wringing, some new initiatives in peacekeeping were begun, but the issue was again more moral than strategic. Indeed throughout the 1990s, the United States continued to sharply restrict the number and size of UN peacekeeping missions in Africa, despite crises in Liberia, Sierra Leone, the Democratic Republic of the Congo (DRC), and the Central African Republic.

The situation began to change in the years since 2000. European Command (EUCOM) began to focus more attention on Africa, into which it had been drawn for several humanitarian or rescue missions, and saw in the weakness and vulnerability of African states a long-term strategic threat. Policy makers also began to notice that other powers were viewing Africa in a different light, with implications for U.S. interests. In 2006, the Council on Foreign Relations published a Task Force Report, *More Than Humanitarianism: A Strategic U.S. Approach toward Africa.*[1] The report called attention to Africa's growing importance as an oil exporter, its importance in the global war on terror, the costly series of conflicts there fostering not only humanitarian crises but criminality on an international scale, Africa's central role in the fight against HIV/AIDS, and the importance of its voting bloc in such international organizations as the World Trade Organization (WTO), the International Atomic Energy Agency, and other international bodies. But the chapter in the report that captured the most attention, quoted and commented on around the world, described China's growing activities in Africa, in energy and other natural resource investments, its growing business presence, and its political objectives there. The chapter noted that India, Malaysia, and both North and South Korea were following China's lead. As one member of the task force put it, "Africa is in play."

President Bush during this period began placing more emphasis on Africa. Much of it was a continuation of the humanitarian focus, with a major commitment to combating HIV/AIDS, a malaria initiative, more emergency relief funding, and various other developmental initiatives. Altogether Bush more than tripled U.S. aid to Africa after 2001, to $6 billion by 2007, and promised to raise it to $9 billion by 2010. Beyond humanitarianism, the Bush administration, in the wake of 9/11, created the Combined Joint Task Force/Horn of Africa (CJTF/HOA), and stationed some 1,200 American service personnel in Djibouti, the first such deployment on African soil since Somalia in 1993 and far more permanent. Bush proclaimed a $100 million counterterrorism program for East Africa. EUCOM initiated the Pan-Sahel Initiative, a small counterterrorism training program in that sparsely populated desert region, an initiative that would grow into the much larger Trans-Sahara Counterterrorism Partnership by the end of the Bush administration. Finally, the Bush administration created a single African Combatant Command, AFRICOM, to bring together and enhance America's military and related security outreach to the continent.

By the end of the Bush administration, Africa was thus gaining attention in strategic terms. But the underlying problems on the continent—poverty, poor governance, conflict, criminality, and corruption—make it challenging, if not downright difficult, for the United States to respond to these interests effectively. The United States is only at the beginning of that process.

Africa's Energy Role

Today Africa provides some 24 percent of U.S. oil imports. Most of this is from the states in the Gulf of Guinea, mainly Nigeria and Angola, but other producers include Equatorial Guinea, Congo (Brazzaville), Chad, Gabon, and soon Ghana. West African crude is low sulfur and is thus particularly well suited to American refineries. Africa is attractive also because it is one of the few areas of the world that encourages private ownership in the oil fields, whereas elsewhere the trend is decidedly toward national ownership. Because of promising offshore finds in this region, Africa could increase its share of U.S. imports over the coming decade. Gas is also becoming a valuable resource in this region. Liquefied natural gas plants have been built in Nigeria and new ones are going up in Nigeria and Equatorial Guinea. A West Africa gas pipeline is providing Nigerian natural gas to several coastal countries.[2]

China has made its major Africa energy forays in Sudan and Angola. China is a major investor (along with India and Malaysia) in Sudan's oil sector and is one of Sudan's primary markets for its oil. China has provided between $5 and $10 billion in loans to Angola for infrastructure and other development projects, collateralized by future oil deliveries. Angola is now China's largest source of imported oil, just ahead of Saudi Arabia. Some fear that China intends to be able to lock up supply in a possible period of great scarcity through its own investments and collateralized loans like those to Angola. The evidence is not convincing, however. China has obtained several oil blocs in West Africa but by most accounts they are neither promising nor economical. China also lacks offshore technology to compete with U.S. and other majors in this promising aspect of the African oil sector, seeking in the case of Angola a joint venture with BP for this purpose. Finally, it is worth remembering that while Africa supplies China with about 33 percent of China's oil imports, China accounts for only 9 percent of Africa's oil exports, while the United States accounts for 32 percent. While energy and other mineral resources are highly important in China's Africa policy, its goals in Africa are broader, as discussed later in this paper.

The primary problem for U.S. energy interests is not China, but instability in Africa. Nigeria, Africa's sometimes number one, currently number two, producer, is experiencing a seemingly intractable insurgency in the oil-producing delta region. Long neglected by the federal government, environmentally despoiled by the early practices of oil companies, poor, and without many alternative economic opportunities, the delta region has spawned an increasingly militant insurgency. It would be reassuring if the underlying and early causes of this situation could now be addressed through better allocation of resources, development investments, etc. But today the insurgent militias are deep into the criminal business of stealing ("bunkering" in Nigeria terminology) as much as 500,000 barrels a day of oil, selling it in return for arms and plenty of cash. At the same time they have shut down about an equal amount of Nigerian official exports. They are also intimately, if irregularly, allied with local politicians, serving as political thugs and enforcers, and with higher-level Nigerian officials, including military officers, in the oil-bunkering trade. Various attempts at military suppression of the militias have failed, producing instead anger at the Nigerian military's harsh tactics. The combination of underdevelopment, violence, corruption, loss of trust, and ever more sophisticated arms in the region makes this an extremely complex situation.

Several other oil-producing countries in Africa are subject to instability. Equatorial Guinea, the third-largest African producer, is a small country ruled dictatorially and with a poor human rights record. It is frequently subjected to efforts to overthrow the regime from abroad. Chad is beset with civil war abetted by the situation in Darfur, Sudan. Congo (Brazzaville) has had several violent changes of government. None of the countries in the Gulf of Guinea are capable of protecting their offshore oil wells, but are reluctant to cooperate too closely or to confront fully the piracy and other forms of criminality in the area. Ironically, perhaps, after decades of civil war, Angola may be the most stable and predictable oil-producing government in the region.

If securing Africa's energy resources is a significant American interest, the challenges to doing so are formidable. As pointed out later in this paper, various efforts to do so have had only marginal results.

Terrorism[3]

International terrorism raised its ugly head in Africa in 1998, with the bombing of American embassies in Kenya and Tanzania. An attack on Israeli facilities in Kenya took place in 2001. Sudan once hosted Osama bin Laden and various terrorist groups until the late 1990s. It is now clear that al Qaeda cells operate down the east coast of Africa and in parts of southern Africa. Their primary role since the incidents in Kenya appear to be to provide a safe haven for terrorists coming to or returning from other countries, and to raise funds.

Somalia has become an even more intense object of U.S. interest, especially since 2006, when a radical Islamic Courts Movement took over the capital, Mogadishu, and made claims against Somalia's neighbors as well as harbored some of those suspected in the Kenya embassy bombing. An Ethiopian invasion that ousted the Islamic Courts Movement from power succeeded only in creating the foundation for a determined insurgency by the most radical of the earlier movement and returned Somalia to the anarchy and violence of the previous fifteen years.

American response to this threat in the Horn vacillated between, on the one hand, hearts and minds intelligence-gathering efforts spearheaded by the CJTF/HOA, along with training of African intelligence and other security personnel, and, on the other, bombing of suspected terrorists in Somalia and provision of encouragement and support behind the scenes to the Ethiopian invasion. At present, as Somalia's anarchy and violence continue, the radicalism of the insurgents has deepened, and the situation has spawned as well the recently publicized piracy in the Red Sea. American policy seems somewhat paralyzed, or at least divided between those who favor more direct military action and those looking for diplomatic possibilities. There is all at the same time modest support for the efforts of a UN-led peace process aimed at strengthening the latest Somali government, led by a moderate Islamist; a faint but largely doomed effort to get the UN to send in a peacekeeping force; and continued consideration of further military action against suspected terrorists. The Obama administration will need not only to review this policy but, once a cohesive policy is agreed upon, work to get all elements of the United States on the same page.

Elsewhere in Africa, U.S. attention has been paid to the threat from al Qaeda in the Maghreb (AQIM) operating in and out of Algeria. As noted earlier, a small pan-Sahel initiative of EUCOM's has grown into a $100 million counterterrorism program that brings together North African and Sahelian countries with the objective of their gaining greater oversight and control over the vast "ungoverned space" in this region. These programs, however, run into local ethnic and political issues that, unfortunately, pit one set of counterterrorism objectives—countering AQIM—against the historical distrust of enhanced military presence in the region that drives disgruntled local ethnic groups into collaboration with AQIM (primarily to protect their traditional smuggling operations). Moreover, embassy and USAID resources are not sufficient, nor are those of these largely poor African states, to offer meaningful development alternatives to these local ethnic groups.

While U.S. focus is on internationally linked terrorism, Africa is faced with numerous indigenous groups that practice terrorism and other heinous acts. The Lord's Resistance Army of northern Uganda forcibly recruits child soldiers and carries out horrifically violent attacks on villages not cooperating with it. Various militias in the DRC have been carrying out similarly vicious acts for years. Some groups in Kenya have been accused of such practices. The president of Sudan has been indicted by the International Criminal Court for war crimes and crimes against humanity in the Darfur region, and some rebel leaders in Darfur have also been indicted. African nations are understandably more focused on these threats than on the more limited international terrorism coming from radical Islam. The bottom line is that bringing terrorism under control, both that which is internationally linked and that which is indigenous, rests more with remedying the underlying weaknesses and vulnerabilities of African states than with, as important as they may be, improving the purely security capacities of these governments.

Conflicts

Conflicts on the African continent have been enormously costly in lives, rape, pillage, destruction of property, and undermining of development progress. The war in the DRC alone has cost directly and indirectly nearly 5 million lives, the most costly war since World War II. Most of the postcolonial conflicts in Africa, however, have in fact been brought to an end: Angola, Mozambique, Liberia, Sierra Leone, Sudan (in the north-south civil war), Côte d'Ivoire, Rwanda, and Burundi—all of these civil wars have ended, some with relatively secure peace agreements, some with more fragile ones. Yet conflict remains a major source of concern.

In Sudan, the conflict in Darfur, labeled a genocide by the United States, continues after six years. The Comprehensive Peace Agreement (CPA) that ended the north-south civil war in Sudan is considered endangered by most observers, as disagreements continue over the sharing of oil proceeds, borders, electoral principles, and other issues. A resumption of that war could have far-reaching consequences in Africa and the Middle East. The DRC experiences only a fragile peace. The DRC conflict, while not getting the same attention in the United States as Darfur, in fact involves one of the largest countries in Africa, one rich in mineral resources, and whose stability or instability affects countries all through central and East Africa. If ever there was a conflict fed by competition for mineral resources, it is in the DRC.

As critically important as these conflicts are, American experience in conflict resolution in Africa is mixed. Under President George W. Bush, the United States played a major role in brokering the CPA in Sudan. Yet despite the labeling of genocide in Darfur, the United States (along with everyone else, it should be noted) has failed to bring this conflict anywhere near to an end. The United States played an important but rather tardy role in bringing the Liberian civil war to a close, with help from an African peacekeeping force and the leadership of Nigerian president Olusegun Obasanjo. The U.S. role in the DRC has been lower profile, helping with some of the regional diplomacy, but not a major player in either peacekeeping like the EU and South Africa, nor in the high-level diplomacy required. The mixed if not negative U.S. role in Somalia has already been noted.

Overall American policy since the Somalia debacle of 1993 has been to foster African peacekeeping and conflict resolution capacity. American policy in this regard was given a boost with the creation of the African Union (AU) in 2001, which created a Peace and Security Council, and whose members pledged to take concerted action in instances of serious even-internal conflicts. The AU subsequently took several initiatives on its own to put peacekeepers on the ground, e.g., in Burundi, Côte d'Ivoire, Darfur, and Somalia, ahead of the UN being ready to do so. In 2003, the United States pledged to help train five African brigades for peacekeeping purposes.

But African leadership has waned in recent years. Strong leaders in Nigeria and South Africa have exited the scene. Kenya, often a vital regional leader in such issues, is preoccupied with its internal political struggle. Moreover, African peacekeeping in Darfur and Somalia has been underfunded, undermined in the case of Darfur by the Khartoum government, and unable to achieve its primary purpose of protection of civilians, let alone undergirding peace processes. Whether African countries will be willing or able to lead in burgeoning conflict situations in the future is questionable.

A New and Growing Threat: Narcotics Trafficking

In October 2008, the executive director of the UN Office on Drugs and Crime warned that drug trafficking from South America through West Africa to Europe was growing and posing a major threat to the security of the region. At that time it was estimated that at least 50 tons of cocaine a year were passing through this route, worth $2 billion. Small, weak states were particularly vulnerable to the blandishment of drug money, such as Guinea-Bissau, Guinea, and São Tomé, but also larger, better-off countries like Senegal, Ghana, and Nigeria were involved. As normally with the syndicates, local countries are corrupted, addiction is promoted in the local population, and alternative economic developments are abandoned. "Time is running out," warned the official. But there is little evidence that the problem is being addressed. South African syndicates, providing the drugs link with West African syndicates in Europe, together easily bypass or corrupt local attempts at prosecution.[4]

While most of the drugs are destined for Europe, American interests are compromised as criminality undermines efforts at peace and development, and opens the door to broader forms of crime that could directly affect American concerns, e.g., furthering lawlessness and piracy in the Gulf of Guinea. The potential alliance of drug syndicates and terrorist elements, as has occurred in Latin America, is also worrisome.

Political and Commercial Competition

China is not new to Africa, having been involved there since the early 1960s. But with its new wealth, and its rapid growth demanding new sources of raw materials, China has made a political and economic foray into Africa that is unprecedented. The greatest attention has been to China's promises of large loans to Angola, the DRC, Nigeria, and elsewhere. China has been happy to use these loans to provide assistance in particular in infrastructure, which the United States and other Western donors had moved away from since the 1990s, and for which there is a huge need in Africa. China is also rapidly moving up to being Africa's number one trading partner, likely bypassing the United States in this regard in 2010. Its investments go far beyond raw materials, moreover. Some eight hundred Chinese companies now operate in Africa in agriculture, telecommunications, health, tourism, and other sectors.

But whereas some analysts saw China's interest in Africa related primarily if not solely to the need for raw materials, there is more to China's interest on the continent. China has steadily reduced the number of African countries that recognize Taiwan (down to four) and has secured African support in the UN and elsewhere to counter Taiwan's efforts to raise its member status in these bodies. Africa has supported China's resistance to human rights criticisms in the UN's Human Rights Council, and African representatives have voted increasingly with China in the UN Security Council.

Some have seen in China's growing diplomacy and investment in Africa a strategic threat to the United States, in locking up energy and other materials, and gaining greater political and economic influence. It is true that China offers political competition to the United States and commercial competition to American companies. China compliments Africa with numerous high-level visits by its president, prime minister, and other senior officials, whereas one or two American presidential visits to Africa over eight years is considered a major gesture by the United States. China has protected Sudan from harsh UN Security Council measures relating to Darfur, and has been both a political and economic supporter—though recently wavering—of Zimbabwe's Mugabe, both in opposition to U.S. policy.

Moreover, China is free from Organization for Economic Cooperation and Development (OECD) rules that prohibit Western donors from combining aid and commercial offerings. China is thus able to sweeten its bid for oil blocs or similar concessions with accompanying offers of aid projects. India is copying some of China's tactics as it steps up its own attention to Africa, again for a combination of economic and political reasons. All of these suggest that the United

States needs to be aware of the areas of real competition coming from Asia and from elsewhere (e.g., Brazil, and recently Russia), and be prepared to address them. But China is not a behemoth and the United States has many more cards to play. China's investment in Africa likely exceeds that of the United States but is dwarfed by that of Western countries altogether. U.S. aid, all now in grants, exceeds Chinese aid, which is moreover predominantly in loans. As pointed out earlier, the energy market is not so easily controlled, by one purchaser or another, and China has found limited success as a producer there.

As China's role has grown, moreover, it has met the same problems as other donors. It has had to abandon its multibillion railroad offer to Nigeria and its offer to take over a refinery there, due to disagreements with the government; has had to scale back its multibillion offer to the DRC in return for greater access to mining rights because of insecurity; and has experienced political backlash for Chinese mining-company practices in Zambia, the flood of Chinese consumer goods in South Africa and elsewhere on the continent, and the use of largely Chinese companies and workers to carry out its aid projects. It has also met with security problems in the Niger Delta, with Chinese workers kidnapped just as other foreign workers have been, and had eight Chinese mining experts killed in Ethiopia by rebels there.

The official U.S. response to China's role in Africa has thus been low-key. Too many commentators in the United States and abroad have suggested that the creation of AFRICOM was in part to counter China's growing interest. Conspiracy theory abounds.[5] However, the Bush administration went out of its way to paint China's growing role in Africa as potentially positive and to indicate that seeking cooperation rather than competition or confrontation would be the U.S. response. The United States began a diplomatic dialogue with China about Africa beginning in 2003, with Sudan and Zimbabwe prominent topics but many other subjects were covered and both sides praised the tone of the talks.[6] AFRICOM, which unfortunately initially proclaimed too wide a mission in Africa and suffered as a result strong African pushback, has pulled back and focused more and more on its core mission of counterterrorism, peacekeeping, and energy security, with scant if any attention to China.

What may make more sense than confrontation is to engage China more on areas of mutual interest. As Sudan moves toward possible southern independence, China has a stake in the country's overall stability and, if needed, a peaceful separation, if only for preserving its already significant stake there. Problems of security in Nigeria and Ethiopia should open the way toward cooperation on countering militias and criminality in Africa. China is not yet ready to sign on to OECD rules and maxims on the uses and conditions for foreign aid, nor ready to join such entities as the Extractive Industries Transparency Initiative. But China has been studying USAID and other Western donor experience and cannot help but be concerned with governance, corruption, and economic stability as its investments grow. China may also be ready to contemplate more cooperation on principles of corporate social responsibility.[7] All of these areas for dialogue may help channel China's competition into more normal political and commercial channels, as the United States has with its European allies, while enhancing areas of mutual interest. In sum, given the scope of

Africa's problems, mobilizing the maximum amount of cooperation from all external actors would be in the U.S. interest.

Other Strategic Concerns

Health

Africa is the epicenter of the HIV/AIDS pandemic, on which the United States has spent $18 billion over the past five years and for which Congress has authorized $48 billion for the next five years. Much progress has been made in treatment, with more than 2 million people brought under a treatment regime since 2003. But the rate of infections continues to outpace the number of new persons being treated. If infection rates are not checked, the commitment to treatment of all those needing it, a commitment made by the G-8 in 2005, could require such major funding over the lifetime of those being treated as to crowd out much else needed for development assistance in Africa.

Beyond HIV/AIDS, Africa is also the weak link in controlling new pandemics such as avian flu, or in eradicating such diseases as polio. Controlling these epidemics requires well staffed and trained monitoring and reporting systems, fast action capacity (such as slaughtering vulnerable poultry), and national health infrastructure. No African country is fully up to the needed capacity in these areas.

Trade

At the beginning of the Doha round of trade negotiations in 2001, Africa and the United States were more or less in sync with each other on objectives and strategy. But in the intervening years, African countries, under South African leadership, have gravitated to support of India, China, and Brazil in their resistance to opening their markets more in return for greater reduction of EU and U.S. agricultural subsidies. Africa's forty votes in the WTO make it impossible to reach agreement in the Doha round if Africa is not using its votes to help promote compromises on the part of their "South-South" partners. The United States will have to develop a far more comprehensive Africa trade policy, offering incentives in the WTO and other steps, to break this logjam.[8]

The U.S. Response

The Obama administration has the advantage of building on the increased attention and resources earmarked for Africa under the Bush administration. Africa policy in recent years has also been remarkably bipartisan, with broad congressional support for the increases in aid, the commitment to HIV/AIDS, and renewal and improvements in the Africa Growth and Opportunities Act, which opened the U.S. market, quota free for many products, to African countries. Without adding too much new funding, the United States should be able to meet Bush's pledge to reach $9 billion in assistance in 2010.

The Obama administration has appointed several high-level officials with strong Africa credentials: Susan Rice, the U.S. ambassador to the UN, is a former assistant secretary of state for African affairs; Gayle Smith, the national security adviser to the president on foreign assistance, is a veteran of African assignments in both USAID and the National Security Council; Samantha

Power, national security adviser on multilateralism and human rights, has written on genocide and is intimately knowledgeable about Darfur; Michelle Gavin, national security adviser on Africa, was Africa specialist for Senator Russ Feingold and a scholar at the Council on Foreign Relations, where she wrote several Africa policy papers; a high-level envoy has been appointed for Sudan and one for the Great Lakes region of Africa will be announced soon; Johnnie Carson, three times ambassador in Africa and most recently national intelligence officer for Africa, has been appointed assistant secretary of state for African affairs. It is a formidable team portending an active Africa policy.

But obstacles to pursuing American interests will loom large. The greatest is the continuing poverty and the weakness of governance in Africa. Even with increased aid, and gradual improvements in Africa's trade capacity, the World Bank predicts that by 2050, sub-Saharan Africa will be the one region having major populations living below the absolute poverty line of $1 a day. While Africa has experienced steady and impressive growth of elective democracies, basic governance is weak and subject to wide swings in performance. For example, Harvard University's Index of African Governance finds several states improved their overall governance in 2006, but several of those states—like Guinea-Bissau, Nigeria, Uganda, Madagascar, and Kenya—experienced setbacks in subsequent years.[9] Africa was troubled as well by a spate of new coups in the past year, in Guinea-Bissau, Guinea, Mauritania, and Madagascar. Coups were a common happening in the first twenty years of African independence, but many thought they had become a thing of the past. The worldwide economic downturn is predicted to throw millions more into poverty in developing countries, which may produce further instability.

Weak and vulnerable states are less able to utilize resources well, especially in the areas of security, justice, and strengthening the rule of law, all essential elements for making progress on the strategic issues described above. Some key states, moreover, are reluctant to accept assistance in these areas. Nigeria has consistently refused offers of U.S. assistance to help police the Gulf of Guinea and contain the bunkering of oil. The Gulf countries are also reticent to cooperate closely to develop regional security arrangements for their offshore oil facilities. As noted, African states may be less willing or able to undertake new or improved peacekeeping operations. This makes AFRICOM's tasks all the more difficult. It can and does offer technical assistance, and undertakes various training programs, but outside the Horn, its reach and influence are limited.

Finally, the U.S. government is not well structured to manage the several conflicts in Africa nor to respond in a unified way to crises as in Somalia or Sudan. The Africa Bureau of the State Department is one of the smallest in the department, despite covering forty-eight countries. It lacks the capacity for staffing and managing the diplomacy necessary and over sufficient time for such major conflicts as that in the DRC. There are problems in State's ability to work across regional bureaus, even though the situations in Sudan, Somalia, and the Sahel demand diplomatic outreach as much to countries of the Middle East as they do to those in Africa.[10] Finally, given the growing deficits in the United States the Obama administration may find it difficult to obtain the rising appropriations necessary to maintain aid to Africa at the enhanced level of 2010, especially funding beyond the already authorized and major commitment to HIV/AIDS.

Conclusion

Africa has reached a new level of recognition in strategic U.S. policy thinking. No longer a purely humanitarian concern, Africa registers today as a factor in energy security, terrorism, international crime, conflict, global health, and trade. U.S. attention to Africa has thus increased, with major increases in economic assistance, a military presence in the Horn, a new unified Africa Combatant Command, and renewed diplomatic attention reflected in the naming of two high-level envoys and a number of other high-level Africa-wise presidential appointees. The challenge for the United States now is to translate that recognition into the long-range, capacity- and institution-building programs that will enable Africa to become a more effective and reliable partner in all the areas of mutual concern. It will require greater coordination of policy across bureaus and departments, better understandings with Congress about the long-term nature of some of the problems to be addressed, and the support of allies and other countries active on the continent. If this commitment is undertaken, Africa will become not only an integral part of America's strategic perspective but a steadily more reliable partner.

Notes

1. Anthony Lake and Christine Todd Whitman, co-chairs, *More Than Humanitarianism: A Strategic U.S. Approach toward Africa* (New York: Council on Foreign Relations, 2006).
2. For a fuller exposition of Africa's energy production and global role, see David L. Goldwyn, "Pursuing U.S. Energy Security Interests in Africa," in *Beyond the Bush Administration's Africa Policy: Critical Choices for the Obama Administration,* ed. Stephen Morrison and Jennifer Cooke (Washington, DC: CSIS, 2009).
3. For a fuller account of this issue, see Princeton N. Lyman, "The War on Terrorism in Africa," in *Africa-US Relations: Strategic Encounters,* ed. Donald Rothschild and Edmond Keller (Lynne Rienner, 2008).
4. UN News Services, "Drug Crime Poses Serious Threat to Region, Warns UN Official," allafrica.com, October 20, 2008, http://allafrica.com/stories/200810280947.html.
5. A good example is F. William Engdahl, "China's $9bn Hostage in the Congo War," Asia Times Online, http://www.atimes.com/atimes.
6. David H. Shinn, "Chinese Engagement in Africa," in *Beyond the Bush Administration's Africa Policy,* ed. Morrison and Cooke.
7. *Africa-China-U.S. Trilateral Dialogue* (Washington, DC: Council on Foreign Relations, 2007), available at http://www.cfr.org.
8. See Princeton N. Lyman, "U.S. Foreign Aid and Trade Practices in Africa," in *Beyond the Bush Administration's Africa Policy,* ed. Morrison and Cooke.
9. Emily Wood, "2008 Index of African Governance," Policy Brief 9 (Program on Intrastate Conflict and Conflict Resolution, Belfer Center for Science and International Affairs, October 6, 2008).
10. For elaboration of these points, see Princeton N. Lyman, testimony before the Senate Committee on Foreign Relations, Subcommittee on African Affairs, *Hearing on Strengthening U.S. Diplomacy to Anticipate, Prevent and Respond to Conflict in Africa,* 111th Cong., 1st sess., April 21, 2009, available at www.cfr.org.

Great Expectations versus Daunting Challenges: Prospects for U.S. Foreign Policy toward Africa during the Obama Administration

Dr. Peter J. Schraeder
Professor and Graduate Program Director
Department of Political Science, Loyola University, Chicago

Barack Obama's historic election as the forty-fourth president of the United States has raised extraordinary expectations among both Africans and Africanists as concerns the future of U.S. foreign policy toward Africa. In addition to being a product of the African diaspora (his father was Kenyan) and publishing two well-received books in which Africa is referenced (*Dreams of My Father: A Story of Race and Inheritance* and *The Audacity of Hope: Thoughts on Reclaiming the American Dream*), Obama is the first sitting president to have visited the African continent prior to taking office, including traveling during two weeks in 2006 as a U.S. senator to Chad, Djibouti, Kenya, and South Africa. The optimism generated by Obama's election was captured by an African colleague, who noted, "How many U.S. presidential candidates prior to taking office can say that they took an HIV/AIDS test in Nairobi to raise awareness about the HIV/AIDS pandemic in Africa, or walked with President Nelson Mandela on Robben Island in South Africa?" The answer, of course, is none. Optimism has been further fueled by Obama's own statements: "Obviously I've got a personal connection to Africa that makes the trip special," explained Obama in reference to his 2006 trip as senator. "I also have a deep abiding interest in what happens to the African continent as a whole."

If history is our guide, however, Africa will remain the region of least concern within the global hierarchy of U.S. foreign policy, as the Obama administration by necessity focuses on domestic issues and other regions of perceived greater importance. Indeed, the sobering reality of domestic and international challenges unrelated to Africa was already evident during the presidential campaign. "The experience of Barack Obama has raised extraordinary expectations in Africa," explained Whitney W. Schneidman, an adviser to the Obama campaign, "but we need to be realistic about these expectations, especially given the financial pressures in the United States." Richard Holbrooke similarly cautioned in an article in *Foreign Affairs* that the next U.S. president would be confronted with a "daunting agenda" as he "inherits a more difficult set of international challenges than any predecessor since World War II." The question that we have to ask ourselves, and the purpose of this paper, is how U.S. foreign policy toward Africa will fare in an Obama administration, when "extraordinary expectations" are confronted by "daunting" challenges, most of which do not originate on the African continent. The remainder of this paper is divided into two sections: a description of five factors that may potentially constrain the foreign policy maneuverability of an Obama administration as concerns Africa, and a discussion of six emerging trends in Obama foreign policy toward Africa.

Potential Constraints on "Great Expectations"

Five sets of constraints may limit the maneuverability of the Obama administration, potentially reinforcing continuity in U.S. foreign policy toward Africa. Whereas some are related to more recent socioeconomic and political-military trends (crisis in the U.S. economy and inheritance of a residual fear of 9/11), others are indicative of long-term constraints inherent in the U.S. policy-making process as concerns Africa (historic White House and congressional neglect of the African continent and bureaucratic influence in the policy-making process as concerns Africa).

1. Crisis in the U.S. Economy

President Obama's number one priority if he wants to be reelected (and all presidents enter office with the goal of winning a second term of office) is responding to the crisis in the U.S. economy. Not since Franklin D. Roosevelt took office in 1933 has a new president inherited an economy in such disastrous shape, including a credit crisis that has led to a financial bailout of more than $700 billion; decreasing home values and foreclosure crisis in the U.S. housing market (typically the one greatest investment of average Americans); dramatic decline in the stock market that has devastated retirement accounts; rising unemployment that economists argue will not peak until late 2010; a growing health care crisis, in which one out of five Americans does not have health insurance; structural decline in the U.S. auto industry, historically one of the mainstays of the U.S. economy; rising national debt from 5 trillion in 2000 to 10 trillion in 2008; and a serious contraction in the U.S. economy. One statistic in particular stands out: the Congressional Budget Office projected in March 2009 that the United States would experience a record $1.8 trillion budget deficit in 2009.

This domestic crisis will clearly consume a significant portion of the Obama administration's first two years in office, leaving little time, relatively speaking, for foreign initiatives and especially those targeted toward the African continent. Moreover, the financial requirements of responding to this crisis will invariably mean the lack of financial resources to fund new initiatives in Africa and other regions of the world.

2. Inheritance of a Residual Fear of 9/11

A second potential constraint involves the Obama administration's inheritance of a residual fear of another 9/11 attack. Although it has subsided significantly in the last eight years and especially in the face of the recent economic crisis, this fear nonetheless remains within the fabric of U.S. society, as demonstrated by the fear generated by the unannounced low-level flight of Air Force One over New York City during the week of April 27 for a photo opportunity. The Africa dimension of this residual fear is that the Obama administration has inherited a series of national security initiatives, often critiqued as the "militarization of U.S. foreign policy toward Africa," that a president seeking to avoid missteps on the path to reelection may be hesitant to dismantle.

One of the best examples of this inherited national security structure is a set of three regionally based counterterrorism programs in the "Islamic littoral" or coastal regions of the African continent: the Trans-Saharan Counterterrorism Initiative (TSCTI) which includes North Africa; the Combined Joint Task Force Horn of Africa (CJTF-HOA), which is responsible for the "Greater Horn of Africa"; and the East African Counter-Terrorism Initiative (EACTI). These counterterrorism initiatives are buttressed by the creation of a continent-wide Africa Command (AFRICOM), expansion of the International Military Education and Training (IMET) program and other U.S. military aid programs, and the Gulf of Guinea Initiative, the primary goal of which is to gradually build up an effective regional security program capable of ensuring the safe transport of oil resources to the United States.

Together these regional security programs provide useful insights into the evolving nature of U.S. foreign policy toward Africa. First, these programs serve as the core of an evolving foreign policy approach that, during the Bush administration, divided Africa into at least four spheres of variable foreign policy interest: (1) those regions (North and East Africa) destined to receive priority attention due to their proximity to the Middle East, the perceived epicenter of the global war on terrorism; (2) regional powers, typically Nigeria and South Africa, but also including Algeria, Ethiopia, Kenya, and Senegal, that are perceived as crucial to the maintenance of regional stability and therefore as "regional anchors" of counterterrorism efforts; (3) countries deemed important to U.S. economic interests, most notably oil-producing countries in the Gulf of Guinea region; and (4) the remainder of sub-Saharan Africa which remains relegated to the back burner of U.S. foreign policy. Not surprisingly, U.S. foreign policy increasingly focused on those countries in which core foreign policy interests intersect, as in the case of Algeria, a regional power with oil resources that is considered crucial to combating perceived terrorist threats in North Africa. Most important, these are not initiatives that a president seeking to avoid missteps on the path to reelection in 2012 will necessarily dismantle.

3. Historic White House Neglect of the African Continent

A third potential constraint involves historic White House neglect of the African continent amidst domestic and other international priorities. It is typically assumed by foreign observers that presidents and their principal foreign policy advisers will be the most influential and the most activist in terms of U.S. foreign policy toward Africa. Throughout the Cold War and its aftermath, however, presidents traditionally have devoted less attention to Africa compared to other regions of perceived greater concern, most notably Europe (including Russia and the other countries that were once part of the Soviet Union) and more recently the Middle East and South Asia.

Historic neglect of Africa at the highest reaches of the U.S. policy-making establishment is the direct result of a wide variety of factors: a president's typical lack of knowledge and therefore the absence of a deep-felt interest in a region that historically enjoyed few enduring political links with the United States as compared with the former European colonial powers; a tendency to view Africa as the responsibility of those same European colonial powers, especially France, whose leaders were often willing to take the lead in crisis situations; the impracticality of one person monitoring relations with nearly 200 countries worldwide, including fifty-three in Africa, and therefore the necessity of delegating responsibility for handling foreign policy for those regions considered marginal by the White House; and, most important, the necessity of balancing domestic priorities with foreign affairs necessities, especially during a first term in office in which the ultimate priority of all presidents is to assure reelection, with simple electoral logic typically suggesting that Africa is not a priority for the vast majority of the voting public.

Even if we recognize that Barack Obama is different from all previous presidents, especially as concerns his personal connection to the African continent, an Obama White House will be consumed by foreign policy issues in other regions of perceived greater importance, potentially leaving little time for high-level White House attention to Africa. This tendency has already emerged during the first 100 days of the Obama administration. In addition to prioritizing U.S. relations with Europe and the other northern industrialized countries (witness Obama's attendance at the G-20 Summit), the Obama administration has clearly demonstrated that the Middle East is second in the foreign policy hierarchy, as witnessed by the mandatory focus on the war in Iraq, the decision to make the pursuit of a Middle East peace an administration priority, and an unprecedented diplomatic overture to Iran. The third region of foreign policy concern is South Asia, as witnessed by the increase of U.S. troops in Afghanistan and focus on Pakistan. Subsequent regions of concern include Asia and Latin America. The key question that one must ask: where is Africa in this U.S. foreign policy priority? Still presumably last.

4. Historic Congressional Neglect of the African Continent

A fourth potential constraint that may limit the maneuverability of an Obama administration as concerns Africa is historic congressional neglect of the African continent. A variety of constitutionally mandated prerogatives, including the confirmation of presidential appointees, the convening of hearings, and the drafting and voting of key legislation, suggests that Congress theoretically should play an important role in defining U.S. foreign policy toward Africa. Like their White House counterparts, however, members of Congress historically have neglected Africa relative to other regions of perceived greater interest. Reelection pressures and time constraints imposed by terms of office (two years for representatives and six years for senators) force them to select and prioritize the domestic and the international issues which will receive their attention. Since the primary objective of most members is to be reelected, and since most U.S. citizens know or care very little about the African continent, conventional wisdom suggests that it is politically unwise to focus too much time on Africa. As a result, membership on the Africa subcommittees is among the least desired congressional positions in both houses of Congress, and is therefore relegated to relatively junior representatives and senators, such as Barack Obama, who served on the Africa Subcommittee of the Senate Foreign Relations Committee.

An important impact of congressional neglect of Africa is that even highly motivated chairpersons of the Africa subcommittees face an uphill task in pushing African issues to the forefront of congressional debate. In the absence of crisis, partisan and ideological differences within Congress prevent activist groups from achieving congressionally mandated changes in U.S. foreign policy toward Africa. Even during short-term crises when an issue may attract the attention of a significant number of members of Congress, control of the policy-making process naturally flows to the White House and the bureaucracies of the executive branch. In this regard, the resurgence of guerrilla activity in the eastern provinces of Congo-Kinshasa at the beginning of 1999, let alone the involvement of several foreign armies in this conflict (what some policy makers typically referred to as "Africa's first world war"), failed to rise to the level of a policy-making crisis in the nonideological context of the post–Cold War era; a far cry from the crisis atmosphere which prevailed in the 1960s when a guerrilla insurgency within the same region was perceived by U.S. policy makers as threatening to install a pro-Soviet regime under the leadership of Patrice Lumumba.

Most important, the fact that both the Senate and the House of Representatives are dominated by the Democratic Party is in fact a double-edged sword for the Obama White House. On average the Congress is more liberal than Obama, and there is a great deal of pent-up demand stemming from the Bush years for a whole host of domestic programs. Key question: Will Obama be able to take the lead and build working majorities that will support his priorities/initiatives, or will he be confronted by an overly activist Congress intent on pushing its own priorities/initiatives?

5. Bureaucratic Influence in the Policy-Making Process

The net result of White House and congressional neglect of Africa is that U.S. foreign policy toward Africa, perhaps more so than that toward any other region of the world, remains largely delegated to the high-level bureaucrats and political appointees within the bureaucracies of the executive branch. Exceptions of course exist, such as the willingness of both the White House and the Congress to pressure Sudan's government to seek a peaceful resolution of civil conflict in the southern portion of the country, but these are rare occurrences typically due to pressures from grassroots constituencies that have the ear of the president and senior congressional leaders and that, most important, are considered crucial to reelection. In the case of Sudan, for example, a wide array of Christian groups deemed essential to Republican victories in 2004 effectively lobbied the White House to "do something" to stop what they perceived as a genocidal policy that a northern-based Islamic regime was carrying out against a southern-based, predominantly Christian population, including the practice of southern Christians being sold as slaves in northern Sudan. In order to understand U.S. foreign policy toward Africa fully, one must therefore focus on the policies and interactions of the African affairs bureaus of the traditional national security bureaucracies, such as the State Department, the Pentagon, and the Central Intelligence Agency (CIA), as well as their counterparts within the increasingly important economic realm, most notably the Department of Commerce. To be sure, the White House sets the overall parameters of U.S. foreign policy toward Africa, as was the case of its predecessors during the Cold War.

But the unique nature of the U.S. policy-making system ensures that specific policy initiatives often emerge from and are coordinated by the national security bureaucracies with little White House input.

The net result of what can be referred to as "bureaucratic influence" in the policy-making process is that it fosters the continuation of established policies, even when an administration with seemingly different beliefs than its predecessor takes office, such as in the shift from the Bush to the Obama administration. The key to potential change is who leads the State Department, and especially the State Department's Bureau of African Affairs, which traditionally has taken the lead as concerns U.S. foreign policy toward Africa. Toward this end, the nomination of Johnnie Carson as Assistant Secretary of State for African Affairs offers important insights. He most recently served as the national intelligence officer for Africa for the National Intelligence Council and senior vice president for the National Defense University. In the State Department, he served as deputy assistant secretary of state for African affairs, ambassador to Uganda (1991–1994), Zimbabwe (1995–1997), and Kenya (1999–2003), and as a Foreign Service Officer (FSO) in Portugal, Botswana, Mozambique, and Nigeria. He began his service in Africa as a Peace Corps volunteer in Tanzania.

An important reason for citing this long litany of Johnnie Carson's Africa experience is to underscore that it is unclear how much a career FSO will push for significant change in the substance and the priorities of U.S. foreign policy toward Africa. Moreover, the recent and significant nature of his involvement in the intelligence arena makes it unlikely that he will be a strong proponent for significantly changing the recent overemphasis on U.S. military/security policy toward Africa. His extended experience in East Africa (Kenya, Tanzania, and Uganda) and Southern Africa (Botswana, Mozambique, and Zimbabwe) nonetheless bodes well for enhanced U.S. foreign policy attention to these two regions.

New Directions in U.S. Foreign Policy toward Africa?

It is always perilous when one tries to peer into a crystal ball to discern future foreign policy tendencies just months after a new administration has taken office. However, if we look at what Obama has said and done during the campaign as well as emerging foreign policy threads during the first 100 days of his administration, against the backdrop of the five potential constraints already mentioned, we can get an idea of where an Obama administration may be headed as concerns Africa. Six trends in particular stand out.

1. Campaign and Presidential Appointments Demonstrate a Heightened Interest in Africa

The Obama campaign oversaw a massive three-hundred-person foreign-policy campaign bureaucracy, organized like a mini–State Department, to assist a presidential candidate whose lack of national security experience was a concern to voters during the campaign. This team included an unprecedented number of close advisers with impressive backgrounds in Africa. Their transitions to political appointees in an Obama White House provide an early sense of Obama's priorities in Africa. To cite but five examples:

- Susan Rice: assistant secretary of state for African affairs under the Clinton administration, who helped lead Obama's campaign foreign policy bureaucracy, and who currently is the U.S. Representative to the United Nations.
- Mark Lippert: Obama's former Senate foreign policy adviser, who in 2008 returned from a Navy tour of duty in Iraq (Navy SEAL), and who currently is White House National Security Council Chief of Staff.
- Denis McDonough: one of Obama's top foreign policy aides, national security coordinator for the campaign, former foreign policy adviser to Senate Democratic leader Tom Daschle, and current White House deputy national security advisor.
- Jonathan Scott Gration, retired Air Force major general who voted for Bush in 2000, accompanied Obama to Africa, refers to Obama as "America's Mandela," raised as the son of missionary parents in the Democratic Republic of Congo (DRC), speaks fluent Swahili, and who currently is White House special envoy to Sudan.
- Samantha Power, Harvard human rights expert and Pulitzer Prize–winning author (*A Problem from Hell: America and the Age of Genocide*), who resigned her post as foreign policy adviser in the Obama campaign in March 2008 after calling Hillary Clinton "a monster," and currently is Senior Director for Multilateral Affairs at the National Security Council.

Three characteristics common to this core Africa foreign policy team include early opposition to the war in Iraq, a tendency toward liberal internationalism, and an emphasis on the use of "soft power" (diplomacy and economic aid) to advance U.S. interests abroad. Interestingly enough, this foreign policy team also demonstrates Obama's reliance on those with military experience (e.g., military backgrounds of Lippert and Gration), with important implications as to how an Obama administration will deal with the current overwhelming influence of strategic/military initiatives in Africa.

2. Obama's Governing Ideology: Pragmatic Traditional Realism with Strong Tendencies toward Liberal Internationalism

President Obama is not a typical liberal, but rather a pragmatic traditional realist with strong tendencies toward liberal internationalism, sometimes referred to as an "optimistic realist" or a "realistic optimist." As succinctly summarized by one observer during the presidential campaign, Obama

- often praises the foreign policy of the Bush senior administration ("I have enormous sympathy for the foreign policy of George H. W. Bush," he said in May 2008);
- does not speak in the moralistic tones (i.e., "good" versus "evil") of the Bush administration, perceiving countries and extremist groups as

complex, and motivated by power, greed, and fear, as much as by pure ideology;

- does not portray countries/movements as part of a monolithic threat (e.g., differentiates between various trends/tendencies within the Islamic world);
- does not use the soaring rhetoric of Bush's freedom agenda, preferring instead to talk about enhancing people's economic prospects, civil society, and his key word, "dignity";
- rejects the Bush administration's obsession with elections and political rights, and argues that people's aspirations are broader and more basic—including food, shelter, and jobs ("Once those aspirations are met, it opens up space for the kind of democratic regimes we want");
- holds a view of democratic development that is slow, organic, and incremental, usually held by conservatives; and
- talks admiringly about Dean Acheson, George Kennan, and Reinhold Niebuhr, all of whom were imbued with a sense of the limits of idealism and of the power of the United States to transform the world.

The implications of this worldview remain unclear. Those in favor of a foreign policy more firmly based on democracy and human rights are nonetheless increasingly beginning to worry that Obama's worldview will result in little change toward authoritarian regimes in Africa, especially those that are known for advancing liberalization within the nonpolitical sectors of their societies, such as Tunisia and its progressive approach to women's rights. According to these commentators, the outline of Obama's foreign policy toward authoritarian regimes was in many respects demonstrated when he made his much-awaited speech on Islam during June 2009 in Cairo, Egypt—an authoritarian U.S. ally at the intersection of the Middle East and Africa that is considered key to advancing the Obama administration's Middle East peace process.

3. Sober Reckoning as Concerns Africa Programs That Will Require Additional U.S. Financial Resources

President Obama has proposed several new foreign aid initiatives and the expansion of existing foreign aid initiatives as concerns Africa. During the campaign, for example, he noted his intention to double U.S. foreign assistance to Africa (which had already trebled during the Bush administration), as part of a promise to double the annual foreign aid budget from $25 billion to $50 billion by 2012. He in particular underscored a desire to showcase the Bush administration's highly successful President's Emergency Program for AIDS Relief (PEPFAR) program, which he also pledged to increase from $15 billion in total funding during the Bush years to approximately $50 billion by 2012. There clearly has been a sober reckoning in this regard, especially in light of the economic meltdown in the U.S. economy. Although current PEPFAR levels will likely remain sacrosanct, it is difficult to imagine how the Obama administration will be able to "sell" massive foreign aid increases to both the American public and the U.S. Congress when so many Americans are either jobless, losing their homes, confronting catastrophic health care costs, or experiencing all three together. One promising trend: the U.S. Congress recently approved a record $900 million for the Global Fund to Fight AIDS, Tuberculosis and Malaria.

Swift movement nonetheless has been and will continue to be evident as concerns Democratic Party/liberal initiatives that are not dependent on additional U.S. financial resources. One of the best examples of this trend occurred on January 23, 2009, when President Obama officially rescinded U.S. restrictions on international family planning policies officially known as the Mexico City Policy but typically referred to as the "global gag rule." Originally put in place under Ronald Reagan, maintained by George H. W. Bush, lifted by Clinton, and subsequently re-imposed by George W. Bush, the gag rule remains a highly charged ideological policy that serves as a litmus test on both sides of the abortion debate. Specifically, this policy ensured during the Reagan and both Bush administrations that no U.S. family planning assistance could be provided to foreign nongovernmental organizations (NGOs) "that use funding from any other source to: perform abortions . . . provide counseling and referral for abortion; or lobby to make abortion legal or more available in their country."

According to "Population Action International," the gag rule exerted a highly negative impact on numerous African countries, including Kenya, during the most recent Bush administration:

- Kenya's leading reproductive health care providers have suffered serious budget cuts and were forced to close eight clinics, lay off large numbers of staff and scale back programs.
- In most cases, those shuttered clinics were the only source of health care for local communities.
- Community-based outreach services throughout Kenya's rural areas have been greatly curtailed as the country's primary family planning organizations cut back due to a lack of funds. Outreach services are often the only access rural men and women have to contraceptive supplies and education on HIV/AIDS.
- Kenya's leading family planning organizations have been forced to withdraw from a U.S.-funded project to provide comprehensive and holistic reproductive and child health care, as well as HIV/AIDS prevention and treatment, to vulnerable populations in Kenya; the project is consequently losing ground.

This policy, and others like it, has been/will be reversed by an Obama administration intent on demonstrating the fruits of electoral victory to its supporters. An early extended presidential visit to the African continent, almost certainly to include stops in Kenya and South Africa, constitutes another policy initiative that will entail little financial cost but that will generate great enthusiasm among both the Democratic Party base and Africans throughout the continent. Indeed, the countries visited (outside of Kenya) will offer perhaps the strongest indicator of Obama foreign policy toward Africa. In this regard, great enthusiasm was generated by President Obama's decision in July 2009 to make a two-day visit to Ghana—an African model for democratic governance and economic development—as part of a longer trip that included participation in the G-8 Summit in Rome, Italy, and a visit to Russia.

4. Cautious Approach to Conflict Resolution That Will Nonetheless Use the White House as a Bully Pulpit

The Obama administration has demonstrated a cautious approach to conflict resolution that is nonetheless willing to use the White House as a "bully pulpit." President Obama entered office mindful of the fact that the United States has not played enough of a proactive role in recent years to resolve conflict on the African continent. In the case of Rwanda, for example, Obama was critical of the Clinton administration's unwillingness to act, and in fact its refusal to label what was happening in Rwanda as genocide. Obama was similarly critical of the Bush administration's refusal to take a more forceful approach to the Darfur crisis, despite the Bush administration's willingness to describe as genocide the policies of the Bashir regime.

President Obama has underscored his administration's intention to take a more proactive approach to African conflict resolution, and has specifically cited ongoing conflicts in Darfur, Zimbabwe, the eastern Congo, the Niger Delta, and Somalia. Although Obama's knowledge of and reference to these conflicts is laudable, an important policy-making reality to consider is that effectively resolving any *one* of these conflicts would require the sustained attention of the White House and, most important, President Obama himself. It is nonetheless unclear if foreign policy challenges and priorities elsewhere, including Obama's pledge to make the Arab-Israeli peace process a White House priority, will permit the degree of high-level White House attention necessary to make effective conflict resolution in Africa a reality.

5. Heightened Focus on Socioeconomic and Development Issues

President Obama has criticized the fact that U.S. development assistance to the poorest African countries decreased by nearly 50 percent from 2000 to 2008. He similarly has criticized the fact that the percentage of development assistance provided to the best-governed African countries has declined even more (by nearly 66 percent) during the same period. These trends were at least partially due to the militarization of U.S. foreign policy toward Africa during the Bush years, as part of the global war on terrorism. The Obama administration has pledged to reverse this trend, by strengthening a variety of nonmilitary programs designed to promote socioeconomic and political development, including the African Growth and Opportunity Act (AGOA), the Millennium Challenge Corporation (MCC), the Middle East Partnership Initiative (MEPI) (which is inclusive of North Africa), and PEPFAR.

The most noteworthy and far-reaching proposal in this regard is Obama's pledge to remake and restore the United States Agency for International Development (USAID) to a position of preeminence in the foreign aid hierarchy. USAID funding was severely curtailed during the Bush administration, as increasing amounts of aid were channeled through the U.S. military establishment and newly created quasi-independent foreign aid programs, such as PEPFAR. As envisioned by its proponents within the Obama administration, this restructuring process would entail the moving to USAID of foreign aid programs currently under at least twenty-one different executive branch agencies. The key to this process is the belief that USAID, and not the Pentagon, should play the central role in the formulation and implementation of development and other related foreign policy strategies, not only in Africa, but in all regions of the developing world.

6. Unclear Change in the Democratic Deficit in U.S. Foreign Policy toward Africa

The Bush administration talked about the normative good of democracy promotion. But when the normative goal of promoting democracy clashed with the strategic goal of containing terrorist threats, the strategic goal almost certainly won, thereby more closely associating the United States with some of the worst abusers of human rights, such as the Egyptian regime of Hosni Mubarak. In fact, all three of the closest U.S. allies in North Africa—Egypt, Morocco, and Tunisia—are dictatorships. Other countries with which the United States has a strategic partnership in the region also have poor human rights records (e.g., Djibouti).

The case of Tunisia is particularly enlightening. Boasting a dictatorship that is perceived in Washington as a "strong U.S. ally in the Arab world" and a "valuable partner in the war on terrorism," Tunisia was actively courted by the Bush administration. The capital, Tunis, is home to a rising number of U.S. facilities commensurate with its increasingly close relationship with Washington, including a new $42 million U.S. embassy, a Foreign Service Institute for teaching Arabic to U.S. government personnel who are preparing to work in Arabic-speaking countries, and a regional office for coordinating the activities of the MEPI program in North Africa (the only other such regional office is in the United Arab Emirates).

Any question as to whether democracy promotion or national security objectives dominated the U.S.-Tunisian relationship during the Bush administration was best addressed by then–secretary of state Colin Powell's response to a question while on an official visit to Tunis in December 2003. When asked about President Ben Ali's use of repression to silence dissent and ensure his unrivaled control over the Tunisian political scene since assuming power via extraordinary means in 1987, Powell's response was that this was "a matter between him [Ben Ali] and the Tunisian people." My wife and I were living in Tunisia at the time, and Powell's statements were highly discouraging to human rights activists. Two months later in February 2004, Ben Ali's strong support for U.S. counterterrorism initiatives was rewarded with a highly coveted head-of-state visit to Washington, including a White House visit with President Bush, amid a rising chorus of criticism by human rights activists. This visit was significant, in that it was the first such visit granted to Ben Ali since he took power in 1987. Although all of Bush's predecessors (Reagan, Bush senior, and Clinton) denied Ben Ali's request for a head-of-state visit, the Bush administration accepted, because of Tunisia's perceived importance in the global war on terrorism. In short, an overriding preoccupation with terrorist threats led the Bush administration to overlook the authoritarian excesses of African regimes in favor of their willingness to support U.S. national security objectives (i.e., the war on terrorism), just as the United States did during the Cold War. The key dilemma for pro-democracy activists: it is unclear whether Obama's governing ideology—pragmatic traditional realism with strong tendencies toward liberal internationalism—will lead to significant change in the democratic deficit in U.S. foreign policy toward Africa.

Panel VII: Africa

Summary of Discussion

Dr. Stephen A. Emerson
Associate Professor of National Security Affairs
Naval War College

The Ruger Workshop's Africa panel began with opening presentations by Ambassador (Ret.) Princeton Lyman and Dr. Peter Schraeder, followed by some brief observations from Dr. Hussein Solomon. While acknowledging that Africa was unlikely to be elevated to the top of the American foreign policy agenda anytime soon, the panelists nonetheless underscored the growing importance of the continent to U.S. security. This reality will require American policy makers to shift their thinking away from Africa as primarily an area of humanitarian interest to an area impacting core national interests in a post-9/11 world. One panelist, however, cautioned that the high expectations—both in and out of Africa—for U.S.-African policy under the Obama administration are unlikely to be met given the reality of the global financial crisis and other more pressing international problems.

Thus, rather than pursue policy objectives that will require a significant new commitment of resources, policy makers would be better advised to find more efficient ways to address current and emerging problems. From terrorism and energy security to drug trafficking, health, and environmental challenges, the United States must learn how to manage its security and foreign policy agenda in Africa better. Likewise, the increasing need to address Africa's "softer" security issues will require Washington to shift its reliance away from the military tool of national power to an increasingly diplomatic and development-oriented strategy. It also would behoove Washington to rethink its go-it-alone approach that has characterized past U.S. engagement on the continent and rather embrace a more cooperative and nuanced internationalist approach to addressing African ills.

The discussion kicked off with a question concerning the prospects for a changed U.S. agenda in the UN Security Council with respect to Africa and more broadly how the Obama administration's focus might differ from that of the Bush administration. It was noted that given the large number of "Africanists" serving in key administration positions (including Susan Rice at the United Nations, Scott Gration as special envoy to Sudan, and Johnnie Carson at the Department of State) many believed Africa would indeed garner more attention under the new administration. Unfortunately, other than a possible more activist internationalist approach many of the U.S. priorities—Darfur, the Sudanese peace process, Somalia, Nigeria, and HIV/AIDS—are likely to remain the same given the lack of new human and financial resources. In fact, talk of a greater emphasis on diplomacy and development has yet to manifest itself in any concrete manner with regard to Africa.

Several questions were raised concerning the role of U.S. trade and investment in Africa, which provoked an extensive discussion. In particular, several participants wanted to know why there was an apparent disconnect between the continent's rising political and economic importance and the unwillingness of the American business sector to pursue emerging opportunities. What could or should the U.S. government do to promote more African trade and investment? Several respondents highlighted multiple examples of U.S. foreign direct investment, particularly with regard to the energy and mineral sector, and the rise of African investment funds in recent years, but this progress was likely to decline in light of the widening global financial crisis. "Africa is an extremely difficult investment environment" to begin with and the real question, according to one respondent, is how the United States can help mitigate the impact of the worldwide recession on African countries.

This led to a very vibrant give-and-take on the role of aid (short-term humanitarian response) versus development assistance (long-term economic investment). Although humanitarian aid will undoubtedly remain a prominent feature of American assistance to Africa, the continent has made significant progress toward greater economic development with U.S. government programs such as the African Growth and Opportunity Act and the Millennium Challenge Account, as well as private entities, such as the Corporate Council on Africa, that are helping to promote better economic decision-making and improved governance. True economic empowerment, however, is not likely to come from government-sponsored programs, but through structures and mechanisms that facilitate growth from the bottom up. This is where the United States can help the most. But, it was noted, "there will be setbacks"; Kenya nearly lapsed into civil war following a flawed election and South Africa is currently undergoing some momentous domestic changes. All the more important, discussants as a whole believed, was for Washington to stay the course and not get disillusioned.

The bulk of the remaining discussion time was spent exploring issues surrounding the objectives and likely nature of American engagement in Africa over the next several years. Or, as one participant put it, what does the strategy look like? And what should be the key areas or components of that strategy? It was pointed out that over the years there has been much disagreement, both within and outside of the U.S. government, as to the best strategic approach for U.S.-African policy and what should be the key U.S. priorities on the continent. One school believes that the focus should be toward improving education, childhood development, health care, and poverty alleviation, while another believes the emphasis should be on infrastructure development, trade promotion, and improving security. Although the jury is still out as to the exact priorities of the Obama administration, discussants underscored the need to continue the Bush administration's themes of improving governance, reducing corruption, and promoting transparency. Throughout the discussion the group kept returning to governance and trade as priority areas essential to underpinning any American foreign policy initiatives in Africa.

At the end of the day, it was acknowledged that the new administration will almost certainly have difficulty grappling with the myriad of problems facing the

United States in Africa and that Washington's engagement on the continent is still likely to remain crisis driven and will be forced to get by with little, if any, new resources for the foreseeable future. It was recommended that the United States focus its efforts on building "a better policy" that is more clearly defined, carefully developed and implemented in a coordinated and nuanced fashion. Rather than seek to institute a vast array of new programs and foreign policy initiatives for Africa, work on improving the mechanics, coordination, and integration of existing successful ones, such as with HIV/AIDS and the Millennium Challenge Account. Moreover, seek to identify ways to address areas of overlapping development and security needs with African countries to advance a common agenda that provides a win-win situation for both the United States and Africa.

Concluding Remarks

Dr. Richmond M. Lloyd
William B. Ruger Chair of National Security Economics
Naval War College

During this workshop we have explored setting future directions for American foreign policy for each of the major regions of the world. We considered the challenges and opportunities the United States and its allies and friends will face in the future. We then explored the changes that should be made to all elements of U.S. foreign policy including the diplomatic, economic, military, and informational elements. In certain circumstances we suggested continuity of policies. We discussed the varying perspectives of nations within each region concerning U.S. foreign policy and changes in policy that they would desire. Overall, our primary focus has been to provide new directions for U.S. foreign policy that will better support the interests and objectives of the United States, its allies, and its friends.

I want to thank all of you for the extensive research you did in formulating your ideas, preparing your formal papers, and for your thoughtful contributions throughout this workshop. Your papers and the strategic conversations we have had during each panel provide a very rich menu of ideas, insights, and pragmatic suggestions that will be of value as the nation reassesses its future foreign policy. We hope that your preparations and work here will also provide you with the building blocks for further dialogue, research, and future publications.

What's going to happen next? We will quickly produce a monograph that will include all of your papers and summaries of our discussions throughout the workshop. We expect to have the monograph online for the general public in midsummer 2009. (Our website is http://www.usnwc.edu/academics/courses/nsdm/rugerpapers.aspx.)

Several thousand printed copies will be available in September 2009. (Send request for printed copies to the William B. Ruger Chair of National Security Economics at richmond.lloyd@usnwc.edu.) We plan to widely distribute the monograph throughout the national security community and the general public.

We will keep you informed of follow-on workshops and conferences that will be of professional interest to you and will benefit from your participation.

Again, thank you so very much for the extensive work you did in preparing for and participating in this workshop.

Participant Biographies

Dr. Hayat Alvi-Aziz

Dr. Hayat Alvi-Aziz is an Associate Professor in the National Security Decision Making (NSDM) Department at the U.S. Naval War College.

She has served as the Director of International Studies at Arcadia University in Glenside, PA. Professor Aziz also taught political science at the American University in Cairo, Egypt, for four years. Her specializations include international relations, political economy, comparative politics, Islamic studies, and Middle East and South Asian studies. She is proficient in Arabic and Urdu. Her publications include numerous journal articles and books including *Regional Integration in the Middle East: An Analysis of Inter-Arab Cooperation* (Lewiston, NY: Edwin Mellen Press, 2007); *An Introduction to International Studies: Exploring Frontiers* (Deer Park, NY: Linus Publications, 2006); and *The Arabian Nights Reader* (Deer Park, NY: Linus Publications, 2006).

Ms. Deborah Bolton

Deborah Bolton is the Department of State Advisor to the Naval War College in Newport, Rhode Island. She is a Senior Foreign Service Officer with the rank of Minister-Counselor.

Ms. Bolton was born and educated in Philadelphia, Pennsylvania. After graduating from St. Joseph's University in 1974 with a degree in International Relations, she entered the Foreign Service that year. She has held assignments in Latin America, Europe, Asia, and the Department of State.

Her early tours include to the American Embassies in Ecuador, Argentina, Hungary, and also Washington. In 1990 she became Chief of the Consular Section at the U.S. Interests Section in Havana, Cuba. In 1993 she earned a diploma in Security Studies at the U.S. Air Force War College. She then became the Deputy Director for International Security and Peacekeeping Operations in State's Bureau of Political-Military Affairs. After language training, she served in Hanoi, Vietnam, for six months then in August 1997 she was assigned to open the U.S. Consulate General in Ho Chi Minh City where she was Acting Principal Officer until May 1998, then the Deputy until May 1999. In July 1999 she was Deputy Chief of Mission at the Embassy in Valletta, Malta. In September 2001 she was appointed the Chief of Mission in Curaçao, Netherlands Antilles. From September 2004 until July 2007 she served as Political Advisor (POLAD) to the Commander, NORAD and U.S. Northern Command in Colorado Springs. Ms. Bolton is the recipient of four Superior Honor Awards and one Meritorious Honor Award from the Department of State. From the Department of Defense she received the Chairman's Joint Distinguished Civilian Service medal and the Armed Forces Civilian Service medal. Her languages are Spanish, Hungarian, and Vietnamese. She is a member of the American Foreign Service Association.

Dr. Emrys Chew

Emrys Chew is Assistant Professor at the S. Rajaratnam School of International Studies (RSIS) in Nanyang Technological University, Singapore. He was educated at the Anglo-Chinese School, Singapore, and St. Catharine's College, Cambridge. From the University of Cambridge, he obtained both a B.A. with First Class Honours in History (1995) and a Ph.D. (2002). His B.A. dissertation, a study entitled 'The Naning War, 1831–1832: Colonial Authority and Malay Resistance in the Early Period of British Expansion', was awarded the Alan Coulson Prize for Imperial and Commonwealth History and subsequently published (*Modern Asian Studies,* May 1998). His doctoral thesis, entitled 'Arming the Periphery', traces the development and dynamics of arms-trade networks in the Indian Ocean between 1780 and 1914, a critical period of Western imperial and industrial expansion as well as indigenous transformation across the frontiers of Asia and Africa. His publications include an article about the impact of arms transfers on military culture and colonial warfare in Indian Ocean societies, particularly in light of contemporary debates about the international war against terrorism ('Militarized Cultures in Collision', *Journal of the Royal United Services Institute,* October 2003) and a chapter about the long-term militarization of South Asia and its periphery ('Globalization and Military-industrial Transformation in South Asia', in G. Till, E. Chew, and J. Ho, eds., *Globalization and Defence in the Asia-Pacific,* Abingdon and New York, 2009). In addition to his research interests, Emrys has taught undergraduate courses on imperialism and nationalism at the University of Cambridge, examining cross-cultural interactions that have generated and shaped much of the modern world. He is currently involved in the Maritime Security Programme at RSIS, where he also teaches postgraduate courses on the international history of Asia and Cold War history and international politics. Beyond professional commitments at RSIS, Emrys is an advisor and member of the Syllabus Development Committee for History at Singapore's Ministry of Education. An avid orchid enthusiast, he is a committee member of the Orchid Society of Southeast Asia, where he also serves as editor of the *Malayan Orchid Review.*

Dr. Patrick M. Cronin

The Honorable Dr. Patrick M. Cronin is Director of the Institute for Strategic Studies, which is based at the National Defense University in Washington, DC. He took up the post in 2008 after leaving the International Institute for Strategic Studies in London, where he had been Director of Studies since 2005. He is simultaneously Director of the Center for the Study of Chinese Military Affairs, which serves as a national focal point for multidisciplinary research and analytic exchanges regarding China.

He was confirmed by the U.S. Senate as the third-ranking official in charge of the U.S. Agency for International Development. During his twenty-five-year career he has spent his time inside government and academic research centers, spanning defense, security, foreign and economic policy, and foreign assistance. He has taught at Georgetown University, Johns Hopkins University, and the University of Virginia, and he is the co-author or editor of five recent books, including, *America's Security Role in a Changing World:*

Global Strategic Assessment 2009, The Impenetrable Fog of War: Reflections on Modern Warfare and Strategic Surprise, The Evolution of Strategic Thought, Double Trouble: Iran and North Korea as Challenges to International Security, and *Civilian Surge: Key to Complex Operations.*

Dr. Sharyl Cross

Sharyl Cross currently serves as Director of Studies for the Marshall Center's flagship course, the Program in Advanced Security Studies. Professor Cross also directs the Marshall Center project on countering ideological support for terrorism. She has taught in the Program on Terrorism and Security Studies, the Senior Executive Seminar, and offers an elective course on United States–Russia/Eurasia security issues.

Dr. Cross came to the Marshall Center from the United States Air Force Academy, where she had been appointed Visiting Distinguished Professor of Political Science. She is a tenured full professor and former chair of the Department of Political Science at San Jose State University in the California State University system. Dr. Cross serves as a periodic lecturer for the NATO School in Oberammergau, Germany, and has lectured widely at leading academic and policy institutions throughout the international community.

She holds a Ph.D. in Political Science with concentrations in International Relations, Comparative Politics, and American Foreign and Security Policy from the University of California, Los Angeles. She was a resident Fellowship Scholar and consultant completing the graduate program in Soviet/Russian studies at the RAND Corporation in Santa Monica, California, in 1984–1987. Originally from Arizona, Dr. Cross graduated Phi Beta Kappa with Honors in Political Science from the University of Arizona in 1983.

Dr. Cross held a post-doctoral fellowship at the Hoover Institution at Stanford University in 1991–1992. She served as Visiting Associate Professor of Transregional Studies from 1992 to 1994 at the United States Air War College at Maxwell AFB.

In 1999, she was awarded a Fulbright Senior Scholar grant to support affiliations as Visiting Research Scholar and Professor at the Institute of USA and Canada Studies in the Russian Academy of Sciences and Moscow State Institute of International Relations of the Ministry of Foreign Affairs of the Russian Federation (MGIMO). Dr. Cross' distance education project developed in collaboration with faculty at MGIMO and St. Petersburg State University School of International Relations has been featured in the *Chronicle of Higher Education*.

She has been the recipient of three (2001, 2004, 2006–2007) post-doctoral research grants at the Kennan Institute for Advanced Russian Studies at the Woodrow Wilson International Center for Scholars in Washington, DC. In 2005, she was invited to serve as Visiting Professor of Transatlantic Relations at the Institute of Political Studies/Sciences Po in Lille, France.

Dr. Cross has also been awarded fellowships in support of her research from the International Research and Exchanges Board, U.S. State Department (Title VIII), Office of the Secretary of Defense, California State University, NATO-EAPC, USAF Institute for National Security Studies, Director Sponsored

Research at the Marshall Center, American Association for University Women, and University of California, Institute for Global Conflict and Cooperation.

Dr. Cross has consulted for the U.S. State Department, Office of the Secretary of Defense, and United States European Command on terrorism-related and other international security topics. Dr. Cross has co-edited books on contemporary international security issues with specialists from Russia and China, and her publications have appeared in leading academic peer-reviewed journals and books in several countries.

Dr. Stephen A. Emerson

Dr. Stephen A. Emerson is an Associate Professor of National Security Affairs and a member of the National Security Decision Making Department at the U.S. Naval War College in Newport, Rhode Island. He is an African affairs specialist with over twenty-five years' experience working on African political and security issues. Dr. Emerson has taught, lived, and traveled widely throughout most of the continent. Prior to joining the faculty at the U.S. Naval War College, Dr. Emerson worked for the U.S. Department of Defense as a political-military analyst for southern Africa and was Chair of Security Studies at the Africa Center for Strategic Studies. His professional interests include southern African area studies, conflict and political instability, and American foreign and security policy in the developing world. Dr. Emerson is the author of numerous governmental and academic articles and studies on African politics, U.S.-Africa policy, and intelligence issues. He holds a Ph.D. in Political Science/Comparative Politics and M.A. in International Relations from the University of Florida.

Dr. Thomas R. Fedyszyn

Dr. Thomas R. Fedyszyn has been a member of the Naval War College faculty for the last nine years, following a thirty-one-year Naval career, serving in six different cruisers and destroyers. He received a Ph.D. from the Johns Hopkins University in Political Science and taught political science at the U.S. Naval Academy. His most recent military assignments included serving as the U.S. Naval Attaché to Russia and two tours at NATO Headquarters in Brussels. A former surface warrior, he commanded the USS *Normandy* (CG 60) and USS *William V. Pratt* (DDG 44). He served in numerous strategy, policy, and long-range planning billets for the Office of the Secretary of Defense and the Chief of Naval Operations. He was a principal contributor to both the Lehman-era Maritime Strategy and NATO's New Strategic Concept following the Cold War. His contributions appear regularly in the *Providence Journal* and the U.S. Naval Institute *Proceedings*. His most recent work on the future Obama Maritime Strategy has stirred debate in Washington policy circles. He was recently appointed as the Naval War College's Eurasia Regional Study Group Chair. He specializes in NATO, naval strategy, and Russian naval affairs.

Dr. John F. Garofano

John Garofano is Professor, Strategy and Policy Department, U.S. Naval War College, Newport, Rhode Island. He received the Ph.D. in Government from

Cornell University and an M.A. from The Johns Hopkins School of Advanced International Studies (Bologna/Washington).

Dr. Garofano's research interests include military intervention, Asian security, and the making of U.S. foreign policy. His writings include *The Intervention Debate: Towards a Posture of Principled Judgment* (Carlisle, PA: 2002), *Clinton's Foreign Policy: A Documentary Record* (ed., Kluwer, 2003), and articles in *International Security, Asian Survey, Contemporary Southeast Asia, Orbis,* and the *Naval War College Review* among other journals. Prior to joining the War College's faculty, Dr. Garofano has been a Senior Fellow at the Kennedy School of Government and a professor at the U.S. Army War College, the Five Colleges of Western Massachusetts, and the University of Southern California. Currently, he holds the Jerome Levy Chair of Economic Geography and National Security, and is an Area of Study Coordinator for Asia-Pacific electives.

Mr. Peter Hakim

Peter Hakim is president of the Inter-American Dialogue, a Washington-based center for policy analysis and exchange on Western Hemisphere affairs.

Mr. Hakim writes and speaks widely on hemispheric issues, is regularly interviewed on radio and television, and has testified more than a dozen times before Congress. His articles have appeared in *Foreign Affairs, Foreign Policy,* the *New York Times,* the *Washington Post, Miami Herald, Los Angeles Times,* and *Financial Times,* and in many Latin American newspapers and journals. He was a vice president of the Inter-American Foundation and worked for the Ford Foundation in both New York and Latin America. He has taught at MIT and Columbia. He has served on boards and advisory committees for the World Bank, Council on Competitiveness, Inter-American Development Bank, Foreign Affairs en Español, Partners for Democratic Change, and Human Rights Watch. He is a member of the Council of Foreign Relations. Peter Hakim earned a B.A. at Cornell University, an M.S. in Physics at the University of Pennsylvania, and a Master of Public and International Affairs at Princeton University's Woodrow Wilson School.

Dr. Timothy D. Hoyt

Dr. Timothy D. Hoyt has been a Professor of Strategy and Policy at the U.S. Naval War College since 2002. He lectures and teaches there on a range of topics including strategy, terrorism, insurgency, warfare in the maritime domain, weapons of mass destruction, and contemporary conflict, and also teaches an elective course on South Asian security. Dr. Hoyt received his Ph.D. in International Relations and Strategic Studies from Johns Hopkins University's School of Advanced International Studies in 1997. At Georgetown University's School of Foreign Service, from 1998 to 2002, he taught graduate courses on security in the developing world, South Asian security, technology and international security, and military strategy. In October 2003, he testified before two subcommittees of the House Committee on International Relations regarding terrorism in South and Southwest Asia. In addition to teaching at the Naval War College, Dr. Hoyt has worked for the U.S. Army, for the Library of Congress'

Congressional Research Service, and served as a lecturer or consultant at other U.S. military schools and government agencies. His recent publications include chapters and articles on the war on terrorism in South Asia, security and conflict in the developing world, the limits of military force in the global war on terrorism, the impact of culture on Iraqi military performance, the evolution of multigenerational terrorist organizations, Pakistani nuclear doctrine and strategic thought, the impact of nuclear weapons on recent crises in South Asia, case studies of the Irish Republican Army and its use of political violence, and the role of maritime cooperation in U.S.-Indian relations. He was recently named co-chair of the Naval War College's Indian Ocean Regional Study Group. Dr. Hoyt is the author of *Military Industry and Regional Defense Policy* (Routledge, 2007), examining the role of military industry in the national security policies of India, Israel, and Iraq, and is beginning work on *American Military Strategy in the 21st Century* (Polity Press, tentative publication 2009) and a history of the Irish Republican Army from 1909 to 2009.

Dr. Christopher R. Jasparro

Dr. Christopher Jasparro is an Associate Professor in the National Security Decision Making Department and the Naval War College's Africa Area of Study Coordinator. He is a specialist in transnational and irregular threats as well as environmental security. Dr. Jasparro is also an Asia-Pacific regional specialist with a secondary regional background in Africa. He holds a Ph.D. in geography and a Graduate Certificate in Transportation Studies from the University of Kentucky, an M.A. in geography from the University of North Carolina–Chapel Hill, and a B.A. in anthropology and geography from the University of Vermont.

From 2006 to 2008 he served on the faculty of the U.S. Marine Corps Command and Staff College. Prior to that, he spent six years at the Asia-Pacific Center for Security Studies (Department of Defense Regional Security Assistance Center–Pacific Command). Dr. Jasparro also served as an Assistant Professor in the Geography Department of Framingham State College and has also taught for Hawaii Pacific University, the University of Kentucky, and Harvard University.

Dr. Jasparro is a former U.S. Naval Reserve intelligence officer. He also has extensive experience as a field archaeologist and has worked in economic development, cartography, and town/transportation planning.

Dr. Michael T. Klare

Michael T. Klare is the Five College Professor of Peace and World Security Studies, a joint appointment at Amherst, Hampshire, Mount Holyoke, and Smith Colleges and the University of Massachusetts at Amherst, and Director of the Five College Program in Peace and World Security Studies (PAWSS), positions he has held since 1985.

Professor Klare has written widely on international security affairs, U.S. military policy, the arms trade, and global resource conflict. His most recent books are *Rogue States and Nuclear Outlaws* (1995), *Resource Wars* (2001), *Blood and Oil* (2004), and *Rising Powers, Shrinking Planet: The New Geopolitics of Energy* (2008).

Professor Klare is the Defense Correspondent of *The Nation* magazine and is a Contributing Editor of *Current History*. He has written for these journals and for *Foreign Affairs, Foreign Policy, International Security, Newsweek, Scientific American,* and *Technology Review*. He also serves on the board of directors of the Arms Control Association and the National Priorities Project.

Dr. Heidi E. Lane

Heidi E. Lane holds a Ph.D. in Islamic Studies from the University of California, Los Angeles, a B.A. from the University of Chicago, and earned her M.A. at UCLA. She has conducted field research in the Middle East over the past two decades. She has held visiting research affiliations with the Truman Institute for the Advancement of Peace at the Hebrew University of Jerusalem and the George Washington University Elliot School of International Affairs/Security Studies Program.

Dr. Lane has also been a recipient of a U.S. Fulbright Grant (Damascus, Syria), a National Security Education Program (NSEP) award, and a fellowship from the Institute for International Global Conflict and Cooperation (IGCC). Her areas of specialization are ethnic conflict, religious nationalism, terrorism/counterterrorism. She is trained in Arabic, Persian, and Hebrew. She is currently conducting research on international counterterrorism programs and what impact these have had on democratic reform and political liberalization in the Middle East. She is a member of Naval War College Strategy and Policy Department and Regional Studies Chair, Greater Middle East.

Ambassador (Ret.) David C. Litt

Ambassador Litt served for thirty-four years as a career U.S. diplomat, specializing in the Middle East and Southwest Asia. In 2005–2006, he was the third-ranking officer at the U.S. Embassy in Baghdad, Iraq, with the title of Political-Military Counselor, providing policy advice to the U.S. Ambassador, and serving as liaison between the embassy and the Multi-National Force–Iraq.

His final assignment as a Foreign Service Officer, prior to retirement in 2008, was as the Associate Director for International Liaison at the George C. Marshall European Center for Security Studies in Garmisch-Partenkirchen, Germany. Ambassador Litt entered the Foreign Service in 1974. He served as the U.S. Ambassador to the United Arab Emirates (UAE) (1995–1998) and as Consul General in Dubai ten years prior. Ambassador Litt was Political Advisor to U.S. Central Command and U.S. Special Operations Command at MacDill Air Force Base, Florida (1998–2004). While at the Department of State, Ambassador Litt served as the Director of the Office of Northern Gulf Affairs (Iran and Iraq), and also as Desk Officer for Saudi Arabia. In addition to a tour as economic/commercial officer in Kabul, Afghanistan, in the late 1970s, he served twice as political officer in Damascus, Syria. Just prior to his recent service in Baghdad, he was the State Department's Diplomat-in-Residence at Duke University in Durham, North Carolina.

Other assignments included Deputy Chief of Mission (DCM) at the U.S. Embassy in Niamey, Niger, and as a consular officer in Palermo, Italy. Among several other languages, he speaks Italian, French, Arabic, and Afghan-Persian (Dari). Ambassador Litt received the Secretary of Defense Meritorious Civilian Service Award in 2004 and USSOCOM's Outstanding Civilian Service Award in 2002. He also received the State Department's Superior Honor Award in 2002 and 2004 for his work with the U.S. military and in 2000 for his proposals to improve strategic planning at the Department of State. In addition, he earned a Superior Honor Award for his service as Ambassador to the UAE, and the Meritorious Honor Award as DCM in Niger.

Ambassador Litt was born on December 27, 1949, in Pittsburgh, PA, and grew up in Miami, FL. He received a bachelor's degree with majors in history and French from the University of North Carolina at Chapel Hill in 1971, and a master's degree in International Relations from the Johns Hopkins School of Advanced International Studies (SAIS) in Washington, DC, in 1973. At SAIS, he specialized in European affairs and international economics. He attended Harvard University's Program for Senior Executives in National and International Security at the John F. Kennedy School of Government in 2000.

Dr. Richmond M. Lloyd

Dr. Richmond M. Lloyd is a Professor of National Security Affairs at the Naval War College and holds the William B. Ruger Chair of National Security Economics. Previously, he served as course director for the Security, Strategy, and Forces course and as director of the U.S. Naval War College's Latin American Studies Group, which coordinates all college activities in Latin America. His research and teaching interests include strategy and force planning, national security and economics, defense and international economics, and logistics. He is the editor of the William B. Ruger Chair of National Security Economic Papers and coeditor of nine textbooks for the Naval War College on strategy and force planning. He lectures on contemporary national defense topics at various sites throughout the United States and South America. He chaired the Naval War College's self-study efforts that led to Congressional authorization for the college to award a M.A. degree in National Security and Strategic Studies and to the accreditation of this degree. He received a Ph.D. in business administration and a B.S. in mechanical engineering from the University of Rochester, and an M.B.A. from the University of Chicago.

Ambassador (Ret.) Princeton N. Lyman

Ambassador Princeton N. Lyman is an adjunct senior fellow for Africa policy studies at the Council on Foreign Relations (CFR). He is also an adjunct professor at Georgetown University.

Ambassador Lyman's career in government included assignments as deputy assistant secretary of state for Africa (1981–1986), U.S. ambassador to Nigeria (1986–1989), director of refugee programs (1989–1992), ambassador to South Africa (1992–1995), and assistant secretary of state for international organization affairs (1996–1998). He served as director of the U.S. Agency for

International Development (USAID) in Addis Ababa, Ethiopia from 1976 to 1978.

From 2003 to 2006, he was the Ralph Bunche Senior Fellow for Africa Policy Studies at the Council on Foreign Relations. From 1999 to 2003, he was executive director of the Global Interdependence Initiative at the Aspen Institute. Ambassador Lyman is a member of several boards, including the American Academy of Diplomacy, the Fund for Peace, the George Washington University Africa Center for Health and Human Security, and the board on African science academy development for the National Academy of Sciences. He is also a member of the African Advisory Committee to the United States Trade Representative.

Ambassador Lyman has a Ph.D. in political science from Harvard University. He has published books and articles on foreign policy, African affairs, economic development, HIV/AIDS, UN reform, and peacekeeping. He has published op-eds in the *Washington Post, Wall Street Journal, Baltimore Sun, Miami Herald, Los Angeles Times,* and *International Herald Tribune*. His book, *Partner to History: The U.S. Role in South Africa's Transition to Democracy* (U.S. Institute of Peace Press), was published in 2002. He was co-director of the Council on Foreign Relations Task Force Report, *More Than Humanitarianism: A Strategic U.S. Approach toward Africa,* issued in 2006, and coeditor of *Beyond Humanitarianism: What You Need to Know About Africa and Why It Matters* (Council on Foreign Relations), published in 2007. His article "Obama and Africa: Matching Expectations with Reality," co-authored with Katy Robinette, will appear in Columbia University's *Journal of International Affairs* in spring 2009.

Dr. Marc Lynch

Marc Lynch is associate professor of political science and international affairs at the Elliot School of International Affairs, George Washington University. Professor Lynch received his B.A. in political science from Duke University and his M.A. and Ph.D. in government from Cornell University. He teaches courses on Middle Eastern politics and international relations. He is the author of *Voices of the New Arab Public* (2006) and writes frequently on Arab media and public opinion, Islamist movements, and Middle East politics. He is also the author of the widely read Middle East politics blog Abu Aardvark at *Foreign Policy* magazine (http://lynch.foreignpolicy.com).

Captain David G. Manero

Captain Dave Manero is the proud son of Carmen and Rosemary Manero and a native of Highland Park, New Jersey. A 1988 graduate of the University of Pennsylvania with a B.S. in Electrical Engineering, he joined the ranks of Naval Aviation earning his Wings of Gold in July 1989.

Captain Manero has made numerous operational deployments to the Middle East, Asia, and South and Central America. A Developmental Test Pilot, he has accumulated over 3,800 flight hours in over thirty-five different aircraft types, and sixty-nine combat missions during Operation Desert Storm. In 2006, he completed his Command Tour with the VT-2 "Doerbirds" where his squadron earned the Chief of Naval Air Training (CNATRA) 2006 Training Excellence Award, the 2006 Vice Admiral Goldthwaite Award for the best training squadron in the Navy (out of twenty-seven), the 2006 Chief of Naval Operations Aviation Safety Award, and the 2006 Meritorious Unit Commendation.

Captain Manero's previous staff experience includes one tour as Flag Lieutenant for Commander, Carrier Group ONE where he was a member of the fly-away JFACC in the Pacific AOR. He also served as Legislative Fellow and later as Military Legislative Assistant to United States Senate Minority Leader Trent Lott.

In addition to graduate-level flight training at the United States Naval Test Pilot School, Captain Manero has earned advanced degrees with emphasis in international affairs and strategy from Harvard University and the Air War College, where he graduated with academic distinction. He is also a designated Acquisition Professional (APC). His personal awards include the Meritorious Service Medal, Air Medal (two Strike Flight), Navy Commendation Medal (six, two with Combat V), Navy Achievement Medal, and the Combat Action Ribbon.

Dr. Daniel Markey

Daniel Markey is a senior fellow for India, Pakistan, and South Asia at the Council on Foreign Relations (CFR). His areas of specialization are security and governance in South Asia, international conflict, theories of international relations, and U.S. foreign policy.

From 2003 to 2007, he held the South Asia portfolio on the Policy Planning Staff at the U.S. Department of State. His responsibilities included analysis and planning for the Secretary of State on regional and global policy issues, participation in departmental and interagency South Asia policy formulation, articulation of regional policy for senior-level speeches and print media, and acting as a liaison with academic, think tank, and diplomatic communities.

Prior to government service, Dr. Markey taught courses on U.S. foreign policy and theories of international relations in the Politics Department at Princeton University and served as the executive director of Princeton's Research Program in International Security. In 2000 and 2001, he was a postdoctoral fellow at Harvard University's Olin Institute for Strategic Studies. His academic research focused on prestige politics and international conflict.

He received a Ph.D. from Princeton University's Department of Politics and a B.A. in international studies from Johns Hopkins University.

Professor Laurence L. McCabe

Professor Laurence L. McCabe teaches the Security, Strategy, and Forces (SSF) and Strategy & Theater Security (STS) courses at the U.S. Naval War College. A recently retired Navy surface warfare officer, Professor McCabe was assigned to cruisers and destroyers in Hawaii, California, South Carolina, and Pennsylvania. He has deployed to every ocean in the world as well as the Caribbean and the Gulf of Mexico. Ashore Professor McCabe served as Flag Secretary to a Commander of a Carrier Battle Group on the aircraft carrier USS *Constellation*. He also served in the Pentagon as Special Assistant to the

Secretary of the Navy for Resources and Programs. Professor McCabe also served as the Military Group Commander, U.S. Embassy in Santo Domingo, Dominican Republic, for three years. Currently, as Chair of the Latin America Regional Study Group, he has presented lectures in Mexico, Jamaica, Dominican Republic, Chile, Argentina, Peru, Bolivia, Uruguay, Costa Rica, and Brazil. In the Africa region, Professor McCabe has lectured in Cameroon, Seychelles, Mauritius, and Madagascar. Professor McCabe lectures on global security and economic development, maritime security, national and military strategy, and globalization.

Dr. Rocky R. Meade

Colonel Rocky R. Meade is the Colonel General Staff of the Jamaica Defence Force (JDF). He has responsibility for the Force's policies on operations, training, intelligence, communications, information systems, civil/military relations, and publications.

He previously served as Commanding Officer 1 Engineer Regiment (JDF), Commanding Officer of the Support and Services Battalion, Commandant of the Caribbean Junior Command and Staff Course, and in two staff officer appointments (operations and administration) in the JDF's Headquarters. He is the JDF's representative on Jamaica's National Security Strategy Committee and worked on the Strategic Defence Review of the JDF. He chairs several committees to include those of the National Training Agency (HEART/NTA) Projects, the Jamaican Military Museum and Library, and the editorial board of the Society for Caribbean Linguistics. He holds B.A. (Hons) and M.A. degrees from the University of the West Indies, a Master of Military Arts and Science (M.M.A.S.) degree from the U.S. Army War College, Fort Leavenworth, and a Ph.D. in Linguistics from the University of Amsterdam. He lectures credit courses at the University of the West Indies and has delivered invited lectures on security, strategic planning, and linguistics topics to audiences in Jamaica and internationally to include Africa, Canada, Germany, the Netherlands, the United Kingdom, the United States, and several Caribbean countries. He has edited the JDF's *Alert* and *All Arms* periodicals and has several publications to include *The Acquisition of Jamaican Phonology* and *The Relevance and Optimal Structure of the Military in Jamaica in the Current and Emerging Geo-Security Environment.*

Dr. Evan S. Medeiros

Evan S. Medeiros is a senior political scientist at the RAND Corporation in the Washington, DC, office. He specializes in research on the international politics of East Asia, China's foreign and national security policies and policymaking, U.S.-Chinese relations and Chinese military issues. He recently served for a year as the Policy Advisor to the Special Envoy for China and the U.S.-Chinese Strategic Economic Dialogue (SED) at the Treasury Department, regularly briefing Secretary Paulson on China affairs. He was sponsored by a Council on Foreign Relations International Affairs Fellowship.

Dr. Medeiros has written on a broad range of Asian security issues. In 2008 he published a RAND study called *Pacific Currents: The Responses of*

U.S. Allies and Security Partners in East Asia to China's Rise. In 2007, he published an independent book called *Reluctant Restraint: The Evolution of China's Nonproliferation Policies and Practices, 1980–2004* (Stanford University Press). Other recent publications include journal articles on China's "Peaceful Rise" strategy and Chinese nuclear doctrine.

He has published several RAND studies including: *China's International Behavior: Activism, Opportunism, and Diversification* (forthcoming 2009), *Dangerous Thresholds: Managing Escalation in the 21st Century* (MG-614-AF); *A New Direction for China's Defense Industry* (MG-334-AF, 2005); *Chasing the Dragon: Assessing China's System of Export Controls on WMD-Related Goods and Technologies* (MG-353, 2005), and *Modernizing China's Military: Opportunities and Constraints* (MG-260-AF, 2005).

Prior to joining RAND, Dr. Medeiros was a Senior Research Associate for East Asia at the Center for Nonproliferation Studies at the Monterey Institute of International Studies. During 2000, he was a visiting fellow at the Institute of American Studies at the China Academy of Social Sciences (CASS) in Beijing and an adjunct lecturer at China's Foreign Affairs College.

He holds a Ph.D. in International Relations from the London School of Economics and Political Science, an M.Phil in International Relations from the University of Cambridge (where he was a Fulbright Scholar), an M.A. in China Studies from the University of London's School of Oriental and African Studies (SOAS), and a B.A. in analytic philosophy from Bates College in Lewiston, ME. He travels to Asia frequently and speaks, reads, and writes Mandarin Chinese.

Dr. R. Craig Nation

R. Craig Nation has been Professor of Strategy and Director of Eurasian Studies at the U.S. Army War College in Carlisle, PA, since 1996, where he has held the College's Elihu Root Chair in Military Studies. He has also been a Fellow with the Clarke Center for the Study of Contemporary Issues at Dickinson College in Carlisle, PA, and a member of the Academic Council of the Institute for East-Central European and Balkan Studies at the University of Bologna, Italy, and served as Professor of European Studies with the U.S. Marine Corps War College at the Marine Corps University in Quantico, Virginia. Professor Nation is a native of Philadelphia, PA, and holds an undergraduate degree in History and Political Science from Villanova University and a Ph.D. in Contemporary History from Duke University. He has taught history, strategy, and international relations with the University of Southern California School for International Relations, the U.S. Army Russian Institute, Cornell University Department of Government, and the Johns Hopkins University Paul H. Nitze School of Advanced International Studies. Professor Nation's professional interests focus on strategic interaction in the Russian, Eurasian, and Southeastern European regions. Major publications include *War on War: Lenin and the Origins of Communist Internationalism* (1989); *Black Earth, Red Star: A History of Soviet Security Policy 1917–1991* (1992); *War in the Balkans 1989–2002* (2003); and *Security in the West* (forthcoming).

Dr. Shannon K. O'Neil

Shannon O'Neil is the Douglas Dillon fellow for Latin America studies at the Council on Foreign Relations (CFR). Her expertise includes political and economic reform in Latin America, U.S.–Latin American relations, and Latin American immigration to the United States. She recently directed CFR's Independent Task Force on U.S.–Latin America Relations: A New Direction for a New Reality. She is currently working on a book on Mexico, analyzing the political, economic, and social transformations Mexico has undergone over the last two decades, and the significance of these changes for U.S.-Mexican relations.

In addition to her work at CFR, Dr. O'Neil has taught in the political science department at Columbia University, and she publishes LatIntelligence (http://www.latintelligence.com), a blog analyzing Latin American politics, economics, and public policies. She is a frequent commentator on major television and radio programs.

Prior to joining CFR, she was a justice, welfare, and economics fellow and an executive committee member and graduate associate at the Weatherhead Center for International Affairs at Harvard University. She was also a Fulbright Scholar in Mexico and Argentina. Prior to her academic work, Dr. O'Neil worked in the private sector as an equity analyst at Indosuez Capital Latin America and Credit Lyonnais Securities. She holds a Ph.D. in Government from Harvard University, an M.A. in International Relations from Yale University, and a B.A. from Yale University.

Commander Victor M. Ott

Commander Ott graduated from the College of Charleston in Charleston, S.C., and was commissioned in September of 1987 after graduation from Officer Candidate School in Newport, Rhode Island. Upon graduating with distinction from Surface Warfare Officers School, Pacific at Naval Amphibious Base Coronado, CDR Ott was stationed aboard the USS *Tripoli,* (LPH-10), at Naval Station San Diego, as the gunnery and communications officer. After deploying to the Western Pacific and qualifying as a Surface Warfare Officer (SWO) on board *Tripoli,* he proceeded to flight training in Corpus Christi, Texas, and Pensacola, Florida, in 1988. He was designated a Naval Aviator in September 1991 and assigned to Helicopter Anti-Submarine Squadron TEN (HS-10) in San Diego, California, for H-60F/H FRS training. Following FRS training, he was assigned to HS-6, as the personnel officer, aircraft division officer and NATOPS officer. While there he made two Arabian Gulf deployments on USS *Abraham Lincoln* (CVN-72) in support of Operations Desert Storm, Southern Watch, and Continue Hope in Somalia. He additionally served on detachments on board USS *Constellation* (CV-64) and USS *Ingraham* (FFG-61).

Beginning in 1995, he served as a flight instructor at HS-10, holding jobs as Scheduling Officer, Assistant Operations Officer, and Standardization Officer. In 1998, he reported for temporary duty at the Combined Air Operations Center in Vicenza, Italy, as the Deputy Director of the Combined Rescue Coordination Center in support of Operations Joint Guard and Noble Anvil in Bosnia-Herzegovina and Kosovo. In 1999 he reported to HS-4 serving there as Safety Officer, Tactics Officer, and Operations Officer, deploying to the Arabian Gulf, once again on board USS *Abraham Lincoln*. While at HS-4 he also was a member of a deployed detachment operating in support of SEAL Team THREE on board USS *Cromelin* (FFG-37).

Upon departure from HS-4 in 2001, CDR Ott entered the Naval War College in Newport, Rhode Island. He graduated in November of 2002 with a Master's degree in National Security and Strategic Studies. In April of 2003 he reported to U.S. Special Operations Command (SOCOM) Tampa, Florida. While at SOCOM he served as the Branch Chief, Current Air Operations (J33 Air) and was selected to become the first Commanding Officer of Helicopter Sea Combat Weapons School, Pacific (HSCWSP).

He assumed command in June of 2005 of HSCWSP where he merged two communities' (HS and HC) tactical training programs and processes, Weapons and Tactics Instructors (WTIs), and weapons load programs into one unit responsible for the wing's advanced weapons and tactics training, standards, and evaluations. Upon completion of O5 command he was assigned to the U.S. Department of State, Bureau of Political-Military Affairs, Directorate of Defense Trade Controls. He is now assigned to the International Engagements division of the staff of the Chief of Naval Operations (OPNAV N52). CDR Ott was selected for promotion to Captain in April of 2008 and will be promoted in July of 2009.

CDR Ott is a graduate of the Joint and Combined Warfighting School taught at the Joint Forces Staff College in Norfolk, Virginia, and a fully qualified Joint Services Officer.

His personal decorations include the Defense Meritorious Service Medal, Meritorious Service Medal (two awards), Navy Commendation Medal (three awards), Navy Achievement Medal (two awards) and various other service and campaign awards including the Battle E (three awards).

Ambassador (Ret.) Mary Ann Peters

Ambassador Peters became the fourth Provost of the Naval War College on September 18, 2008. Previously, she held the position of Dean of Academics of the College of International and Security Studies at the George C. Marshall European Center for Security Studies in Garmisch-Partenkirchen, Germany. Prior to becoming the Dean of the College, Ambassador Peters served as Associate Director for International Liaison at the Marshall Center.

Before joining the Marshall Center in 2003, Ambassador Peters spent more than thirty years as a career diplomat with the U.S. Department of State. From 2000 to 2003 Ambassador Peters served as the U.S. Ambassador to Bangladesh, leading the Mission's efforts in support of the war on terrorism and other key U.S. foreign policy goals. She received a Presidential Meritorious Service Award in 2003 for her work in Bangladesh. Prior to her posting in Dhaka, Ambassador Peters was the Deputy Chief of Mission at the United States Embassy in Ottawa, Canada, responsible for the management of the Embassy and supervision of the six U.S. Consulates General in Canada.

From 1995 to 1997, Ambassador Peters served in the White House as Director for European and Canadian Affairs at the National Security Council. Among other portfolios in this position, Ambassador Peters worked on the diplomatic and security aspects of the search for peace in Northern Ireland. From 1993 to 1994, Ambassador Peters served as Deputy Assistant Secretary of State with oversight responsibility for U.S. relations with nineteen Western European countries and Canada. In this capacity she acted as the U.S. Chair of the U.S.-Canadian military coordination body, the Permanent Joint Board on Defense.

A senior diplomat, fluent in six foreign languages, Ambassador Peters has also served in Sofia, Bulgaria, as Deputy Chief of Mission; in Moscow as Economic Counselor; and in Mandalay, Burma, as Principal Officer. Prior to her assignment in Moscow, she studied Russian at the U.S. Army Russian Institute in Garmisch, Germany. From 1988 to 1990, Ambassador Peters was the Deputy Director of the Office of Pakistan, Afghanistan and Bangladesh Affairs in the U.S. State Department. She began her career as a Vice-Consul in Frankfurt in 1975.

Ambassador Peters holds a Bachelor of Arts degree from Santa Clara University and a Masters in International Studies from the School of Advanced International Studies at The Johns Hopkins University. Her formal education also included course work in Paris, France, and Bologna, Italy. She is married to Timothy McMahon. They have two children: Maggie and Blaise.

Dr. Jonathan D. Pollack

Jonathan D. Pollack is Professor of Asian and Pacific Studies and Chair of the Asia-Pacific Regional Study Group at the U.S. Naval War College, in Newport, Rhode Island. Between 2000 and 2004 he also served as Chairman of the College's Strategic Research Department. Prior to joining the War College faculty in 2000, Dr. Pollack was affiliated with the RAND Corporation in Santa Monica, California, where he served in a wide range of research and management positions. His major research interests include Chinese national security strategy, U.S. foreign and defense policy in Asia and the Pacific, Korean politics and foreign policy, and nuclear weapons and international politics. He is a member of the International Institute for Strategic Studies (IISS), the Council on Foreign Relations, the National Committee on U.S.-China Relations, and an emeritus member of the Committee on International Security and Arms Control, a standing committee of the National Academy of Sciences.

Dr. Pollack's recent publications include three major edited volumes on U.S. strategy and policy in the Asia-Pacific region: *Strategic Surprise? U.S.-China Relations in the Early 21st Century* (2004); *Korea: The East Asian Pivot* (2006), and *Asia Eyes America: Regional Perspectives on U.S. Asia-Pacific Strategy in the 21st Century* (2007); and guest editor and contributor to "The Major Powers and the Two Koreas: Stability at Risk?" (*The Korean Journal of Defense Analysis,* March 2009). He has also authored numerous journal articles, book chapters, and strategic commentaries, with particular emphasis on Chinese military development, U.S.-Chinese relations, the North Korean nuclear issue, and U.S. defense strategy in East Asia. During 2008–2009, he is undertaking research on "Rethinking Korean Nuclearization," supported by a grant from the John T. and Catherine D. MacArthur Foundation. A book derived from this project, *No Exit: North Korea, Nuclear Weapons, and International Security,* will be published by the IISS in 2010 as one of the inaugural volumes in the institute's new book series.

Dr. Terence Roehrig

Terence Roehrig is an Associate Professor in the National Security Decision Making Department, at the U.S. Naval War College. He is a coauthor of a forthcoming book entitled *South Korea since 1980: Democratization, Economic Struggle, and Nuclear Crisis* (Cambridge University Press) with Uk Heo. In addition, he is the author of two books, *From Deterrence to Engagement: The U.S. Defense Commitment to South Korea* (Lexington Books, 2006) and *The Prosecution of Former Military Leaders in Newly Democratic Nations: The Cases of Argentina, Greece, and South Korea* (McFarland Publishers, 2002), and a coeditor of *Korean Security in a Changing East Asia* (Praeger, 2007). Professor Roehrig has published articles and book chapters on North Korea's nuclear weapons program, Korean and East Asian security issues, deterrence theory, the U.S.–South Korea alliance, human rights, and transitional justice. He received his Ph.D. in political science from the University of Wisconsin–Madison and is the current President of the Association of Korean Political Studies.

Ambassador (Ret.) Teresita Schaffer

Ambassador Teresita Schaffer is Director, South Asia Program at the Center for Strategic and International Studies (CSIS). She came to CSIS in August 1998 after a thirty-year career in the U.S. Foreign Service. She devoted most of her career to international economic issues and to South Asia, on which she was one of the State Department's principal experts.

From 1989 to 1992, she served as deputy assistant secretary of state for South Asia, at that time the senior South Asia position in the department; from 1992 to 1995, she was the U.S. ambassador to Sri Lanka; and from 1995 to 1997, she served as director of the Foreign Service Institute. Her earlier posts included Tel Aviv, Islamabad, New Delhi, and Dhaka, as well as a tour as director of the Office of International Trade in the State Department. She spent a year as a consultant on business issues relating to South Asia after retiring from the Foreign Service.

Her publications include "Sri Lanka: Lessons from the 1995 Negotiations," in *Creating Peace in Sri Lanka* (Brookings, 1998); two studies on women in Bangladesh; and "Kashmir: Fifty Years of Running in Place," in *Grasping the Nettle* (USIP, 2004). Her CSIS publications include *Kashmir: The Economics of Peace Building* (2005), *Pakistan's Future and U.S. Policy Options* (2004), *Rising India and U.S. Policy Options in Asia* (2002), and several reports on the HIV/AIDS epidemic in India. Schaffer has taught at Georgetown University and American University. She speaks French, Swedish, German, Italian, Hebrew, Hindi, and Urdu, and has studied Bangla and Sinhala.

Dr. Peter J. Schraeder

Peter J. Schraeder (born June 2, 1961) is a tenured professor and graduate program director in the Department of Political Science at Loyola University Chicago, where he also teaches as part of the Black World Studies, International Studies, and Islamic World Studies programs. He received his M.A. (1986) and Ph.D. (1990) in International Studies from the University of South Carolina, after completing a double major in International Studies and French at Bradley University (1979–1981) and a degree in French Civilization and Language at the Sorbonne in Paris, France (1981–1982). His administrative, teaching, and research skills are the direct result of wide-ranging overseas experiences, including having lived, lectured, or carried out research in twenty-nine African countries in all regions of the African continent, most recently North Africa. In addition to teaching at the University of Tunis in Tunisia (2002–2003) and at Cheikh Anta Diop University in Senegal (1994–1996) as part of the Fulbright scholar exchange program, Schraeder has held visiting appointments at the John Felice Rome Center in Italy (2003–2005), Somali National University (1985), the U.S. Embassy in Djibouti (1987), the French Institute of African Research in Zimbabwe (1996), and the University of the Antilles in Guadeloupe (1999). He teaches in Tunisia every January, and leads a twenty-two-day summer travel course for American students to Tunisia each June. Schraeder's research interests span four major areas: comparative foreign policy theory, United States and European foreign policies toward Africa and the Middle East, African politics and foreign policy (including North Africa), and intervention in world politics and international democracy promotion. His research has been published in such diverse scholarly journals as *African Affairs*, *The Journal of Modern African Studies, The Journal of Politics, Middle East Journal, Politique Africaine,* and *World Politics*. He is the author or editor of ten books, including *Globalization and Emerging Trends in African Foreign Policy: A Comparative Perspective of Eastern Africa* (2007), *African Politics and Society: A Mosaic in Transformation* (2nd ed., 2004), *Exporting Democracy: Rhetoric vs. Reality* (2002), *United States Foreign Policy toward Africa: Incrementalism, Crisis, and Change* (1994), and *Intervention into the 1990s: U.S. Foreign Policy toward the Third World* (1992). He is currently working on two books, "African Foreign Policy: Democratization and Its Impact on Policy Formulation and Implementation" and "The Cross, the Crescent and the Ballot Box: Catholic and Islamic Perspectives on the Rule of Law and Democracy Promotion." Fluent in English and French, Schraeder is also actively involved in a wide number of international research networks, most notably in France and francophone Africa. He resides in Arlington Heights, Illinois, with his wife, Catherine Anne Scanlon, and three children: Maximilian (ten years old), Marianne (seven years old), and Patrick (four years old).

Dr. Hussein Solomon

Dr. Hussein Solomon holds a D.Litt. et Phil. (Political Science) from the University of South Africa. Currently he is Professor in the Department of Political Sciences. His previous appointments include being Director of the Centre for International Political Studies, University of Pretoria; Research Manager at the

African Centre for the Constructive Resolution of Disputes (1998–2000); Senior Researcher: Institute for Security Studies (1996–1998); and Research Fellow: Centre for Southern African Studies, University of the Western Cape (1993–1995). In 2007 he was Visiting Professor at the Global Collaboration Centre at Osaka University in Japan and in 2008 he was Nelson Mandela Chair of African Studies at Jawahrlal Nehru University in New Delhi, India. Currently he is Visiting Fellow at the MacKinder Centre for the Study of Long-Wave Events at the London School of Economics and Politics Science in the United Kingdom.

In addition, he is a Member of the Board of the Centre for the Study of Violence and Reconcilitation; Deputy Chair of the Pugwash Conferences on Science in World Affairs (South African Chapter); Member of the Security Council of the Gerhson Lehrman Group of Companies; Member of the International Steering Committee of Global Action to Prevent War; Research Associate of the Centre for Defence Studies, University of Zimbabwe; and a Research Associate of the South African Institute for International Affairs. He is also a Member of the International Advisory Council of the Toda Institute for Global Peace and Policy Research in Hawaii. He is also Vice President of the African Studies Association. Moreover, Member of the Editorial Board of *Politeia;* Special Advisor to the Editor of the journal *Globalizations;* Contributing Editor to the *Journal for Contemporary History;* and a member of the Editorial Boards of *Africa Insight, Alternatives,* and *Scientific Journals International* as well as Editor of *Africa Insight* (India). He also sits on the Board of Trustees of All Africa Women for Peace. Hussein Solomon also holds the rank of a Captain in the Reserve Force of the South African Air Force.

His research interests include conflict and conflict resolution in Africa; South African Foreign Policy; international relations theory; religious fundamentalism and population movements within the developing world. His publications have appeared in South Africa, Nigeria, the United States, the United Kingdom, Switzerland, the Russian Federation, the Netherlands, Norway, Denmark, Egypt, Ethiopia, Israel, Lebanon, India, Bangladesh, Spain, and Japan.

His most recent publication is an edited book, *Challenges to Global Security: Geopolitics in an Age of Transition* (2008), that has been published by IB Tauris Publishers in London.

Professor Sean C. Sullivan

Sean C. Sullivan is an Assistant Professor of National Security Affairs at the U.S. Naval War College in Newport, Rhode Island. He is assigned to the National Security Decision Making Department teaching the Policy Making and Process and the Contemporary Staff Environments sub-courses. He is a subject matter expert on defense planning and the Department of Defense Formal Resource Allocation processes. Professor Sullivan coordinates all curriculum development on Defense Resource Allocation and is the author of numerous related articles, readings, and case studies on formal defense planning processes.

A retired naval officer, Sean Sullivan served in the U.S. Navy for twenty-three years. He served at sea for over fifteen years in various surface combatants, amphibious ships, and afloat staffs. He deployed five times to the western Pacific and Arabian Gulf and once to the southeastern Pacific Ocean.

Sean Sullivan attended the Naval War College, graduating in March 1999 with a Master of Arts Degree in National Security and Strategic Studies. He also holds a Bachelor of Arts degree in Political Science from the University of Rochester.

Ambassador (Ret.) Paul D. Taylor

Ambassador Taylor is a Senior Strategic Researcher in the Strategic Research Department of the Center for Naval Warfare Studies of the Naval War College and former chair of the Latin American Studies Group. Currently also teaching elective courses on Latin America and the United States and on international economics, he spent a career in the U.S. Foreign Service and four years as an international business executive responsible for establishing the Latin America subsidiary of a multinational satellite broadcasting firm. He served as U.S. Ambassador to the Dominican Republic. Ambassador Taylor was assigned in 1992 as Professor and International Affairs Adviser to the President of the Naval War College and taught the core curricula in the Strategy and Policy Department and in the National Security Decision Making Department.

As a Foreign Service Officer, he also was Deputy Assistant Secretary of State responsible for overall relations with Mexico and economic relations with Latin America and the Caribbean. Ambassador Taylor was assigned diplomatic duties for periods of two to three years each in Ecuador, Thailand, Brazil, Spain, and Guatemala. He served for two years on the staff of the Peace Corps in Washington and in Ecuador and for three years on active sea duty as a commissioned officer in the U.S. Navy. His languages include Spanish and Portuguese. He studied economics at Harvard, where he received an M.P.A., and politics at Princeton leading to an A.B. *magna cum laude*.

He has written articles on strategy, economic sanctions, and other subjects of international affairs and edited *Perspectives on Maritime Strategy: Essays from the Americas,* Newport Paper 31 (Newport: Naval War College Press, 2008); *Latin American Security Challenges: A Collaborative Inquiry from North and South,* Newport Paper No. 21 (Newport: Naval War College Press, 2004); and *Asia and the Pacific: U.S. Strategic Traditions and Regional Realities* (Newport: Naval War College Press, 2001).

He is currently editing *Perspectivas sobre estrategia marítima: Ensayos de las Américas, la nueva estrategia marítima de EE UU y comentario sobre Una Estrategia Cooperativa para el Poder Naval en el Siglo XXI,* the first volume to be published by the Naval War College Press in a language other than English.